SEASHORE LIFE
BETWEEN THE TIDES

CORAL REEF SCENE NEAR KEY WEST

Pelagia cyanella (upper middle); *Stomolophus meleagris* (right middle);
Octopus americanus (lower left); *Oreaster reticulatus* (lower middle);
Centrechinus antillarum (lower right).

SEASHORE LIFE
BETWEEN THE TIDES

By

WILLIAM CROWDER

Illustrated by the Author

DOVER PUBLICATIONS, INC.

NEW YORK

Published in Canada by General Publishing Com-
pany, Ltd., 30 Lesmill Road, Don Mills, Toronto,
Ontario.
Published in the United Kingdom by Constable
and Company, Ltd., 10 Orange Street, London
WC 2.

This Dover edition, first published in 1975, is an
unabridged republication of the work originally
published by Dodd, Mead & Company, New York,
in 1931 under the title *Between the Tides*. The
frontispiece, originally in color, is here reproduced
in black and white.

International Standard Book Number: 0-486-23221-2
Library of Congress Catalog Card Number: 75-16036

Manufactured in the United States of America
Dover Publications, Inc.
180 Varick Street
New York, N.Y. 10014

18073

THIS WORK IS DEDICATED
WITH GRATEFUL APPRECIATION
TO THE FOLLOWING NATURALISTS

ROBERT R. COLES
PHILIP O. GRAVELLE
ROBERT HAGELSTEIN
RAWSON J. PICKARD
CHARLES P. TITUS

WHOSE KINDNESS ALONE
HAS RENDERED POSSIBLE
ITS PREPARATION

TABLE OF CONTENTS

I

INTRODUCTION

AN INDOOR SALT WATER AQUARIUM

CHAPTER I

INTRODUCTION

THIS is a popular reference book and guide describing the lower sea-shore animals of the Atlantic coast of North America. But while the book embraces to more or less a degree all the forms occurring in the western Atlantic seas, it is designed with special reference to the commoner inhabitants of temperate waters. To this end a most careful selection of material has been made.

The illustrations will, it is hoped, be as informative as the text. For it can not be gainsaid that despite its primitiveness, the method of referring to pictures still remains the easiest, if not precisely the surest, whereby the identity of animal forms may be determined. Every animal described in the text is represented graphically as well.

Technical terms not generally familiar are usually italicized the first time they are used in each chapter. Moreover, a glossary is furnished at the end of the volume, to which the reader may refer when necessary. Also there is a bibliography containing a list of the most important works for general reference relating to each phylum of marine animals.

All measurements are given in the metric system; this by reason of its efficiency and its almost universal use in natural science literature. Moreover, its adoption in this work will familiarize the inexpert student with its use; consequently, this will not be without profit should he perchance consult a technical reference in search of more detailed and specific information. A metric scale is represented in the glossary.

It should not surprise the reader to discover that certain specific names listed in the text differ somewhat from those found to obtain in other or older works. The truth is, if one compares the accounts of authors writing at different dates, one will find that they often differ not only in regard to specific names, but also in regard to the systematic position of various animals. The reason for this is that the earlier naturalists had not the information available to succeeding investigators. That even modern knowledge is far from complete is attested by the fact that today authorities are not all agreed as to the exact number and rank of the subdivisions of the various phyla.

The most casual reader can not have failed to observe that the scientific names of all organized beings are Latin. Latin or latinized names are given to both plants and animals because Latin is a classical language, a language presumably (but not necessarily actually) understood by natural scientists throughout the world. Accordingly, when a new plant or animal is discovered in any one country, the name given to it is understood and recorded by the workers of all other nations; and when this new name appears in scientific literature, it is recognized by all alike to mean a certain specific form only: there is no later confusion as to its identity, as to precisely what plant or animal is meant.

Whenever an animal listed in this book has a popular, or common name, this also is usually given; though it is evident from the preceding paragraph that such a name is not of great aid to the student in his search through the various literatures. Only a very small proportion of the total number of animals bear a popular name in addition to the one given by science. Popular names are for the most part local. For example, along some parts of the Atlantic coast, *Limulus polyphemus* is called the horseshoe crab; elsewhere it is called the horsefoot crab; still elsewhere, the bait crab. As a matter of fact, it is not a crab: more, it is not recognized by most naturalists even as a crustacean; it seems, in truth, more closely related to the spiders and the scorpions than to the crabs.

This *binomial,* or two-name, system of nomenclature was introduced by Linné, known usually by his Latin name Linnaeus, a Swedish naturalist; and it is called the Linnaean system. This was near the middle of the 18th century. Linnaeus grouped species into classes and orders, but he did not employ the term *family.* Since his time the general scheme of classification has been greatly expanded, and now international rules govern the naming of plants and animals. It sometimes happens that two or more workers will discover the same species of animal independently; each worker will give it a specific name of his own devising which is different from that of the others; yet each will be in agreement with the others regarding the genus to which the animal belongs. When these independent discoveries are made, the animal usually retains the name it received from the person who first records in published form his scientific description of it. The fact that independent discoverers can arrive at a determination of the correct genus is because classification is based on definite rules of natural relationships and structural similarities.

It should not be assumed by the reader, however, that one form is lower or higher than another merely by virtue of precedence or succession of presentation in the present text: the order of no classification can show in linear arrangement the proper relationship between lower and higher animals.

One word more. Of all the haunts of life, the seashore is the most interesting; no other portion of our globe is so rich in numbers of varieties and individuals; yet this region has received the least attention from the people generally and from the popular exponents of natural science. The author of this work feels that much of this inattention is owing to a certain confusion that confronts the attempts of the average person when he essays to identify the various creatures he meets along the shore. For one thing, the lack of unscientific, or common, names of the vast numbers of animals he encounters is apt to deter the rambler from taking more than a momentary interest in their presence. A psychological reaction is here involved. It is a peculiar circumstance that although a plant or an animal may have been totally unfamiliar to us and although we may have known absolutely nothing regarding its nature or capacity, once we have learned its name its strangeness is gone. When we meet with it again it assumes the status of a somewhat familiar, if not exactly an old, acquaintance. It is with this end in view, that is, to help form such acquaintances, that this book is designed. The reader will have introduced to him by name the various inhabitants that compose the population of the shallow sea and the shore. It is hoped, too, that the following account will enable him to realize a closer further acquaintance with this extraordinary group of animals, a group whose cultivation and understanding have too long been the prerogative of specialists. Once this has been done, the reader can easily be relied upon to attend and interpret for himself the great drama that is constantly in progress between the tide marks and beneath the waves. . . .

II

SEASHORE LIFE

PLATE I

WHARF PILE GROUP

Sagartia leucolena (upper middle); *Balanus balanoides* (upper center); *Pelagia cyanella* (left center); *Aurelia aurita* (upper right); *Ciona intestinalis* (middle right); *modiolus modiolus* (lower middle); *Beroë cucumis* (lower left); *Microciona prolifera* (lower right).

CHAPTER II

SEASHORE LIFE

No proper appreciation or understanding of the structural features and the habits of animals can be possible without a knowledge of the environmental factors that relate to those features, and have to do with the various forms of adaptation. It is the difference in structural features that distinguishes one species from another; and in a certain sense, it may be said that even variations in a species are merely expressions of an adaptative response to a changing environment during the history of the race or a group of individuals.

Now, of all the physical changes that have been and are now taking place on the surface of the earth, the sea and its shores have been the scene of the greatest stability. The dry land has seen the rise, the decline, and even the disappearance, of vast hordes of various types and forms within times comparatively recent, geologically speaking; but life in the sea is today virtually what it was when many of the forms now extinct on land had not yet been evolved. Also, it may be parenthetically stated here, the marine habitat has been biologically the most important in the evolution and development of life on this planet. Its rhythmic influence can still be traced in those animals whose ancestors have long since left that realm to abide far from their primary haunts. For it is now generally held as an accepted fact that the shore area of an ancient sea was the birthplace of life.

Still, despite the primitive conditions still maintained in the sea, its shore inhabitants show an amazing diversity; while their adaptative characters are perhaps not exceeded in refinement by those that distinguish the dwellers of dry land. Why is this diversity manifest? We must look for an answer into the physical factors obtaining in that extremely slender zone surrounding the continents, marked by the rise and fall of the tides.

It will be noticed by the most casual observer that on any given seashore the area exposed between the tide marks may be roughly divided into a number of levels each characterized by a certain assemblage of animals. Thus in proceeding from high- to low-water mark, new forms constantly become predominant while other forms gradually drop out. Now, provided that the character of the substratum does not change, these differences in the types of animals are determined almost exclusively by the duration of time that the individual forms may remain exposed to the air without harm. Indeed, so reguarly does the tidal rhythm act on certain animals (the barnacles, for instance), that certain species have come to require a definite period of exposure in order to maintain themselves, and will die out if kept continuously submerged. Although there are some forms that actually require periodic exposure, the number of species inhabiting the shore that are able to endure exposure every twelve hours, when the tide falls, is comparatively few.

With the alternate rise and fall of the tides, the successive areas of the tidal zone are subjected to force of wave-impact. In certain regions the waves often break with considerable force. Consequently, wave-shock has had a profound influence on the structure and habits of shore animals. It is characteristic of most shore animals that they shun definitely exposed places, and seek shelter in nooks and crannies and such refuges as are offered under stones and seaweed; particularly is this true of those forms living on rock and other firm foundations. Many of these have a marked capacity to cling closely to the substratum; some, such as anemones and certain snails, although without the grasping organs of higher animals, have special powers of adhesion; others, such as sponges and sea squirts, remain permanently fixed, and if torn loose from their base are incapable of forming a new attachment. But perhaps the most significant method of solving the problem presented by the surf has been in the adaptation of body-form to minimize friction. This is strikingy displayed in the fact that seashore animals are essentially flattened forms. Thus, in the typically shore forms the sponges are of the encrusting type, the non-burrowing worms are leaflike, the snails and other mollusks are squat forms and are without the spines and other ornate extensions such as are often produced on the shells of many mollusks in deeper and quieter waters. The same influence is no less marked in the case of the crustaceans; the flattening is either lateral, as in the amphipods, or dorso-ventral, as in the isopods and crabs.

In sandy regions, because of the unstable nature of the substratum, no such means of attachment as indicated in the foregoing paragraph will suffice to maintain the animals in their almost ceaseless battle with the billows. Most of them must perforce depend on their ability quickly to penetrate into the sand for safety. Some forms endowed with less celerity, such as the sand dollars, are so constructed that their bodies offer no more resistance to wave impact than does a flat pebble.

Temperature, also, is a not inconsiderable factor among those physical forces constantly operating to produce a diversity of forms among seashore animals. At a comparatively shallow depth in the sea, there is small fluctuation of temperatures; and life there exists in surroundings of serene stability; but as the shore is approached, the influence of the sun becomes more and more manifest and the variation is greater. This variation becomes greatest between the tide marks where, because of the very shallow depths and the fresh water from the land, this area is subjected to wide changes in both temperature and salinity.

Nor is a highly competitive mode of life without its bearing on structure as well as habits. In this phase of their struggle for existence, the animals of both the sea and the shore have become possessed of weapons for offense and defense that are correspondingly varied.

Although the life in the sea has been generally considered and treated as separate and distinct from the more familiar life on land, that supposition has no real basis in fact. Life on this planet is one vast unit, depending for its existence chiefly on the same sources of supply. That portion of animal life living in the sea, notwithstanding its strangeness and unfamiliarity, may be considered as but the aquatic fringe of the life on land. It is supported largely by materials washed into the sea, which are no longer available for the support of land animals. Perhaps we have been misled in these considerations of sea life because of the fact that approximately three times as many major *types* of animals inhabit salt water as live on the land: of the major types of animals no fewer than ten are exclusively marine, that is to say, nearly half again as many as all land-dwelling types together. A further interesting fact is that despite the greater variety in the form and structure of sea animals about three fourths of all known *kinds* of animals live on the land, while only one fourth lives in the sea. In this connection it is noteworthy that sea life becomes scarcer with increasing distance from land; toward the middle of the oceans it disappears almost completely. For example, the central south Pacific is a region more barren than is any desert area on land. Indeed, no life of any kind has been found in the surface water, and there seems to be none on the bottom. In those remote regions deep-sea dredges bring up nothing living; instead, the hauls contain manganese nodules, the ear-bones of whales and quantities of sharks' teeth, some of the latter belonging to sharks that have been extinct since the Miocene. During all the intervening ages since that far-off time these teeth have been lying on the ocean bed uncovered and undisturbed; proving the sterility of the waters above, from which there descends no rain of organic materials which would tend to cover the red clays that compose that ocean floor.

In passing it may be well to mention another interesting fact: sea animals are largest and most abundant on those shores receiving the most copious rainfall. Particularly is this true on the more rugged and colder coasts where it may be assumed that the material from the land finds its way to the sea unaltered and in greater quantities.

We shall now proceed to those details of shore life with which every naturalist should become familiar before he attempts to make a systematic study of its inhabitants, details as to *where* the various forms are to be looked for; that is, where to collect. As to the *methods* of collecting, discussion of these will be reserved for the following chapter.

To the untrained eye, the seashore does not usually at first sight present itself as an area teeming with life. Particularly is this true of sandy reaches. Those common signs of animal life, such as one is accustomed to recognize elsewhere, are few, indeed. Nevertheless, no observant person can remain long in association with any particular shore locality without detecting unmistakably the presence of an exceedingly dense population. As to the range of species such a population may sometimes include, an idea may be gained from the fact that one observer recorded no fewer than 1268 species in the vicinity of Buzzard's Bay.

Just as different levels of the tidal zone are characterized by their varying local forms, so the nature and situation of a geographical area will indicate in a general way the types and kinds of forms that predominate there. Thus the Atlantic Coast of the United States may roughly be divided into three regions, or ranges; the northern range extending from the Bay of Fundy to Cape Cod, the central range from Cape Cod to Cape Hatteras, the southern range from Cape Hatteras to

Key West. These divisions, although somewhat arbitary, are not without an important justification; for, while it is obvious that many forms found in one range may also be native to another, each region is the haunt of certain groups of animals not characteristic of the others. In the main, however, the dominant species occurring in one range seldom extend far beyond its limits. Again, it is due to the physical characters of the seashore and to temperature that the animal life exists in such diversity as to make these divisions not only possible but expedient as well. Take, for example, the rock-bound coast of Maine, together with the colder waters that prevail in that area, which harbor a fauna peculiarly their own. In the multitudinous tide pools and covering the rocks the green sea urchin (*Strongylocentrotus drobachiensis*) holds forth in numbers that it reaches nowhere else. Then, too, in the central range, certain squids (*Loligo pealei,* var. *pallida*) are confined almost exclusively to Long Island Sound, although these animals are good swimmers and fully capable of venturing into other waters; and the same statement will apply to that other lover of temperate climes, the large jellyfish (*Cyanea capillata,* var. *fulva*). And everybody knows that the warm water of the Gulf Stream off the coast of Florida is peculiarly instinct with lives innumerable, while in the shallower depths nearer to the shore dwell the fixed sub-tropic corals, the sea fans, and that strange phosphoric phantom of the sea, the "Girdle of Venus," or the ctenophore *Cestum.*

It is, however, with physical, rather than with geographical, areas that we are to concern ourselves here. But before proceeding further, let us acquaint ourselves with certain terms which will be used occasionally in the course of this book. Classes of animals can be defined according to the depth of water they inhabit: this is called a *bathymetrical* (depth-measurement, or depth-level) division. Those living near or upon the shore are *littoral* species, those inhabiting the open sea are *pelagic,* while those frequenting great depths are *abyssal.* Other terms are used to indicate their mode of life. The forms that float at or near the surface and drift with the currents, such as protozoa, larval and other minute organisms, and jellyfishes, are called "plankton." Strong swimming animals that roam at will are called "nekton." Fixed forms like sponges, and bottom-dwelling forms like starfishes and crabs, which are able to move about, are known as "benthos."

It must be borne in mind that the season of the year has much to do with the kinds and numbers of species to be found at the seashore. Many animals during the colder months migrate to deeper waters; others, such as certain jellyfishes, die on the approach of winter, leaving their larvæ to settle upon the bottom where they await the approach of spring before further development; nor can such delicate forms as prevail among the hosts of hydroids thrive in situations close to the tidal zone which are subject to frosts and the scouring action of ice. Unquestionably the best time for the seashore collector without special equipment is during the summer months; even during this season there occur times that are more productive than others; these are after periods of stormy weather, and at the time of the fortnightly low spring tides. After a storm, forms from the deeper waters are frequently dislodged and washed up high on the beach; low spring tide reveals an area rich in a fauna not ordinarily accessible. Spring tides can always be determined by consulting the local tide tables or by noting the periods of the new and of the full moon, with which periods they are coincident.

SALT MARSHES

Salt marshes (see Plate II) are characterized not only by their vegetation, which is composed of almost a pure culture of thatch (*Spartina*), but also by the countless millions of fiddler crabs which honeycomb the loose and soggy soil of the hummocks. The species peculiar to these localities is the gray slaty-colored one matching the mud, *Uca pugnax*. Here, too, lining the banks of the runnels and wider waterways thrive hordes of horse mussels (*Modiola plicatula*); and in occasional places those other two common dingy mollusks, *Littorina litorea* and *Nassa obsoleta*. But while the forms just mentioned, particularly the fiddler crabs, dominate the territory almost to the exclusion of all other larger animals, the salt marshes can be profitably searched for species not commonly identified with these localities; for in the deeper pockets or depressions of the water courses creatures may sometimes be found which the flood tides have brought in from the open waters, and which lingered too long to escape with the ebb.

MUD FLATS

On mud flats (see Plate II) and shores generally, fiddler crabs and the mud snail *N. obsoleta* will also be found in profusion. The dominant vegetation here, however, is the eelgrass (*Zostera marina*), and on its long slender blades may be found many curious and interesting forms. Crawling over the submerged plants are small gasteropod mollusks, while the eggs of the various species may be seen attached thereon. These plants are also the base whereon such forms as bryozoa, compound ascidians and the spiral-shelled, mollusklike worm *Spirorbis* are found affixed. Around their roots resort several varieties of *Panopeus,* a group of mud crabs; others, besides, seem to be as numerous here as elsewhere, such as the spider crab *Libnia dubia,* and the young of that crablike creature which is not a crab (*Limulus polyphemus*), the so-called horseshoe crab. In the muddy substratum numerous worms, some of remarkable structure and of considerable beauty, bury themselves, exposing only their long filamentous tentacles into the outer medium. These include such annelids as the tube-building *Amphitrite ornata,* the blood worm (*Polycirrus eximius*), and others too numerous for mention here.

WHARF PILES

On wharf piles (See Plate I), bridge-piles, the bottoms of old barges, old hulls (see Plate II) and wood structures generally, forms may often be found which, though common elsewhere, seemingly on occasion elude the most diligent search. Such fixed animals as the sea squirts, sea anemones, branched hydroids, moss animals and the beautiful tubularian hydroids not infrequently establish themselves in these likely places in large colonies. Barnacles and mussels are numerous, and occasionally a sponge gains a foothold. One creature, however, usually outnumbers all the others, but it is so inconspicuous that only the trained eye usually detects its presence. This is the ship worm (*Teredo navalis*), a boring mollusk, wormlike in form, whose communication with the outer world from its retreat within the wood is through a minute hole through which it thrusts its slender paired siphons.

SANDY SHORES

The animal population of sandy shores (see Plate III) lives for the most part under the surface; except for the dead shells and perhaps the perforations made on the sand by fiddler crabs and the soft clam (*Mya arenaria*) the superficial and unknowing observer would never suspect the presence of a vast and varied number hidden out of sight. One kind of fiddler crab, whose species may be told by the oven-like structures at the entrance to its burrow, is the pretty *Uca minax*.

Of the mollusks, there are so many to be looked for that mention here of all will be impossible. Deserving of notice, however, are the razor clam (*Ensis directus*), a facile burrower, and a good swimmer also; the moon snail (*Natica heros*), hidden under a little mound of sand where the retreating tide has left it exploring the substratum in search of the hard-shelled clams (*Venus mercenaria*) which it is so fond of eating; and forms occasionally stranded, like the dog whelk (*Fulgur carica*) and its curious cocoon.

Large marine worms are numerous. Various annelids, such as the sand worm (*Nereis virens*), and that monster flatworm *Meckelia ingens,* may be found by digging.

Sandy beaches are the favorite resorts of horseshoe crabs during the spawning period. Nor must the collector overlook the sea wrack and other débris cast up by the tides. Often this material will be found to harbor a multitude of smaller living forms, and not infrequently the remains of species in a good state of preservation; small shells, sponges, the skeletons of bryozoan colonies, etc.

ROCKY SHORES

The fauna of a rocky region (see Plate III) is to a considerable extent associated with the plant life. Here the rockweed (*Fucus*) (see Plate III) grows in profusion, and its tough branching fronds form an ideal anchorage for many small fixed forms as well as offer support to a numerous host of free living creatures. Among the likely fixed animals are sertularian hydroids, compound ascidians, bryozoan colonies and the egg capsules of mollusks. A pocket magnifier is essential to detect the gross differences between the various species blonging to

PLATE II

1—Examining an old hull for marine life.   2—Salt marsh.   3—Mud flat.

PLATE III

1—Sandy shore. 2—Seaweed *Fucus* covering the rocks.

3—At the edge of a rock pool. 4—Rocky shore.

the same group, since many of them, like the hydroids and bryozoa, for example, have a close superficial resemblance to each other. For more exact identifications, the microscope may often be necessary. As a rule the seaweed abounding in the more sheltered places among the rocks will be found to harbor the greatest number of forms. It is on the vegetation in such places that one is most apt to come across the most beautiful of the commoner hydroids, such as the sea plume (*Obelia commissuralis*) or the pink passion-flower (*Thamnocnidia spectabilis*), and clinging to or making their way over these last-named forms, may often be seen one or more representatives of those two groups containing the strangest minute grotesqueries of the sea, the skeleton shrimps (*Caprellidae*) and the sea spiders (*Pycnogonidae*).

Of course the most conspicuous animals of a rocky shore are the mussels and the barnacles; but there are many species of these latter, and it is an interesting diversion to determine the number of species and the variety of different forms of the same species occurring in a small restricted area. Other crustaceans besides barnacles also abound among the rocks. In clefts and crannies and under stones are secreted the larger crabs together with a crowd of smaller forms like the little amphipod *Gammarus*. On the bare rocks themselves, near or just below the low-water level, are the ever-present starfishes or sea urchins; in this situation, also, are to be seen sponges, often of the red varieties, and occasionally a creeping chiton. Of the other smaller mollusks, there are the limpets, the periwinkles and the forms like the oyster drill, to mention but a few.

TIDE POOLS

Perhaps the most picturesque, and therefore the most alluring, of spots along the seashore, whether the region is largely of rock or of sand, are the tide pools (see Plate III). These natural aquaria with their submarine gardens of greenery are usually rich with a varied fauna, a fauna comprising floating free and fixed forms easy of observation and which are at once a source of pleasure as well as of profit to the seashore naturalist.

While many types of animals seem to prefer the tide pools, because of their quiet and protected waters, there will be found in these precincts not a few that have been marooned during the intervals of the ebbing tides. Among such wanderers are ophiurans, certain shrimps and prawns, jellyfishes and comb jellies, or maybe sepiolas and squids. Sponges, too, will sometimes be found growing in tide pools, whose normal dwelling-place is in the deeper waters offshore.

Although it is certainly true that no animals are typically tide pool dwellers, it is equally true that tide pools contain with unfailing frequency certain types of animals. And such forms, while largely sedentary, compose an extensive list. It is needless here, however, to mention but a few. First there the calcareous sponges including the simple sycon sponge *Grantia* and the still simpler ascon sponge *Leucosolenia;* next there are certain tube-building worms like *Nerine agilis* and a species (*Pectinaria gouldii*) that builds a pretty conical tube of sand grains which it carries about; then there are those two attractive specimens of nudibranchs, or naked mollusks, *Æolis* and *Dendronotus;* and last such lower crustaceans as the isopods *Cirolana concharum* and *Idotea marina,* and the higher forms like *Palæmonetes* the prawn.

OFFSHORE WATERS

No one can maintain an interest in the lower animals of the seashore without wishing ere long to acquaint himself more extensively with those groups of plankton, nekton, or benthos creatures, individuals of which are at times discovered in the tide pools or stranded dead or alive on the beach. Such a one, however, has usually passed the neophyte stage of the seashore naturalist, and his larger experience has familiarized him with the use of the compound microscope.

Indeed, the smaller plankton cannot possibly be studied without the aid of a microscope of at least a moderate power; the protozoans particularly are to be identified only by employing this instrument.

First and foremost among the pelagic protozoans to engage the marine naturalist's attention will be the rhizopods known as *Foraminifera* and *Radiolaria;* not because these are the most ubiquitous and numerous— for they are not,—but, because of their exceeding beauty and marvelous structure, they have been the most widey advertised. Foraminifers prevail in far greater numbers and over much wider areas than do radiolarians; but both forms flourish in tropical seas more commonly than they do in temperate waters. Actually, the first type of protozoan likely to be snared in the tow-net is one of the dinoflagellates, a group which contributes largely to the phosphorescence of the sea, and of which *Noctiluca* is its most famous member.

Other plankton animals there are besides the protozoans which require a high magnification for study; these are numerous copepods and other minute crustaceans, and crustacean larval forms that in season always swarm the seas. Eggs and the larvæ of various invertebrates will often be brought in with the same haul that captures the adult forms of such animals as free-swimming colonies of hydroids, pelagic worms and amphipods.

To the deeply submerged tow-net there will occasionally come such creatures as swimming-crabs and lobsters, depending on the range in which this device is used; but for obtaining true benthos forms, nothing excels the dredge. This apparatus, working as it does by literally scraping up the floor of the sea, collects every sessile and slow-moving object in its path. Sponges, sea cucumbers, starfishes, worms, crustaceans, mollusks and other forms that live beyond the limit of the lowest tides are usually easily obtained by dredging when all other methods fail.

There are some active creatures, however, that live on the bottom, and these require special devices to capture them; such are certain crabs and smaller animals attached to long fronds of seaweed which when torn loose from their moorings float out of reach of the dredge.

A description of the various collecting devices and of their use will be found in the following chapter.

III

COLLECTING

PLATE IV

MINIATURE MARINE ANIMALS

Pleurobrachia pileus (upper left); *Calocalanus pavo* (upper right); *Aeginella longicornis* (lower left); *Molgula arenata* (middle right).

CHAPTER III

COLLECTING

ALTHOUGH not all persons who have occasion to refer to this book will
be more than casually interested in the animals of the seashore, it is assumed that those who are especially attracted by the contents of this
chapter desire to become intimately acquainted with the manifold mar-
vels and modes of living exhibited by these lowly creatures; therefore,
for the benefit of the enthusiastic student, the various appliances required
for the proper prosecution of his work will be detailed in full.

Nevertheless, the truth must be admitted that one whose occupation
on the beach is merely that of collector, one whose work with one's
collections is merely that of mounting, arranging and labeling, can
hardly be considered a naturalist. The true naturalist is one who delights
in the study of living forms more than in the contemplation of their
preserved remains, who observes their development, growth and habits,
and takes account of their wonderful adaptations to their environments.
Such creatures as one uses in the home or laboratory in furtherance of
these studies, one will keep alive as long as it is expedient or possible
to do so; when one has no further use for them, one will return them
to their natural home, the sea. This is also to say that the naturalist is
a *real* lover of lower animal life, and is therefore a conservationist.

On the other hand, there do occur instances when in the otherwise
pleasurable pursuit of natural science, in order better to understand the
living, it becomes regrettably necessary to kill and to study the corpse.
On such occasions, the beginner will do well to err on the side of gather-
ing insufficient material for his purpose, rather than try to obtain a
sufficiency or oversupply. For as most forms destined to such sacrifices
on the altar of science are put directly into the killing-solution as soon
as collected, there is obviously no way of liberating surplus specimens.

Before proceeding in regular order with the details of paraphernalia,
a suggestion here may not be out of place regarding personal outfits. The
season of the year, as well as individual taste, indicates the type of clothing to be worn. Most modern women collectors—and many men col-

lectors, too—adopt the abbreviated style of breeches known as "Boy Scout shorts"; in summer, however, a swimming-suit will serve for either sex. (Plate V.) Footwear should always consist of rubber hip-boots or tennis-shoes; few collectors have feet so hardened that they can wade or walk barefoot with impunity over mussel beds or through dense crab-infested vegetation.

A good pocket lens (Fig. 1) of about 10- or 15-power magnification should be one's constant companion; it should be of the sturdy metal folding type; this may be attached to a lanyard of reasonable length and

Fig. 1—Pocket magnifier.

Fig. 2—Sheath-knife.

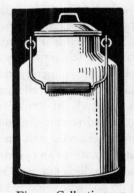

Fig. 3—Collecting can.

suspended around the neck, which not only prevents losing the lens but also keeps it in readiness to be easily and momentarily used. Such a lens is especially serviceable in the examination of hydroids and other forms of minute detail. A sheath-knife (Fig. 2) carried on one's belt, or, in lieu of this, a common sharp case-knife, is a useful accessory for detaching anemones and mollusks or for cutting away from timber the bases of fixed forms.

Such objects as shells, dried sponges and other non-living material may be collected in a basket; but for living animals, a receptacle containing fresh sea water is essential. The writer has found that the most generally efficient container used in his own work both on the shore and in offshore waters is an ordinary covered white porcelain milk can (Fig. 3). The capacity of the container will be determined by the amount and proportions of the material to be collected; however, the portability of a two-gallon can recommends its use for most work. Caution must be used in putting animals that are inimical to each other into the same receptacle. For smaller forms, mason-jars, bottles or vials will suffice; but care here also must be taken lest the animals succumb from lack of oxygen; the receptacles should be left unsealed until they are actually in transport from the beach.

LITTORAL COLLECTING

When arriving at the beach, it is perhaps the best practice to start by turning over the stones of various sizes nearest the low-tide level, or by examining the crevices and sheltered nooks of such rocks as may occur in the neighborhood, which are left uncovered for only a short period. Here the collector must be on the alert. Some forms will elude him by their agility; others by their protective resemblance to their surroundings. Indeed, some conform so closely in color and markings to that of their retreats that they will often escape the notice of even the trained observer. Such forms include all manner of anemones, worms and crustaceans which remain in hiding until the return of the next tide. Then, following this employment, attention may be given to the forms that live in the sand; the line of sea wrack (see Plate V) and detritus that marks the level of high tide should be examined last. Aside from perhaps a spade, no special equipment for obtaining specimens is necessary.

In the meantime, however, probably the most productive work is to be accomplished in the shallow water below the tidal zone and in the tide pools; and it is here that the real business of collecting begins.

1. *Indirect methods.* Numerous macroscopic (visible to the naked eye) organisms, although comparatively small, are discerned with difficulty in the shore water; most of them, however, will be found associated with seaweeds, and if these latter be collected and allowed to remain for a few days in water in small dishes (white porcelain photographic developing trays are ideal for the purpose), small cœlenterates, annelids, crustaceans and mollusks will appear in abundance. The rockweed (*Fucus*) (see Plate V) is a particularly likely plant for such forms. The procedure here, of course, is merely to gather the material

by hand. Objects such as old submerged shells, shell fragments and waterlogged driftwood also often repay careful scrutiny.

2. *The dip-net.* The handiest article in the wader's equipment is the dip-net (Fig. 4). This is used to great advantage in capturing large

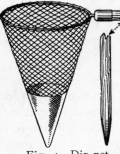

crustaceans, especially the active swimmers such as certain crabs and shrimps or prawns; large jellyfishes and comb jellies, too, are best caught and transferred from the sea to the containers; nor is it of mean advantage to its possessor when one chances to come upon a roving squid or a school of little sepiolas.

The dip-net should consist of a conical netted bag measuring about a foot in diameter at the opening and eighteen inches in depth, which is attached to a stout ring of brass or galvanized

Fig. 4—Dip-net.

iron, and is fixed to a wooden handle about seven feet in length. A lining of fine-woven cheesecloth sewed within the lower third will retain the more minute forms. The distal end of the handle might advantageously be pointed rather than blunt; this enables the net handle to be forced into the substratum to act as a marker in the event of some necessity for leaving a likely spot to obtain additional accessories, or it may be to leave both hands free for the work of examination.

3. *The seine.* For collecting invertebrates, the seine (Fig. 5) is by no mean indispensable. Still, by reason of the large area which these

Fig. 5—Seine.

nets are able to sweep, certain strong swimmers like squids and other cephalopods are caught which sometimes elude the smaller dip-net. Seines are made in lengths ranging from about twelve feet to more than a hundred; for the collector, however, the most convenient, and yet efficient, length is one measuring about forty feet. Seines such as are usually supplied have a depth of four feet, and are weighted with lead at intervals along the bottom margin, while at the upper margin is a line of cork or wood floats to keep the seine upright in the water. When received, a seine should be reinforced at each end with a strong light wooden bar or pole, wherewith the net may be maneuvered more easily. The meshes should number not fewer than four to the inch.

Seines are used to the best advantage on sandy bottoms. Two persons are required for the proper operation of a seine (see Plate VII), each of whom holds the bar at his end in such a way that the weighted bottom is kept in contact with the substratum while the floats are at or near the surface, with the net forming a trailing arc between the two operators. At the end of a haul, a landing should be made quickly and in such manner that the seine sags troughlike between the margins carrying the weights and the floats. Such material as is not to be retained by the collector should be returned to the water immediately.

All nets should be washed in fresh water and cleaned of seaweeds and other organic refuse after use, thus preventing premature decay. Before storing they should be thoroughly dry. Nets not treated to fresh water are, by reason of the salt they contain, hygroscopic, and the dampness thus resulting hastens their decay.

OFFSHORE COLLECTING

The term "offshore collecting" as used in this book is meant to indicate that method of collecting near the shore beyond the depths where wading is possible; it does not include collecting done in the outer reaches of the open sea, or in the abyssal regions of the deep sea. For such collecting as is discussed here, either a row-boat or small power-boat is fully sufficient; while the apparatus to be described will be of the simplest type, not beyond the ability of many persons to make, and within the means of most to procure.

Fig. 6—Water-glass.

1. *The water-glass.* Bottom-dwelling forms in shallow areas can best be observed by using a water-glass (Fig. 6), particularly when the surface is rough. This device consists of a box-like frame about one foot square and one foot deep with a pane of plate glass fitted into one of the open ends and the whole made water-tight.

With its glass end let into the water, the water-glass floats; and by its use objects below stand out as clear and distinct as if seen through the glass of an aquarium.

2. *The seaweed grapple*. Submerged vegetation, with its attendant animals, which is not easily obtained by other methods may frequently be secured by trailing behind the boat an apparatus known as "Pieter's plant grapple" (Fig. 7). It is made by passing eight or more galvanized steel wires through a one-foot length of 1½-inch galvanized pipe, bending back the wires at one end in the form of hooks, and twisting the wires at the other end so as to form a loop or eye for the attachment of a line. The inside of the pipe should be filled with lead to give added weight and to consolidate the several parts.

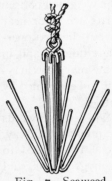

Fig. 7—Seaweed grapple.

3. *The tow-net*. Animal plankton is collected with a tow-net (see Plate V). This consists of a muslin cone attached to a hoop at its mouth and having at the apex of the cone a small glass receptacle (see Plate VI); the net is trailed behind the boat on a line which is tied to a number of leaders affixed to the hoop. The size of the tow-net varies according to the towing-power available; but the dimensions here given are for a tow-net that may be used with an outboard motor-boat or only slightly less conveniently and efficiently with a row-boat.

This very valuable apparatus is probably the most inexpensive and easiest to make of all that compose the marine naturalist's equipment. The hoop is preferably made of wood—an ordinary wooden barrel hoop will serve excellently—and should be about fifteen inches in diameter. To this at frequent intervals around the circumference should be fastened by tie-strings the muslin cone measuring approximately fifty inches in length. At the apex of the cone, an opening large enough to pass around the neck of a test-tube or a small, wide-mouthed bottle should be made by cutting off the tip; around the circumference of this small opening a draw-string should run, leaving ample length to the draw-string for tying a bow-knot when the bottle is secured in place. The leaders may be made from clothes-line and should number not fewer than six, equal in length and each about a yard long.

When making a towing, cast the net so that the bottle end falls clear of the bag and does not foul the line or leaders, and let it trail behind the boat (see Plate VI) at whatever distance is compatible with the depth from which towings are desired. The speed of the boat will determine the length of rope to be let out; for surface towings, not more than a few feet of line are necessary. Haulings should be made by lifting the net clear of the water and allowing the contents to settle to the bottle. This may then be untied and the organisms transferred to a separate receptacle.

4. *The tangle.* This ingenious device (Fig. 8) is used as a fouling snare to collect echinoderms and occasional spiny crustaceans living on the bottom. A simple form of it may be made out of a common household mop such as is composed of long, loose cotton strands. The strands should be frayed out, the handle removed and a weighted line attached for trailing. The method of using it is as simple as its construction: it is merely trailed along the bottom: it captures such spiny creatures as it may come into contact with by entangling them in the strands.

Fig. 8—Tangle.

5. *The dredge.* For collecting bottom-dwelling forms, nothing exceeds the efficiency of the dredge (Fig. 9). There are several types and sizes of this apparatus in use by various workers, but the one here described will meet all the requirements of the average collector.

As the bottom over which the dredge travels is oftentimes rough, it is absolutely essential that the materials entering into its construction be of the stoutest and most durable that can be conveniently employed. A triangular frame is made for the mouth of a heavy tow-net out of three lengths of 75° angle-iron, each measuring about a foot long and 1½ inches over the face of each angle and bearing a number of perforations along the face of one angle for tying the net. These angle-irons are bolted together in such a way that they are connected with one end of the three supporting rods, each about forty inches long. The opposite end of each rod is secured to its fellows with a narrow piece of strip-iron. From the illustration it will be seen that such a dredge is constructed so that no matter what side rests on the bottom the triangular framework at the mouth presents a broad scoop prepared to bite into the substratum when the dredge is trailed.

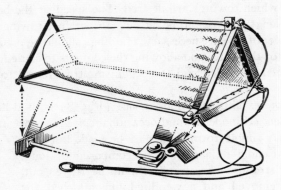

Fig. 9—Dredge.

As the dredge operates by literally scooping off a layer of the floor together with such plants and animals as may rest on the bottom or burrow into it, some means must be provided for sorting out the contents of a haul without dumping the material directly on deck or on the bottom of the row-boat, as the case may be. For this purpose nothing is better than an ample square of waterproof canvas or a piece of tarpaulin, which can be emptied overboard and trailed for washing after each haul.

INDOOR AQUARIA

To the reader who has kept only fresh-water aquaria, the salt-water tank will be a revelation. Not only are many of the forms that are amenable to confinement unsurpassed in beauty, but also their ways and habits are so diversified that observation of them has a never-ending charm. It can not be denied, however, that the successful maintenance of a sea-water aquarium is to some extent a matter of experience; that is to say, a sea-water tank is not so easy for the beginner to keep as is one of fresh water. But on the other hand, the keeping of one is not hard to learn, and when once the knack has been acquired such trifling attention as such a tank requires will be regarded as a diversion rather than a task. Nor is there any good reason why the inland dweller should lack the privilege of continuing his acquaintance with certain of the creatures met with on a visit to the seashore. In one's own home far removed, one may continue to keep many forms transported thence in small containers without further resort to the sea for water after the initial supply is obtained.

Sea water is certainly to be preferred for the aquarium when it can be secured; but in place of this it can be artificially prepared. If a known weight of sea water is evaporated to dryness and the remaining solid residue of sea salt is weighed, it will be found that this residue forms about three and one half per cent. of the original weight. Now, if the evaporation has been conducted very slowly, the residue will be crystalline in structure, and a microscopic examination will reveal crystals of various sizes and shapes, but the greater number of these will be cubical in form: these cubical crystals consist of common salt (sodium chloride), which composes approximately three fourths of the entire residue, while the remainder of the three and one half per cent. is composed mostly of various salts of magnesium and calcium. Specifically, an average 1000 grams of sea water contain

Sodium chloride (NaCL)	27.213	grams.
Magnesium chloride ($MgCl_2$)	3.807	"
Magnesium sulphate ($MgSO_4$)	1.658	"
Calcium sulphate (Ca SO_4)	1.260	"
Potassium sulphate (K_2SO_4)	0.863	"
Calcium carbonate ($CaCO_3$)	0.123	"
Magnesium bromide ($MgBr_2$)	0.076	"
	35.000	"

Thus, it will be seen from this table that out of every hundred pounds of sea water there is a solid residue of about three and one half pounds; therefore, if sea salt can be obtained, an artificial composition of seawater can readily be made by adding three and one half pounds of this salt to ninety six and one half pounds of water. To determine the weight of the water for the aquaria, it need only be remembered that a pint weighs one pound. If, however, sea salt can not readily be obtained, the following formula may be used, purchasing the different salts separately:

Water .... 96½ pints, or 12 gals.
Sodium chloride (Common pure salt) ...... 43¼ ozs.
Magnesium chloride .... 5¾ "
Magnesium sulphate (Epsom salts) .... 3¾ "
Calcium sulphate (Plaster of Paris) .... 2¼ "

Notwithstanding that certain substances contained in natural sea-water in very small quantities are missing in the foregoing formula, the mixture will serve about as well as artificial sea-water made from pure sea salt, and therefore it may be used whenever sea salt or sea water are obtainable.

An aquarium, to be successful, must have some means of maintaining a constant supply of fresh oxygen as fast as this vital necessity is used up by the inmates. There are several methods of such maintenance: the water may be changed at frequent intervals or it may be replenished from a stock reservoir by a circulating pump; aëration may be accomplished by pumping a stream of air from a fine jet or by introducing a sufficient quantity of seaweeds. As all except the last-mentioned are either troublesome or expensive, and of superior value only in the case of very large tanks, further discussion of them will be omitted.

One reason that the salt-water tank is more difficult to keep than the fresh-water tank is because, as a rule, seaweeds do not supply oxygen to the water in the quantities that fresh-water plants do; moreover there are many seaweeds that will not thrive in small tanks, and there are some whose presence seems positively poisonous. Red seaweeds, generally speaking, should never be selected for this purpose; nor should many kinds of olive or purple plants be employed. One seaweed that may be used with utter reliance is the "sea lettuce" (*Ulva latissima*. See Plate V). This plant with its broad, crinkly, glistening green fronds is ornamental as well as most admirably adapted to the aquarium. It is common on all shores, and it may be collected in sheltered situations and quiet tide pools where it grows into great expansive sheets a yard or more long. For the tank, specimens of such size should be selected as will not unduly crowd their quarters. It will be found that the individual

PLATE V

1—Examining sea wrack at the high tide mark.   2—*Fucus*.   3—*Ulva*.
4—Collecting in a tide pool.

PLATE VI

1—Trailing the tow-net.　　　2—Bringing in the haul.

3—Closeup of tow-net haul.

plants are usually attached to some solid surface; therefore it is best, if one has a liking for the decorativeness of one's aquaria, to use several smaller fronds, each fixed to some such anchorage as a shell or pebble, rather than a large single frond. Such smaller seaweeds lend themselves to a more suitable arrangement, with their fronds lifting gracefully upward; whereas larger seaweeds have a tendency to fold upon themselves unattractively, and in addition interfere with their own light. Although the utility of the following mentioned plants may be doubtful, their beauty alone recommends them to those who desire a variety of forms. These are common, and easy to collect : *Ceramium rubrum* (Plate VIII), a filamentous red seaweed with clawlike ends; *Rhodomela rochei,* (Plate VIII), a delicate feathery red-brown seaweed; *Chondrus crispus* (Plate VIII), a dense-growing purplish seaweed, sometimes called "carrageen" or "Irish moss"; *Enteromorpha intestinalis* (Plate VIII), a graceful green seaweed with tubular fronds.

Good light is necessary to a well-managed aquarium. While direct sunlight is not harmful in itself, a tank remaining in such light soon reaches a temperature that is deleterious to plants and fatal to animals. And as the capacity of the water to absorb oxygen is partly dependent on the temperature, an aquarium should always be kept in the coolest place possible that is yet consistent with good lighting. Mention of the absorbent capacity of water leads us directly to another and important matter regarding the proper maintenance of the indoor aquarium. It is known that a very salty solution limits the capacity of water for absorbing oxygen much in the same way as does temperature; consequently, it is plain that the normal evaporation of water in the salt-water aquarium should be compensated for by the addition, not of more salt water, but of fresh water only.

Aquaria should be rectangular in construction; they may consist entirely of glass, molded in one unit, or they may be composed of separate sheets of glass reinforced with metal corners. The latter type is preferable, as the molded tanks, although admittedly leak-proof, are practically worthless for photographic and microscopic purposes, the uneven surface of the glass distorting the image of small details. The globular aquaria commonly sold in stores should never be used; besides restricting the surface area of the water, thus limiting the absorption of atmospheric oxygen, this type does not lend itself well to visual and optical work. For containing very small creatures to be studied, cylindrical glass jars or finger-bowls will serve. (See Plate VIII.)

With the plants properly in place, a sprinkling of gravel and shell fragments should be strewn very sparsely over the bottom, with perhaps the addition of one or two sizable stones placed in the tank for the support of anemones or the shelter of crustaceans. It is important to remember that the substratum of a salt-water aquarium, unlike that of a fresh-water aquarium, cannot be composed of a thick layer of material without soon becoming black and unattractive and even foul with sulphuretted hydrogen gas throughout the entire layer. Therefore the material used should be so thinly distributed that the light can penetrate the interstices to the bottom. Care, of course, must be used when stocking the tanks, so that there will be no animals disproportionate in size or numbers to the volume of water. As a rule, not more than three animals the size of the little hermit crab, *Pagurus longicarpus,* should be kept in less than a gallon of water. Nor should destructive or predacious species be confined with less ruthless ones. Nor, again, should any unconsumed food be allowed to remain in the tank; cleanliness is one of the great secrets of keeping a healthy aquarium.

Further details regarding the selection of animals for the aquarium must be left to the discretion of the reader. Indeed, it is not possible by written advice to guide the beginner infallibly in the matter of choosing and maintaining an aquarium colony; only experience can combat the various causes of loss and failures that may from time to time occur. If, however, the few hints given here be faithfully acted upon, there should be no other losses than those accruing from an injudicious selection of animals; for it must not be forgotten that there are many creatures which are utterly unable to live more than a few hours in the indoor aquarium; but the knowledge of these will increase with acquired experience and familiarity with their various habits.

PERMANENT COLLECTIONS

Specimens composing permanent collections are usually either dried or preserved in ethyl alcohol or in formalin. Starfishes, chitons, crabs and many other hard or spiny-skinned creatures are best preserved dry; soft-bodied animals invariably require immersion in some appropriate preserving fluid. But before any type of animal can be preserved by one of the preceding methods it should be properly killed by first using one of the following killing-agents: weak alcohol, corrosive sublimate in hot or cold saturate solution, acetic acid, or some such solution as Perenyi's fluid, the formula for which is

Nitric acid 10% .... 4 ozs.
Chromic acid ½% .... 3 "
Alcohol 90% .... 3 "

Many organisms, however, contract violently on coming into contact with these reagents, and they must first be rendered insensible by being anæsthetized before being introduced into the killing-fluid. The anæsthetizing agents ordinarily used are chloral hydrate, cocaine, chloroform, weak alcohol and chloretone.

Prolonged immersion in the preserving fluids now commonly used has a tendency to harden the tissues. In some instances this is advantageous, as collapsible specimens present an unattractive appearance in cabinet jars; moreover, certain soft tissues are troublesome for anatomical purposes, and are harder to study and to dissect.

The most generally useful, and indeed the most indispensable, of reagents is alcohol. As pure ethyl alcohol is practically unobtainable by the average collector, the denatured product, containing 10 per cent. of methyl alcohol, will serve as a fairly efficient substitute. For preserving delicate, transparent animals, however, it is advisable to use only purified spirit which has been filtered and diluted with distilled water. Frequent changes of alcohol are often necessary on account of discoloration; changes should be continued, though, until the fluid no longer shows a tendency to discolor. To prevent contraction of large soft-bodied animals, it is advisable to immerse them first in weak alcohol and transfer them every few days to mixtures of increasing strength; that is to say, start with about 30 per cent. alcohol and make the transfers to 40, 50, 60 and 70 per cent. solutions in turn. Alcohol of 70 per cent. is the most desirable for permanent preservation of specimens. A stronger solution is not needed, and can actually be harmful by hardening the objects too much and rendering them so brittle that they may be broken by ordinary handling.

Formalin, or formaldehyde, is a good temporary preservative, but is not to be recommended for use with permanent specimens. It is much less costly to employ and is less bulky than alcohol, and therefore its use is advisable on long journeys or extended seaside excursions away from the home or laboratory. For gelatinous animals like the large jellyfishes, the solution should be about three per cent. strength; and for animals somewhat greater in consistency, such as ascidians or squids, a six per cent. solution is sufficient. The solutions may be made either of fresh or of salt water, but the sea water solution better preserves the transparency of gelatinous organisms than does the other. Non-contractile forms may be killed by putting them directly into the solution, but contractile forms must first be narcotized. Echinoderms, shell-bearing mollusks, and other calcareous organisms should not be kept long in formalin, since all calcareous parts are attacked by the free acid, although this free acid may be neutralized to some extent by sodium carbonate. Colors keep their brilliancy for a longer time in formalin than in alcohol, but those that are fugitive in the one will eventually disappear in the other also.

Chromic acid is a useful reagent, serving especially well for killing and hardening gelatinous and soft-bodied animals. Specimens should not be allowed to remain in this fluid longer than is necessary, however, as they are likely to become too deeply tinged as well as rendered fragile. Solutions may be made with ordinary fresh water in the following proportions:

Chromic acid solution of 1 per cent. . . . . 20 parts
Concentrated acetic acid . . . . 1 part

The mixture will not keep long; when it has turned green, it is not suitable for use. After objects have been treated, they should be well washed before transferring to a permanent preservative; otherwise a precipitate will form and the animals will acquire a greenish hue. Care should be taken in making up the solution, not to allow any chromic acid crystals to touch the fingers since this acid is a powerful irritant poison.

Like the keeping of a salt-water aquarium, success in treating subjects for permanent preservation depends in the final analysis on experience, on the acquired skill as well as the native ability of the preparator.

METHODS IN MICROSCOPY

The employment of the microscope and the preparation of material to be studied by its aid are certain soon or late to engage the attention of the zealous collector. Although the microscope is by no means indispensable to the enjoyment of seashore life, that enjoyment is undoubtedly enhanced and its sphere extended by the occasional use of this instrument. But good microscopes are expensive. And the technical knowledge necessary to prepare permanent microscope slides is so patently within the province only of the specialist that their mention should properly have no place in this book. Yet, for the benefit of many beginning naturalists, certain hints here for their present or future guidance in this matter will be very much in order.

1. *The microscope.* This should, as a first consideration, be an instrument made only by one of the firms having a high, long-standing and an international reputation. The principal parts of the microscope (Fig. 10) are the *stand,* which includes all the mechanical features, and the lenses, which compose the *optical system.* For such persons as may find it more convenient to start with a less expensive type, it is advisable first to purchase a good stand to which later may be added such accessories and optical parts as one's purse will permit. Such a stand will consist of the heavy foot, or *base,* usually horseshoe-shaped; the upright *pillar,* having a joint that allows the instrument to be inclined; the *handle arm,* which supports the *body tube* and the focusing adjustments consisting of the *rack and pinion adjustment* and the *micrometer adjustment;* the *stage,* containing screw threads for attaching a *substage condenser;* and the *mirror,* provided with a plane and a concave surface. The optical parts for this preliminary outfit may consist of a 10x (ten-power) *ocular lens,* or *eyepiece,* at the top of the body tube; and an objective lens of 10x with an equivalent focus of 16 mm, which screws into the lower end of the tube; this combination giving a magnification of 100 diameters. With such an outfit for a beginning, the intending marine microscopist will have

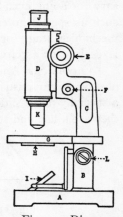

Fig. 10—Diagram of a compound microscope. A, base; B, pillar; C, handle arm; D, body tube; E, rack and pinion adjustment; F, micrometer adjustment; G, stage; H, substage screw threads; I, mirror; J, eyepiece; K, objective lens; L, tilting joint.

a good working start. After one has familiarized oneself with the performance of this optical system, the higher-power lenses that will afterwards be added will present no difficulties in operation. These added lenses should consist of two objectives, a 4 mm objective and an "oil immersion" objective of about 2 mm equivalent focus, together with perhaps two oculars of 5x and 12½x respectively. Thus fully equipped, the microscope will give a range of magnifications extending from fifty to about twelve hundred diameters or more, the latter being the limit of practical magnification in present-day instruments. For the perfect performance of these higher-power objectives, a substage condenser (which is a lens system for transmitting light) is essential. Although not necessary for doing even the best work, certain conveniences and refinements such as special nosepieces for the rapid exchange of objectives and a mechanical stage for the facile examination of slides, may be added later if desired or as circumstances permit. (See Plate VII.)

2. *Technique.* Animals to be examined under the microscope are either observed living, usually by being placed in cells such as a watch-crystal or a specially made slide (Fig. 11) containing a ground-out depression, or they are killed and stained and mounted on slides for permanent preser-

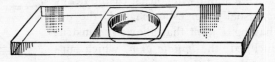

Fig. 11—Microscope slide with cell and coverglass.

vation. As to observing living protozoans, hydroids, minute crustaceans, etc., the type of animal will usually suggest the best method of handling to the manipulator; but regarding the preparation of mounted material, recourse must be had to the special literature and various manuals on the subject, some of which are listed in the bibliography at the end of this book. The subject is so vast that only a general outline of common methods can be given here.

Many solid organisms that have passed through the various preparatory stages have their tissues rendered so transparent that the details of both external and internal structures cannot easily be made out under ordinary observation, so microscopists have made use of certain stains to make such structures visible. The stains more commonly used are made from hæmotoxylin, anilin dyes, or coal-tar products. Some stains are selective in that they will stain certain tissues in an organism, leaving other tissues unaffected; thus it is possible to stain some objects in a way that will render its various tissues each in a different color. Stains are also either aqueous or alcoholic. Specimens fixed in alcohol are put directly into the alcoholic stain, but when placed in aqueous stains they are first washed with water. In either case they are again dehydrated in alcohol and finally placed in xylol or an oil that will mix with alcohol on the one hand and with some mounting medium, such as Canada balsam, on the other. The xylol or oil also acts as a clearing solution. This is then drained off and a slide covered with a drop of Canada balsam is prepared on which the object is placed. A clean cover-glass, consisting of a small extremely thin piece of glass, is then gently placed in contact with the balsam-immersed object, and the whole is then set aside to harden.

Soft-bodied organisms like infusorians, hydroid jellyfishes, etc., are not mounted ordinarily in Canada balsam, but are preserved in a medium that remains more or less viscid or actually liquid. To contain such media, and to prevent their evaporation, the cover-glasses are "ringed" with a rapidly hardening material.

UNDER-WATER PHOTOGRAPHY

Even without professional equipment, there are some ways in which the amateur collector can make satisfactory photographic records of those creatures he wishes to portray in their under-water surroundings.

Objects situated in clear, unrippled, shallow water only a few inches deep would commonly appear to present no difficulty to the photographer; yet when an exposure is made by the inexperienced person, it will often be discovered that the developed plate or film shows no trace of the desired image. This absence of an image in the negative is owing to the fact that the reflection of the sky has registered more strongly than have the more subtle details apparently so pronounced to the eye. To overcome the obscurity of such details, it is only necessary to erect some sort of screen that will shut off all incident counter-light and reflections. The area to be photographed must lie within the reflected image of the screen, not within its shadow. Such a screen (Fig. 12) may be made by utilizing a dark-colored cloth or a tarpaulin which is supported along one edge by a pole, this in turn being upheld by an upright at each end; the opposite edge of the screen may be staked or weighted with stones, inclining the cloth at an angle much in the manner of a lean-to shelter tent.

Fig. 12—Photographic screen to eliminate reflection.

For taking pictures in shallow water, either a small hand camera or a view camera will suffice; but it is obviously impossible to take close-up photographs unless one has an outfit with a very long bellows extension or is using a lens of very short focal length. Extreme close-up and low-magnification photographs are best produced in the laboratory, where there are good artificial lighting facilities and where there is more convenience for suitably arranging and grouping the subjects.

To photograph objects which are at a depth of several feet, in places where the water is clear and the bottom plainly visible, the water-glass is used in conjunction with the camera. Although a small hand camera can be used, the most convenient type is a reflecting camera. With the latter, especially when working in a drifting boat, the image can be better framed and more accurately focused and watched up to the moment of exposure; moreover, this type permits the operator to work with greater freedom.

The method of photographing consists merely in adjusting the camera within the frame of the water-glass so that the lens is as close as possible to the glass, thus avoiding sky reflections. Except in the shallower areas, the operation of photographing is limited to the making of long-shots and vertical views. For nearer and horizontal views taken at appreciable depths, a submerging camera and a diving equipment are required; but as these are properly part of the professional worker's outfit, a description of them will not be given here. However, it should be borne in mind that the value of most pictures made in deeper water levels is artistic rather than scientific; that is to say, owing to the refraction of

PLATE VII

1—Common types of salt water tanks.   2—Seining near the shore.
3—Low power binocular microscope.

PLATE VIII

1—*Ceramium rubrum*; 15x.   2—*Rhodomelia rochei*; 18x.
3—*Chondrus crispus*; 1x.   4—*Enteromorpha intestinalis*; ¾x.

the water and the quality of the light, all finer details are either diffused, distorted, or lost entirely; the method of photography that can render sharp and distinct objects removed a dozen feet or more from the camera has not yet been devised.

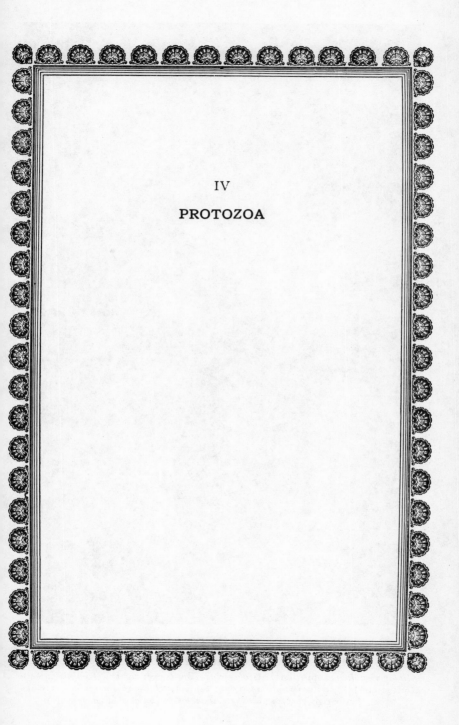

IV

PROTOZOA

PLATE IX

MARINE PROTOZOONS

Globigerina dubia (upper left); *Globigerina rubra* (upper right); *Orbulina universa*
(left center); *Noctiluca scintillans* (right center); *Heliosphaera inermis*, a radiolarian (lower right); *Amphilonche messanensis*, a radiolarian (lower left).

CHAPTER IV

PHYLUM PROTOZOA

(SINGLE-CELLED ANIMALS)

THE marine *Protozoa* are minute animals, each consisting of a single cell. Their distribution is world-wide; they swarm the littoral seas and are found elsewhere from the surface to the greatest depths that have yet been dredged. So small are they, however (indeed, they are for the most part microscopic), that their presence is seldom noticed except when their numbers in the case of certain species seem to turn the water red by day, or when other species cause the phosphorescence of the night.

In spite of the fact that the body of a protozoan consists of a single cell, the animal performs all the essential functions that characterize higher animals. This body, like any other animal cell, contains as its chief constituent a mass of protoplasm with one or more nuclei; and although special organs, as these are commonly understood, do not exist, there are present specialized structures that perform definite functions. For instance, in the simpler forms locomotion is achieved by thrusting out temporary extensions of the cell-body; these extensions, called *pseudopodia,* are projected in the direction of movement. For a similar purpose, higher forms develop permanent projections from the outer covering, or *ectosarc,* such as *cilia,* which are short and lashlike, or such as *flagella,* which are long and whiplike. In these latter instances, locomotion is accomplished by the vibration of the cilia or the flagella.

Nutrition takes place in the internal region called the *endosarc.* All sorts of organic matter compose the dietary of the protozoans. Among the lower forms, solid particles of food are simply taken in through the ectosarc at any region of the body; in certain higher forms, however, as in most infusorians, a well-developed mouth is present which communicates with the endosarc by a distinct gullet. Ingested food is usually surrounded by a fluid which forms the *food-vacuole,* the digestible material being absorbed and the remainder excreted. In some of the lower protozoans excretion is performed in the reverse manner of ingestion, that is, any point of the body surface will serve as a vent; but in the higher forms, indigestible portions are cast out through a permanent anal opening. Besides the food-vocuole, many protozoans possess a structure known as the *contractile vacuole,* the function of which is the extraction of waste fluids and gases. Numerous shell-bearing forms are able to raise or lower themselves in the water, owing probably to the presence of their distinct gas vacuoles; for they seem capable of altering the amount of carbon dioxide in these vacuoles and thereby of changing their specific gravity.

Respiration in the *Protozoa* is carried on by an interchange of gases through the ectosarc; thus, oxygen is taken in from the surrounding water, and carbon dioxide is transmitted to it by osmosis. Yet, unquestionably the symbiotic relationship that seems to exist between some protozoans and certain algæ has for its main ends functions that are respiratory as well as nutritive, the animal deriving oxygen and carbohydrates from the plants, and the plants getting carbon dioxide and nitrogenous waste from the animals.

Conjugation by the temporary or permanent fusion of individuals,—a reproductive process that in certain respects resembles the method of fertilization in the *Metazoa,* or many-celled animals,—is common among all classes of the *Protozoa;* but still more common is that simple mode of reproduction whereby the individual divides into two equal parts. In this process, the nucleus divides first, being followed shortly afterward by the remainder of the cell-body, thus forming two new individuals from the single parent-cell. A modification of the process of simple division is brought about in certain other forms by the production of spores; a mode that characterizes the *Sporozoa,* and which occurs occasionally in some other protozoans.

Although there are a number of forms among the marine *Protozoa* that may be seen and their identity determined with the aid of a good hand lens, a proper study of the group entails the use of a compound microscope. Owing perhaps to the fact that special technical apparatus is essential for the observation and study of protozoans, naturalists and writers generally have seldom attempted to popularize them. The truth is, this important division of marine animals, exclusive of certain forms that have attracted collectors mainly by reason of their beauty or other physical attributes, has received comparatively little attention from the professional zoölogists themselves; in short, with the exception of such notable shell-bearers as the foraminifers, radiolarians and dinoflagellates, the marine *Protozoa* have been perhaps the most neglected group in the animal kingdom. Here, then, is one department in which the amateur investigator can do much original research; and this with the certain knowledge that one's labor is not likely to be a duplication of that of many predecessors. At any rate, protozoan forms merit more than passing attention; constituting as they do a considerable element in the economy of the seashore, certainly no one who aspires to a complete acquaintance with seashore life will neglect to learn as much as convenience permits of their manifold ways and wonders.

There are about 8000 known species contained in the phylum *Protozoa,* and these include those of both fresh and salt water. The majority of species are marine; and the majority of these are radiolarians. The main groups of the *Protozoa* may be arranged according to the following classification:

PHYLUM PROTOZOA

CLASS **Sarcodina**   SUBCLASSES ..
{ Rhizopoda
Heliozoa
Radiolaria

CLASS **Mastigophora**   SUBCLASSES ..
{ Flagellidia
Dinoflagellidia

CLASS **Infusoria**   SUBCLASSES ..
{ Ciliata
Suctoria

CLASS **Sporozoa** (Parasitic forms; not within the scope of this book).

CLASS **Mycetozoa** (Terrestrial forms; not within the scope of this book).

CLASS **SARCODINA**

The marine *Sarcodina* are protozoans in which the protoplasmic body is usually without definite shape, and which move about by the protrusion of protoplasm in the form of lobed and fingerlike or tenuous and raylike processes, termed *pseudopods* (meaning false feet). Numerous species possess a rigid skeletal structure that acts as a support for the protoplasmic body and the pseudopods. Unlike the fresh-water *Sarcodina,* the marine forms are usually without a contractile vacuole. The entire group is characterized both by conjugation and by encystment.

By reason of their possession of pseudopods, together with having the utmost simplicity of structure and the greatest generalization of life processes, the *Sarcodina* are considered to be the most primitive of all animals. Most of them are so minute that they cannot be seen with the unaided eye. Those consisting of naked masses of protoplasm usually tend to a globular form when first placed under the microscope, but soon thereafter they resume their characteristic shapes. Nearly all of them are herbivorous, subsisting chiefly on bacteria, diatoms, and other algal material; some, however, are known to be carnivorous, as they have been seen feeding on closely related species. There are three subclasses.

Subclass **RHIZOPODA**

This division contains those sarcodinas that possess pseudopods having no central axial supporting filament such as distinguish the heliozoans; the pseudopods are usually lobose or slender and anastomosing. Both naked and shelled forms occur. There are two orders recognized. Upwards of 1300 species of rhizopods are found in salt water.

ORDER **AMOEBIDA**

Rhizopods without a shell and having simple pseudopods.

Family **AMOEBIDAE**

Genus *Amoeba*

A. guttala. (Fig. 13.) A minute club-shaped protozoan that moves about with the bulbous end to the fore. It is without distinct pseudopods,

and is extremely hyaline in appearance, having only a few large granules
in the swollen region. Common in de-
composing marine vegetation. Length
about 0.04 mm. Known to occur in
Vineyard Sound; probably frequent in
other ranges.

Fig. 13—A. guttala.

ORDER **RETICULARIIDA**

In this order are included the foraminifers, famed among microscopists and collectors of the minute. Members of the order are characterized by having branching and anastomasing pseudopods, and being either
with or without a shell.

FAMILY **GROMIIDAE**

GENUS *Gromia*

G. lagenoides. (Fig. 14.) This protozoan is pear-shaped, and its
body is covered with a tight-fitting, plastic, chitin shell, which in turn is

Fig. 14—G. lagenoides.

covered with a thin layer of protoplasm. The shell is hyaline and has
an opening at the larger end where
it turns inward, forming a tube.
The walls of this tube are thicker
than the rest of the shell. From the
thin layer of investing protoplasm,
fine, branching pseudopods are given off in all directions; while at the
shell-opening a mass of protoplasm collects, from which a dense network of pseudopods extends. These latter seem to be used as a snare
to catch smaller protozoans and diatoms. The densely and evenly-granular
protoplasm contains a spherical nucleus. Length of shell about 0.25 mm;
greatest diameter 0.12 mm. Found on the fronds of different kinds of
seaweeds. Woods Hole region; but probably occurs in other localities.

Family LITUOLIDAE

Genus *Haplophragmium*

H. canariense. (Plate X.) A free, shelled form in which the test is planospiral and composed of several coils, each coil being composed of a number of chambers. The shell wall is constructed of minute grains of sand and appears somewhat smooth owing to the cementing material. Shell about 0.75 mm across planospiral. Relatively common in offshore waters. Cape Cod to Cape Hatteras.

Family TEXTULARIDAE

Genus *Verneuilina*

V. polystropha. (Plate X.) This foraminifer has an elongate, tapering shell composed of small grains of sand smoothly cemented. The shell is spirally arranged and consists of a series of chambers. At the margin of the last-formed chamber is the aperture. The test is reddish brown with markings of irregular dots. Test about 0.7 mm long. Not uncommon. Offshore waters. Cape Cod to Cape Hatteras.

Family LAGENIDAE

Genus *Uvigerina*

U. angulosa. (Plate X.) A foraminifer with an elongate, spiral test consisting of numerous chambers. The wall is calcareous, smooth and hyaline, with perforations. The aperture is a short tubular neck at the end of which is a phialine lip. Length of test about 0.5 mm. Offshore waters. All ranges.

Family GLOBIGERINIDAE

Genus *Globigerina*

G. dubia. (Plate IX.) This seems to be a bottom-living species, as it is a common inhabitant of the mud dredged up in several seas. It has a calcareous shell composed of numerous inflated chambers arranged in a trochoid (wheel-like) coil. The chambers usually number about five or six, with reticulate walls. Diameter about 0.75 mm. A widely distributed species; particularly common in the offshore areas of the Florida coast.

G. rubra. (Plate IX.) In this species the test is composed of several inflated chambers which form an elongate trochoid spire. There are about three volutions, each with three chambers having reticulated walls and bearing spines. The animal is both pelagic and a bottom-dweller; in the latter habitat, however, the spines are less well developed. The test is pink in color in specimens taken in the southernmost ranges; this color is generally wanting in those farther north. Measures about 0.75 mm in length and 0.50 mm in diameter. Rather widely distributed, being particularly abundant in the path of the Gulf Stream.

Genus *Orbulina*

O. universa. (Plate IX.) When young this species has a test somewhat like that of *G. rubra* and other typical globigerinas, and might easily be taken for a member of that genus; later, however, in the adult form it develops a globular chamber about this early globigerinalike structure. The surface of the shell is finely reticulate with a small pit at the bottom of each reticulation, and it bears long radiating spines. About 1 mm in diameter. The species is pelagic. Common. Off the Florida coast.

Family **ROTALIDAE**

Genus *Truncatulina*

T. lobatula. (Plate X.) A calcareous species with a test coiled in such a way that the whole of the segments are visible on the superior side; the chambers, however, are seen from both sides. The walls are smooth and somewhat hyaline. About 0.8 mm over greatest diameter. A littoral species; very common in all ranges.

Genus *Rotalia*

R. beccarii. (Plate X.) This species has a shell composed of numerous chambers arranged in a flattened spire. All chambers are visible from the superior side, only those of the last-formed coil being visible below. The chamber walls are ornamented with raised bosses. Beneath the inner angle of the last-formed chamber is a narrow slit which forms the aperture. Measures about 1 mm over greatest diameter. This species is a littoral form and is probably the most abundant of all the foraminifers. It lives in the sand and mud. Common in all ranges.

FAMILY NUMMULITIDAE

GENUS *Polystomella*

P. crispa. (Plate X.) In the species represented here, the test is composed of numerous chambers arranged in a regular, bilaterally symetrical involute spire. These chambers extend back to the central, or so-called "umbilical region." The last-formed whorl is made up of twenty chambers. The test is calcareous and has the appearance of glazed white china. It has a diameter of about 1 mm. Not uncommon. Found in sand and mud in littoral and offshore waters. Cape Cod southward.

FAMILY MILIOLIDAE

GENUS *Spiroloculina*

S. limbata. (Plate X.) An imperforate form; that is, having no apertures in the test except the typical one in the last-formed chamber. The shell is oddly composed of chambers arranged planospirally, all being visible from opposite sides of the test and the whole having a roughly lenticular shape. At one end of the last chamber is a short necklike extension in which is the round aperture. The test is calcareous and resembles dull white porcelain. Its measurement is about 1 mm long and 0.75 mm across the chambers. Found frequently in shore sands, but occurs more commonly on deeper bottoms. Cape Cod southward.

GENUS *Orbitolites*

O. duplex. (Plate X.) This beautiful species has a discoidal test composed of numerous chambers that in specimens subjected to the abrasive action of the waves resemble a delicate lacelike structure of glazed china. The chambers are divided into chamberlets, each with an aperture on the rim. The early chambers are minute, but as the test increases in size, the chambers likewise become larger and are arranged in a gradually widening spiral of exquisite proportions. Diameter about 2 mm. This species is fairly abundant; it occurs on shore vegetation and in the mud and sand of offshore waters. Cape Hatteras to Florida.

SUBCLASS HELIOZOA

The protozoans included in this group may be either naked or shell-bearing, but they are of a typically spherical form, having little or no tendency to change their shapes by amœbalike movement. The pseudopods are fine and raylike, seldom changeable, are generally supported with axial filaments, and radiate from all parts of the body. The axial filament, as well as the skeleton in those forms containing shells, is composed of silica. The heliozoans reproduce either by cell-division or by spore formation; in the latter case, the spores are flagellate and lead an active swimming life; later the flagella are lost and the animal assumes the adult form. There are upwards of fifty known species, but most of them are found in fresh water, some in moist earth, the marine forms being comparatively few. Some species are known to occur in both fresh and salt water, the marine forms differing from the fresh-water forms only in the absence of contractile vacuoles.

SUBORDER APHROTHORACIDA

This suborder comprises the naked heliozoans. They are without a skeleton, but are provided with pseudopods that have axial filaments.

GENUS *Actinophrys*

A. sol. (Fig. 15.) A heliozoan that inhabits both fresh and salt water. It is characterized by having no distinction between the ectosarc and the endosarc. The body is spherical, and the pseudopods show granule-streaming. Diameter of body 0.04 mm; length of pseudopods about 0.13 mm. Not uncommon. Found in

Fig. 15—A. sol.

tide pools and among littoral seaweeds. Cape Cod to Cape Hatteras.

SUBORDER **CHLAMYDOPHORIDA**

Heliozoans with a soft or felted fibrous body-covering. Body usually spherical.

GENUS *Heterophrys*

H. myriapoda. (Fig. 16.) In this species, the body has a slight differentiation between the ectosarc and the endosarc. A thick gelatinous

coating envelops the body, between which coating and the body is sometimes another covering of bright yellow cells; the body also frequently contains chlorophyl cells, chlorophyl being the green coloring matter peculiar to plants. Locomotion is performed in a peculiar rolling manner. The body proper has a diameter of 0.02 mm; with the gelatinous covering about 0.035

Fig. 16—H. myriapoda.

mm; while the pseudopods are about twice as long as the body diameter. Rather common on shore algæ. Cape Cod southward.

SUBCLASS **RADIOLARIA**

No species of radiolarians are known to occur along the Atlantic Coast of the United States. They are found mostly in the deep sea. Still, as this book already contains an occasional reference to them, and as the uninformed reader may wish to gain some idea of their appearance, the siliceous skeletons of some common forms are pictured here. (Plate XI.) The living shelled animals are often of comparatively large size, of comparative complexity, and always of great beauty (Plate IX). These protozoans are truly the jewels of the sea. The body is divided into two regions, the central chitinous capsule and the surrounding extra-capsular skeleton of silica, the latter being a transparent glass-like structure usually of exquisite delicacy. They are similar to the heliozoans in having raylike pseudopods; and like *Heterophrys* individuals, they often contain yellow unicellular algæ (*zoöxanthellæ*). The beauty of the living animals is sometimes enhanced by the appearance of the extra-capsular region, which is frequently vacuolated, forming a spherical mass of bubblelike chambers. They have long been favorite objects of study with European microscopist-collectors. There are more than 700 genera containing nearly 4500 named and described species.

CLASS **MASTIGOPHORA**

Such protozoans as are included in this class are chiefly characterized by the possession of a motile apparatus in the form of one or more flagella. The body is almost invariably provided with a covering such as a shell or cuticle or membrane; but in many instances the latter is exceedingly delicate, and, as a consequence, the body is more or less amœbalike. No well-marked division into ectosarc and endosarc is usually present. The protoplasm frequently contains granules of green (chlorophyl), yellow or brown (diatomine) coloring matter. The organisms are mostly microscopic.

The method of food-taking differs among certain forms; in some the food, like that of green plants, consists of products derived from simple compounds by the protozoan itself; in others, the food, like that of fungi, consists of dissolved organic material; while in others still, the food is composed of solid proteid particles and other such assimilable substances.

This class was formerly known as the *Flagellata,* a name that still lingers to some extent in modern writings. The animals are inhabitants of both fresh and salt water, and numerous forms are parasitic in higher animals. Upwards of 350 species are known, which are grouped in two subclasses, each subclass having marine representatives.

SUBCLASS **FLAGELLIDA**

Mastigophorans composing this subclass are small organisms having a body with a distinct cuticle, thus giving them a definite shape; however, certain forms have so thin a covering that changes in shape often take place. The body has a definite anterior and posterior region; from the anterior end extend one or more flagella. Of the eight orders recognized in this group, four are treated in this work.

ORDER MONADIDA

Members of the *Monadida* are characterized by a simple structure, a body frequently amœbalike with one or two flagella at the anterior end. No distinct mouth-opening exists, but food particles are ingested in a localized area at the base of the flagella.

FAMILY CODONECIDAE

GENUS *Codoneca*

Fig. 17—
C. gracilis.

C. gracilis. (Fig. 17.) A small colorless creature contained in an urnlike structure that is borne on a filamentous stalk. The contained animal is variable in shape, sometimes assuming a spherical, and sometimes an ovoid form. It does not fill the cup-urn, nor is it attached to it by any filament. A single flagellum is contained at the end of the constricted part of the body passing through the "neck" of the urn. Total length of urn and stalk about 0.02 mm; length of urn alone 0.01 mm. Found attached to shore algæ. Vineyard Sound.

ORDER CHOANOFLAGELLIDA

These are the "collar" flagellates, the individual members of the group having one or more collarlike structures about the base of the flagellum.

FAMILY CRASPEDOMONADIDAE

GENUS *Monosiga*

M. ovata. (Fig. 18.) The animal is usually unstalked, but is sometimes provided with a very short stalk less long than the body length.

PLATE X

1—*Haplophragmium canariense*; 40x.   2—*Verneuilina polystropha*; 40x.
3—*Uvigerina angulosa*; 40x.   4—*Truncatulina lobatula*; 40x.
5—*Rotalia beccarii*; 40x.   6—*Polystomella crispa*; 40x.
7—*Spiroloculina limbata*; 40x.   8—*Orbitolites duplex*; 22x.

PLATE XI

1—Plankton from the gulf stream, with species of *Calanus* (top)
and of *Lucifer* (bottom); 50x.

2—Radiolarian shells; 70x.

Body spherical or ovate; broadest at base, and tapering at top. A funnel-shaped, hyaline collar surmounts the top of the tapered portion of the body, encircling the base of the flagellum. The collar length is about 0.03 mm; body length 0.05 mm. Occasionally found on shore algæ. Southern New England waters.

Genus *Codonosiga*

C. botrytis. (Fig. 19.) Individuals of this species are either single or colonial. They are provided with a long stalk measuring about 0.014 mm. In other respects they resemble *Monosiga* rather closely. Found on red shore algæ. Southern New England waters.

Fig. 18—M. ovata.

Fig. 19— C. botrytis.

ORDER **HETEROMASTIGIDA**

Flagellates with two or more flagella: one being directed forward and used in locomotion, the others directed backward.

Family **BODONIDAE**

Genus *Bodo*

B. caudatus. (Fig. 20.) The body of this organism is usually flattened and pointed posteriorly, but it may vary in shape, being somewhat amœbalike. It has two long flagella; the forward flagellum being a little longer than the one trailing in the rear. Length of body about 0.015 mm. Quite common among shore waters and tide pools. Southern New England waters.

Fig. 20—B. caudatus.

ORDER EUGLENIDA

The flagellates in this group are generally large with spiral markings on the outer surface.

FAMILY ASTASIIDAE

GENUS *Astasia*

Fig 21—A. contorta.

A. contorta. (Fig. 21.) This creature has a flexible, transparent, colorless body with an obliquely striated cuticle, giving it a twisted appearance. The shape of the body is variable, but is usually tapered posteriorly, while the anterior region is produced into a necklike extension which bears the flagellum. Length of body is 0.06 mm; greatest diameter 0.03 mm. Frequent in decaying seaweeds. Southern New England waters.

FAMILY PARANEMIDAE

GENUS *Anisonema*

A. vitrea. (Fig. 22.) An ovate form of flagellate that has eight furrowed surfaces running somewhat spirally from end to end. It possesses a flagellum that is directed forward from an oral opening, and one that trails in the rear.

Fig. 22—A. vitrea.

Length 0.05 mm; diameter 0.02 mm. Common in decaying vegetation. Southern New England waters.

SUBCLASS DINOFLAGELLIDIA

In the majority of forms, these flagellates possess a hard outer covering, or shell, having two grooves; one of these runs transversely, and in it lies a flagellum; the other runs longitudinally, and from it extends a flagellum that is directed away from the body. Some forms may be naked or protected only by a firm membrane.

ORDER ADINIFERIDAE

Those dinoflagellates without a transverse groove are included in this order. The shell usually consists of two valves.

FAMILY **PROROCENTRIDAE**

GENUS *Exuviella*

E. lima. (Fig. 23.) In this species the shell is ovate and thick, and is swollen posteriorly. The anterior border of both valves is slightly indented, from which region the flagella extend. The animal is dark brown in color. Moves through the water slowly. Length 0.04 mm. Southern New England waters.

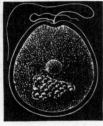

Fig. 23—E. lima.

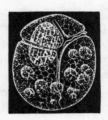

Fig. 24—G. gracile.

ORDER **DINIFERIDAE**

In this order are included those dinoflagellates with both a transverse and a longitudinal groove.

FAMILY **GYMNODINIIDAE**

GENUS *Gymnodinium*

G. gracile. (Fig. 24.) A naked dinoflagellate in which the body is divided by the transverse groove into a shorter anterior and a longer posterior region. The body, except for a slight flattening dorsoventrally, is almost spherical. Each groove carries its characteristic type of flagellum. The endosarc is evenly granular and usually contains a number of large ingested food bodies which can be plainly seen. The color is brownish. Length of body 0.068 mm. Common. Southern New England waters.

Family NOCTILUCIDAE

Genus *Noctiluca*

N. scintillans. (Plate IX.) This naked dinoflagellate is perhaps the most widely known protozoan extant. It is one of the commonest causes of the nightly phosphorescence of the sea. It is also one of the largest protozoans. The body is somewhat kidney-shaped to spheroidal. The endosarc is considerably vacuolated, with strands of protoplasm extending from the ectosarc to the central mass containing the nucleus. A large deep oral groove runs around the circumference for about a fourth of the distance. From the groove arises a tentacle, more or less mobile, which is directed backward. *M. scintillans* is the only species of the genus. This organism is sometimes classed as a cystoflagellate bearing the name *M. miliaris.* The size is variable, measuring from 0.2 mm to 2.0 mm in diameter. One of the most cosmopolitan of marine organisms. Its distribution is world-wide. Common in all ranges.

Family PERIDINIIDAE

Genus *Peridinium*

Fig. 25—P. digitale.

P. digitale. (Fig. 25.) A dinoflagellate characterized by the numerous pits distributed over the surface of the spherical shell. The shell has an oblique furrow separating the anterior from the posterior region. The former region bears two spines of unequal size; the latter, a single spine. The visible nucleus is situated in the posterior half of the shell. Length about 0.07 mm; diameter 0.66 mm. Not infrequent. Southern New England waters.

Genus *Ceratium*

C. tripos. (Fig. 26.) A fairly large species with a body as broad as it is long. The body is covered with conspicuous pores, is somewhat triangular, and bears three strong horns, two of which are shorter than the third and slightly curved upward. The two short horns are pointed and closed at the tips, the other is blunt and open. The species is very variable in form, the relative shape and proportion of the horns not being constant. Length about 0.3 mm. Very common in littoral waters and the open sea. All ranges.

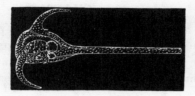

Fig. 26—C. tripos.

C. fusus. (Fig. 27.) Another very common species causing nightly phosphorescence of the water. The organism is very elongate owing to the alignment of the two horns at the ends of the body. Length about 0.29 mm. Color, yellow. Found frequently in shore algæ as well as in the open water. All ranges.

Fig. 27—C. fusus.

CLASS INFUSORIA

Protozoans included in the marine *Infusoria* are the most highly specialized of the entire group. Locomotion and capture of food is effected by special organs, the cilia, which are minute processes or hairlike extensions of the ectosarc through the cuticle. The cilia differ from flagella in their relatively shorter length and greater number; they also differ in their mode of movement, which is rapid, vibratory and often rhythmic. The body is invariably limited by an ectosarc; and there are always two nuclei, specialized in size and in function, which are termed respectively the *micronucleus* and the *macronucleus*. The *Infusoria* subsist on food composed of organic material of all kinds; however, there are some whose principal food is animal matter, some live upon plants, some are scavengers of decaying material, and some are parasites. All forms, with the exception of those that are parasitic, possess a mouth and gullet; the anal opening is usually temporary.

Reproduction is accomplished by simple division, by budding, or by conjugation.

Two subclasses are generally recognized.

SUBCLASS CILIATA

In this subclass are those infusorians that are provided with cilia during the adult as well as during the embryonic stage. Typical reproduction is by simple transverse division. A mouth and gullet are present in most of the marine forms, of which there are over 400 species.

ORDER HOLOTRICHIDA

Ciliate infusorians in which the cilia are usually similar and distributed evenly over the surface of the body; those near the mouth usually being longer.

FAMILY ENCHELINIDAE

GENUS *Lacrymaria*

L. lagenula. (Fig. 28.) A flask-shaped infusorian with a body two or three times as long as broad, bearing at one end a slightly elastic, conical proboscis. The surface of the body is obliquely striated with about fifteen well-defined lines. The cilia are uniformly distributed on the body; those at the base of the proboscis are longer. The anterior region of the endosarc is usually filled with larger granules, while the posterior half contains numerous smaller granules. Size variable; length about 0.16 mm; greatest width 0.06 mm. Fairly frequent in decaying seaweeds. Southern New England waters.

Fig. 28—L. lagenula.

GENUS *Trachelocera*

T. phoenicopterus. (Fig. 29.) This curious form has an extremely long and ribbonlike body capable of considerable extension and re-

Fig. 29—T. phœnicopterus.

traction. It is a variable form in that the anterior end in some specimens is square, in others it may be cylindrical; again, in certain individuals the posterior end is pointed, while in others it is rounded. The body is striated with fine, longitudinal markings. The nucleus appears fragmented with the parts distributed throughout the protoplasm. The macronucleus may be either a round unit, or in two parts, or many scattered parts. Length 1.7 mm. Found occasionally in plant débris. Southern New England waters.

Family TRACHELINIDAE

Genus *Loxophyllum*

L. setigerum. (Fig. 30.) A species in which the body is flattened, and the general outlines somewhat irregular. The anterior end is obtusely pointed; the posterior end is rounded. Ventral surface only is ciliated, the dorsal surface merely being striated longitudinally and faintly marked radially. Numerous

Fig. 30—L. setigerum (side view).

setæ, or strong bristle-like processes, surround the margin except at the anterior end. Viewed from the side this organism has a pronounced hump rising from a flattened peripheral or marginal extension. The body is very plastic, folding the marginal extension over the objects on which it creeps. Length about 0.1 mm; breadth 0.05 mm. Sometimes found in decaying vegetation. Southern New England waters.

Genus *Lionotus*

L. fasciola. (Fig. 31.) The body of this infusorian is somewhat sigmoid in shape with a tapering anterior end, the extremity of which

Fig. 31—L. fasciola.

is turned dorsally; the posterior region is plumper than the rest of the body, and is bluntly pointed. Cilia are present only on the under side; however, a row of large cilia mark the long oral opening along the lower left side. Color, brown or bright yellow. Length, variable; up to 0.6 mm. Occasionally found on shore algæ. Southern New England waters.

FAMILY **CHLAMYDODONTIDAE**

GENUS *Nassula*

Fig. 32—N. micro-
stoma.

N. microstoma. (Fig. 32.) A species distinguished by its small mouth. The ciliated body is plump and somewhat cylindrical, being rounded at both ends. The small mouth is marked by the longer cilia at a slight depression on the surface. The cuticle is very firm and decorated with longitudinal, spirallike rows of cilia and *trichocysts,* or hair-cells. Color, yellowish brown. Length 0.05 mm; diameter 0.03 mm. Found on shore algæ. Southern New England waters.

FAMILY **CHILIFERIDAE**

GENUS *Uronema*

U. marina. (Fig. 33.) A small ellipsoidal animal about twice as long as it is broad; longitudinally striped and covered with cilia. Mouth located in upper half of body. An undulating membrane, with longitudinal striæ and covered with vibratile cilia, is present at the mouth. The distinguishing feature of this species, however, is the long, delicate bristle at the posterior end of the body, being about two thirds of the body-length. Length of body 0.05 mm; width 0.02 mm. Common on decaying algæ. Southern New England waters.

Fig. 33—U.
marina.

FAMILY **PARAMECIDAE**

GENUS *Paramecium*

P. aurelia. (Fig. 34.) The slipper animalcule. An infusorian characterized by an elongate body, the forward end rounded and the hinder end bluntly pointed. A long, deep oral groove leads to the mouth, which is in the middle or posterior region of the body. Two micronuclei are present. Length about .15 mm. Common. Found in decomposing organic matter. Cape Cod southward.

Fig. 34—P. aurelia.

FAMILY OPALINIDAE

GENUS *Anoplophrya*

A. branchiarum. (Fig. 35.) A cylindrical or pear-shaped body, the surface of which is distinctly marked with somewhat spirallike longitudinal striations that take the character of depressions, are the chief distinguishing features of this protozoan. Long cilia are present over the entire surface; but these are relatively wide apart and inserted along the longitudinal stripes. Nucleus longish, somewhat curved, and coarsely granular; micronucleus lies in concavity of curvature. Length of body 0.1 mm; diameter 0.04 mm. Found among algæ. Southern New England waters.

Fig. 35—A. branchiarum.

ORDER HETEROTRICHIDA

The infusorians included in this order are characterized by a uniform covering of cilia and an *adoral zone* (region near the mouth) along the oral groove consisting of short cilia fused together into *membranelles,* or diminutive membranes.

FAMILY BURSARIDAE

GENUS *Condylostoma*

C. patens. (Fig. 36.) A species with an elongate, somewhat sac-like body five or six times as long as it is broad. The body is plastic, frequently containing brightly colored food granules. The forward end is largely taken up by the triangular *peristome* (around the mouth),

Fig. 36—C. patens.

the mouth itself being at the sharper angle of the triangle. Longitudinal striations mark the cuticle, the markings running somewhat spirally and having fine cilia inserted in them. Along the right edge of the peristome is an undulating membrane. Locomotion is effected by the membranelles which are ranged along the left edge of the peristome and the front edge of the body. The nucleus is a long, singular structure resembling a string of beads extending the full length of the left side. Arranged in the same region are a number of micronuclei. Length 0.4 mm; diameter 0.1 mm. Common on decomposing algæ. Cape Cod southward.

FAMILY TINTINNIDAE

GENUS *Tintinnopsis*

T. beroidea. (Fig. 37.) A medium-sized ciliate possessing a colorless, chitinous house, or protective case, with embedded sand grains. The case is thimble-shaped and rounded at the hinder end, and the animal is attached to the bottom by a peduncle. At its forward end the animal has two complete circlets of cilia, one of which, the outer, forms the adoral zone and is composed of twenty-four thick, tentacle-like membranelles; the other circlet is composed of shorter cilia within the adoral zone. The entire body is ciliated, but, unless the animal can be made to leave the house, this is not observable owing to the opacity of the case. The body is cylindrical. Length 0.05 mm; diameter 0.04 mm. A free-swimming, sand-loving form. Vineyard Sound.

Fig. 37—T. beroidea.

ORDER HYPOTRICHIDA

This order includes those ciliates that have the cilia only on the bottom side of a dorso-ventrally flattened body.

FAMILY EUPLOTIDAE

GENUS *Euplotes*

E. harpa. (Fig. 38.) An oval, rigid ciliate, quite flat on the ventral surface, but decidedly arched and longitudinally ridged dorsally. On the frontal and median ventral surface are ten large *cirri,* or filamentous appendages; in the rear are five similarly stout anal cirri together with a number of smaller and finer marginal cirri. A long, broad peristome is present, bearing an adoral zone consisting of a row of membranelles arranged in a continuous curve extending from the mouth to the extreme right frontal margin. It is a rapid swimmer, but it is also given to creeping, especially when it pauses to go over

Fig. 38—E. harpa.

organic material in search of food particles. Body length 0.1 mm; width 0.05 mm. Found on shore algæ. Southern New England waters.

GENUS *Diophrys*

D. appendiculatus. (Fig. 39.) This form is somewhat like *E. harpa* except for the character and number of its cirri and a considerable indentation at the hinder end on the right side. There are 7 great cirri in the frontal median region; the anal cirri are 5 in number and are huge and powerful, extending a considerable distance beyond the posterior of the body. The peristome is defined by the adoral zone and bears on its right border a row of cilia, a similar row running along the base of the membranelle. Locomotion the same as that of *E. harpa.*

Fig. 39—D. appendiculatus.

Length 0.05 mm; width 0.03 mm. Found occasionally on algæ. Southern New England waters.

GENUS *Uronychia*

U. setigera. (Fig. 40.) A ciliate readily distinguished from the two preceding forms by the absence of anterior cirri and the strikingly larger cirri at the posterior end, these being relatively enormous.

The body is ovoid, the hinder region being somewhat wider. The flat

under side of the animal has two hollows in the posterior end, the right one of which is the larger and contains 5 large cirri of unequal size : the left hollow contains 2 cirri, also unequal. Above the 5 right cirri are 3 curved cirri with their points to the left. A wide and open peristome with a small hollow on its left border indicates the region of the mouth. Length 0.04 mm ; width 0.03 mm. This is a very common species in decomposing algal débris. Southern New England waters.

Fig. 40—U. setigera.

ORDER **PERITRICHIDA**

The animals in this group are cup-shaped or cylindrical. Except for the cilia forming the adoral zone at the oral groove (a few forms may have a secondary row of cilia at the hinder end), the body is bare of ciliary processes. Most members of this order are *sessile,* or fixed.

FAMILY **VORTICELLIDAE**

GENUS *Vorticella*

V. patellina. (Fig. 41.) A bell-flower animalcule. The chief characteristics of this and all other members of the genus are the belllike body supported at the end of a contractile stalk which is attached to some solid object, and the absence of colony formation. In the present species the body is distinctly campanulate, or belllike, widest at the rim, from which it tapers to the pedicel, or stalk. The rim is marked by a row of rapidly vibrating cilia (the adoral zone). A singular feature of the pedicel is the manner of its contraction when the animal is subjected to shock or irritation, which resembles a coil or spiral spring released from tension. The stalk contains a contractile thread, which is very visible in the living animal. The macronucleus is long and looped, with attached micronucleus. Length of

Fig. 41.— V. patellina.

body 0.05 mm. Common. Found attached to algæ. Cape Cod southward.

V. marina. (Fig. 42.) This form has all the general characteristics of *V. patellina,* except that the bell is more conical and that it is transversely marked with a series of rings. Length of body 0.03 mm. Found on algæ. Cape Cod southward.

Genus *Zoothamnium*

Z. elegans. (Fig. 43.) A colonial form; however, the individuals are few in number, usually three or four. The bodies are belllike, but variable. The peduncle, which is smooth, slender and transparent, branches rather sparsely at the distant end. The so-called "ciliary disk" extends beyond the rim of the bell. Length of body 0.08 mm. Found on algæ. Cape Cod southward.

Fig. 42—
V. marina.

Fig. 43—
Z. elegans.

Fig. 44—
C. crystal-
lina.

Genus *Corthurnia*

C. crystallina. (Fig. 44.) In a general way this form resembles *Vorticella* species; however, it differs in certain specific details such as the possession of an elongate shape and the fact that it occupies a cuplike house. This protective covering is variable, sometimes being cylindrical or thimble-shaped or pouch-shaped, or, again, corrugated or smooth on the sides, and wavy and smooth on the border. The house may be attached to an object either directly or by means of a short stalk. By reason of its great contractibility, the animal can extend half of its length out of the cup or retract well within the interior. Length of cup 0.07 mm to 0.2 mm. Found frequently fixed to algal fronds. Southern New England waters.

Subclass **SUCTORIA**

A group of infusorians, usually sessile, in which the adults have no cilia but possess long, hollow tentacles adapted either for sucking or for piercing. Some are predatory on other protozoans to which they attach their tentacles and suck them out; others are parasitic in other infusorians. No orders are recognized; but there are eight families containing about 200 species; two families are represented by the following species.

Family PODOPHRYIDAE

Genus *Podophrya*

P. gracilis. (Fig. 45.) This protozoan has a globular body provided with short, drumsticklike tentacles radiating in all directions; and it is mounted on a stalk of variable length (usually 0.04 mm) but relatively long in proportion to the body. This stalk is extremely slender, bent and has no apparent structure. Diameter of body 0.008 mm. Found attached to algæ and calcareous detritus. Southern New England waters.

Fig. 45
—P.
gracilis.

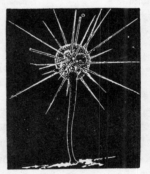

Fig. 46—E. coronata.

Genus *Ephelota*

E. coronata. (Fig. 46.) A form of variable shape, sometimes spherical and sometimes ovoid or pear-shaped. The body bears numerous sharp-pointed tentacles and a few straight uniform tentacles. A comparatively heavy stalk, tapering from its thickest part at the point of insertion in the body to the slenderest part at the base, supports the animal. Length of body 0.09 to 0.2 mm; length of stalk 0.2 to 0.6 mm. Not uncommon. Found on campanularian hydroids and bryozoans as well as on algæ. Southern New England waters.

FAMILY ACINETIDAE

GENUS *Acineta*

A. divisa. (Fig. 47.) The distinguishing feature of this suctorian is that the stalk is surmounted by a cuplike structure within which a saclike membrane is suspended from the rim, on which the body rests. The comparative opacity of the animal when contrasted with its exceedingly transparent cup gives the body the appearance of resting on the rim. Relatively long knobbed tentacles are distributed over the top of the body where they sway back and forth very slowly. Length of body 0.03 mm; length of stalk 0.1 mm; length of extended tentacles 0.06 mm. Occurs frequently on bryozoan colonies. Woods Hole region.

Fig. 47—A. divisa.

Fig. 48—A. tuberosa.

A. tuberosa. (Fig. 48.) This rather odd form is characterized by tentacles arranged in two bunches. The tentacles are capitate, or knobbed and they number about fifteen in each group. The supporting cup for the elongate body is delicate and usually difficult to detect. As the stalk is quite variable, no definite measurement can be stated. Body length about 0.3 mm. Its color is yellow, frequently distributed in a patternlike arrangement throughout the body. Found occasionally on calcareous and algal material. Southern New England waters.

V

PORIFERA

PLATE XII

CORAL REEF LIFE ON THE COAST OF SOUTHERN FLORIDA

Physalia pelagica (upper left); *Cestus veneris* (middle right); *Panulirus argus* (lower middle right); *Tripneustes esculentus* (lower middle); *Moira atropos* (lower right).

CHAPTER V

PHYLUM PORIFERA

(SPONGES)

IN THE preliminary text of the preceding chapter was given an account of the fundamental features that characterize the *Protozoa,* or unicellular animals. The other great subdivision of animal life is known as the *Metazoa,* and it includes all creatures having multicellular bodies. And although the sponges were formerly considered to be colonial protozoans and were once classed with the *Protozoa,* they are now recognized as being the lowest of the *Metazoa* and constitute the first group of that subdivision.

Sponges, notwithstanding their many-celled structure, are lowly organized. In the laboratory they have been shaken apart, and the individual dissociated cells have coalesced to form a new sponge, or have grown independently into other sponges. A common method of propagating commercial sponges is by cutting the living animals to pieces and placing the cuttings in favorable situations to grow. Again, the low organization of these animals is attested by the fact that those of the same species growing side by side often will coalesce when coming into contact; though sponges of different species will not.

All sponges are fixed, aquatic forms; and all, with the single exception of the family *Spongillidae,* are marine. Sponges have specialized organs and tissues, but these are reduced to a minimum. Perhaps the most striking of their structural features are those that are involved in the inorganic skeleton, or framework. It is according to the skeletal characteristics that these animals are classified.

Most sponges when living, especially commercial species like the common bath sponge, would not be readily recognizable by the average person; for in life the individual animal is usually a dark, unprepossessing mass with a slimy exterior perforated with numerous openings. The organic portion, or flesh, is soft and jellylike. The more simply organized sponges, however, usually retain more of their natural appearance after being killed and having the soft parts removed.

There are three main types of sponges: the *ascon type,* the *sycon type,* and the *leucon,* or *rhagon type.* (Fig 49.)

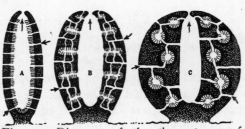

Fig. 49—Diagram of the three types of sponges. A, ascon type; B, sycon type; C, leucon type.

The ascon type of sponge is the simplest in structure. These are usually cylindrical, sometimes colonial, and the exterior surface is covered with numerous pores through which the water flows through the thin body wall directly into the interior chamber called the *cloacal cavity.* From the cloacal cavity the water passes to the outside by the way of a larger orifice known as the *osculum,* located at the summit or free end of the sponge.

In the sycon type of sponge, the structure is somewhat more complex, thus denoting a higher type of organism. The body wall is thicker, and the water flows through numerous cylindrical passages before reaching the cloacal cavity.

Still more complex and still higher in rank is the leucon type of sponge. And it is to this type that the great majority of sponges belong. Here the body wall is thicker than the two previously named types, the water in this instance passing through an involved system of both canals and widened portions of canals, or chambers, before being expelled through the various oscula.

The body wall of all sponges is composed of three layers: the external layer, or *ectoderm;* the internal layer, or *mesoglæa;* the internal layer, or *endoderm.* (Fig. 50.) In the leucon type of animals, the mesoglea constitutes the far greater portion of the body.

What the dermal epithelium, or outer skin, is in higher animals, the ectoderm is in the sponge. This skin is a single covering of flattened cells, sometimes sensitive, and contractile; its contractibility

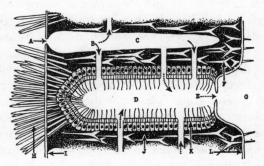

Fig. 50—Section of a sycon sponge. A, incurrent pore; B, prosopyles; C, incurrent canal; D, radial canal; E, apopyle; F, spicule; G, cloacal cavity; H, spicules; I, ectoderm; J, mesoglœa; K, collar cells; L, endoderm. (Dotted arrows show direction of water current through the sponge.)

being due to certain elongated cells, which in addition to their protective function form sphincters around the surface openings and often line the cloacal and other cavities.

It is chiefly in the mesoglœa that are found the distinguishing characters of most sponges. In this layer, which varies considerably in thickness in different animals, according to their primitiveness or higher organization, are contained various cellular bodies and the elements that make up the framework of the different forms. This framework may consist of calcareous or silicious spicules or of horn-like fibers, called *spongin,* or of a combination of silicious spicules and spongin. All sponges, with the exception of a very small group composing the order *Myxo-spongiœ,* possess some kind of skeleton.

Spicules occur in a great variety of forms. (Fig. 51.) The

Fig. 51—Sponge spicules. Megascleres (left) and microscleres (right).

simplest and perhaps the commonest form is that which is needle-shaped and pointed at both ends. In certain species of sponges, this simple form may be accompanied by other forms more complex. As spicules vary from the most simple to the more complex they are usually given the names *uniaxial* (= 1-rayed) *type, triradiate* (= 3-rayed) *type, quadriradiate* (= 4-rayed) *type, sexradiate* (= 6-rayed) *type* and *multiradiate* (= many-rayed) *type.* The complexity and diversity of shapes in the different types make them interesting objects of study with the microscope. Regardless of their shapes, however, they are placed in two general groups known as the *megascleres* and the *microscleres;* the former being larger and elongate or radiate, while the latter are the minuter and variously shaped types. In some sponges the megascleres are distributed so numerously and oftentimes so methodically that they give rise to a continuous loosely matted or felted skeleton; in other sponges microscleres are scattered throughout the mesoglœa, not forming a continuous skeleton; in still other sponges the spicules may project through the ectoderm. In calcareous sponges containing mineral spicules, the latter consist of carbonate of lime, with the crystalline structure and other properties of calcite; in sponges containing silicious spicules the latter consist of colloid silica or of opal; but, although the two kinds may differ widely in composition, there is no important structural unlikeness.

Spongin is familiar to everybody. It is this substance that composes the skeletal framework of commercial sponges. In chemical composition, it is allied to silk and contains a large amount of sodium. There are a few rare instances in which spongin occurs in the form of tri-, quadri- and sexradiate spicules, but most usually the spongin skeleton takes the form of fibers consisting of a central core of soft granular substance around which the spongin is disposed in concentric layers, making a hollow cylinder of variable size, according to the species. Sometimes these tubular cavities enclose sand, silicious spicules, foraminifera and other foreign bodies. (Plate XIII.) It may be added, however, that foreign material may contribute also to the formation of some silicious sponges; indeed, in certain cases such bodies introduced from without may form the entire skeleton, no other hard parts being present or produced. Spongin fibers are of different characteristic growths; they may branch antlerlike or like bushy tree growths, or anastomose to form networks.

The endoderm in all sponges is made up of peculiar cylindrical cells bearing a collar that encircles the base of a flagellum. These collar cells are not unlike in appearance and performance the protozoans described in the preceding chapter, which are known as choanoflagellates. In fact, these cells in the spongial endoderm are known as *choanocytes* (collar cells). The flagella do not act in unison, but each independently of the others. Their function is to lash the water, thus creating currents which bring food particles and oxygen and at the same time carry away waste products. Ingestion is also performed by the collar cells. The circulating water first enters through the pores whence it passes through the radial canals and flagellate chambers, and thence flows into the cloacal cavity, streaming out through the osculum. (Fig. 49.) The canal system is simple or complex, according to the organization of the animal; in the highest type of sponges there are involved systems of incurrent and excurrent canals communicating with the flagellate chambers.

Reproduction in sponges may occur in four different ways: by budding, by fission, by the formation of gemmules, and by sexual methods. Budding is normal growth that produces new oscula, each osculum representing a new individual; in some instances, however, the buds may become separated from the rest of the animal and grow independently. The latter instance is a form of reproduction by a modified method of fission; but true fission occurs in cases where composite units divide into two or several parts autonomously or through the medium of external agencies, such as wave action, the attacks of other animals, etc. Gemmule formation is a reproductive process wherein certain cells within the mesoglœa migrate to one spot in the middle layer, the other cells of which secrete a capsule around them. Gemmules seem to be formed only at the approach of an unfavorable season. On the death of the sponge, the gemmule falls to the bottom to await a more favorable period, when the capsule will burst and the contained cells will develop into a new sponge. In the sexual method of reproduction, both spermatozoa and ova are formed; though no specialized organs are present for this purpose. Both sex elements may sometimes be found in the same individual; but even in hermaphrodites, one or the other element usually is present in greater proportion in different individuals, so that some are predominantly male while others are predominantly female animals. In other sponges, the sexes are distinct. Although some cases are known in which the early embryonic stages are passed through outside of the parent's body (the common boring sponge *Cliona celata* is an example), the vast majority of sponges undergo their early stages of development still enclosed within the parent. The fertilized egg usually is contained in a brood-capsule; after a period of incubation the ciliated larva escapes and swims about in the water, soon or late attaching itself to the bottom, whereupon a metamorphosis occurs and an adult animal develops.

The distribution of sponges is world-wide, and there are about 2500 known species. Commercial sponges, however, and other sponges having skeletons wholly composed of spongin do not occur on our Atlantic shores north of Cape Hatteras, that is to say they grow only within the southern range. Those found north of Tarpon Springs, Florida are usually worthless for domestic purposes, their framework being too harsh and brittle when dry. Key West and Tarpon Springs are the principal markets; but the majority of sponges bought in this country and throughout the world come from the Mediterranean Sea. These are the finest-textured sponges found; and they are lighter in color; however, the extremely light-colored kinds commonly seen in the stores have been bleached with chemicals, which impairs their durability.

Sponges are now usually classified with the *Metazoa* either as a separate phylum or as a division of the *Cœlenterata*. Since sponges, with very few exceptions, possess a skeleton consisting either of mineral spicules or of organic fibers, it is on the characters of the framework that the principal divisions are founded. Thus, the following classification may be adopted:

PHYLUM PORIFERA

CLASS **Calcarea**   ORDERS ......$\left\{\begin{array}{l}\textbf{Homocoela} \\ \textbf{Heterocoela}\end{array}\right.$

CLASS **Hexactinellida** (Exotic
and deep-water forms including
the curious "Venus' Flower
Basket" and other glass sponges;
not within the scope of this
book).

CLASS **Demospongiae**   ORDERS ......$\left\{\begin{array}{l}\textbf{Tetractinellida} \\ \textbf{Monactinellida} \\ \textbf{Ceraospongiae} \\ \textbf{Myxospongiae}\end{array}\right.$

CLASS **CALCAREA**

These are the calcareous sponges and they are mostly small cylindrical forms usually found only in tide pools and shallow water. They are colorless and grow either singly or in colonies. The spicules may be of the uniaxial, triradiate or quadriradiate type consisting, as the class name indicates, of carbonate of lime. To obtain the spicules for study, a solution of potash may be used to eliminate the organic matter.

This is a sharply defined group, no forms being known that in the remotest degree are intermediate between the *Calcarea* and the other classes. Their calcareous spicules, together with their relatively simple structure and other distinctive features of organization, make them comparatively easy to identify.

ORDER **HOMOCOELA**

Included in this order are those calcareous sponges that are the simplest in organization and structure. The individuals are commonly known as "ascon sponges," and they are distinguished by their thin body walls and by having the central or cloacal cavity lined throughout with collar cells. In this group of animals, no intermediate radial canals or chambers are present between the pores and the central cavity.

FAMILY LEUCOSOLENIIDAE

GENUS *Leucosolenia*

L. botryoides. (Plate XIII.) A small, ivory-white species, tubular in form, occasionally branched and sometimes anastomosing. The spicules are uniaxial and triradiate, and faint yellow in color. It is usually colonial, although it not uncommonly occurs singly. Length about 35 mm. It grows in shallow water on stones, wood piling and other solid supports. Gulf of St. Lawrence to Vineyard Sound.

Fig. 52—L. fragilis.

L. fragilis. (Fig. 52.) This species, usually found growing in colonies, is extremely small in height, but the massed individuals may cover a comparatively conspicuous area, the separate individuals being about 1.5 mm long, while the colony may be 10 mm or more in diameter. The color is white or yellowish. Uniaxial and triradiate spicules are present; the former sometimes occurring in arched forms. Found in comparatively shallow water below the low-tide levels. Gulf of St. Lawrence to Long Island Sound.

ORDER HETEROCOELA

In this order are contained calcareous sponges somewhat more advanced than those comprised in the previous order, in that well-defined radial tubes or chambers are present in their thicker body walls, intermediate between the pores and the central cavity. These tubes or chambers are in communication with the pores by means of small water passages, or incurrent canals, and are lined with collar cells. The cloacal cavity is lined with flattened ectodermal cells, and not with collar cells. (Fig. 50.) The sponges in this group are all small and more or less cylindrical in shape. These are known as the "sycon sponges."

FAMILY GRANTIIDAE

GENUS *Grantia*

G. ciliata. (Plate XIII.) The urn sponge. This small sponge is a solitary species, but it is not infrequently found growing in clusters. The individual sponge is urn-shaped or oval, with a large osculum at the summit, rimmed with projecting spicules. The spicules at the osculum are uniaxial and of the same type as the others throughout the body except that the former are longer. Length about 12 mm; thickness 3 mm. Color gray or drab or dull yellow. Common in tide pools and sheltered shallow waters where it may be found attached to stones, pilings and other solid supports. Maine to Cape Cod southward.

CLASS DEMOSPONGIAE

The sponges in this class are primarily noteworthy for their thick walls, which contain small, round, collar-celled chambers connected by branched excurrent canals with the cloacal cavity. They are usually large forms, and the various types make up a quite heterogeneous group. In this class are contained those species that possess a framework of either spongin alone, or a combination of spongin and silicious spicules, or silicious spicules alone, or no skeleton whatever.

ORDER TETRACTINELLIDA

Sponges included in this order usually have a hard external covering containing megascleres which compose the principal skeletal structure, the microscleres being distributed throughout the mesoglœa. (Extra-territorial and deep water forms; not described in this work.)

ORDER MONACTINELLIDA

A group of sponges in which the skeleton consists chiefly of spicules of the uniaxial type, with or without spongin fibers, composes this order. Some forms may have other types of spicules than those that are needlelike; but no species has a framework consisting of spongin alone. Two suborders make up the group, and in these are included the majority of all sponges.

Suborder HADROMERINA

In this division of the *Monactinellida,* only those forms in which the spongin is either absent or very poorly developed are included. The sponges are for the most part massive but are occasionally cup-shaped or stalked, and have compact bodies with a hard, crustlike surface.

Family TETHYIDAE

Genus *Tethya*

T. hispida. (Fig. 53.) The orange sponge. Like the object from which it gets its popular name, this species has the form, color and somewhat the surface-texture of a small orange. The thick-leathery rind of the sperical body is pimpled with slight projections which are produced by the projecting ends of groups of long, uniaxial spicules, the meglascleres, extending radially from the center of the body. Found in offshore waters. Gulf of Maine.

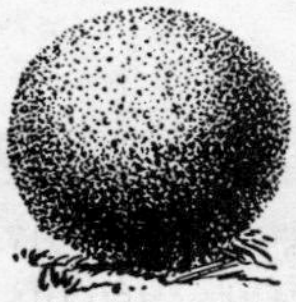

Fig. 53—T. hispida.

Family SUBERITIDAE

Genus *Suberites*

S. compacta. (Fig. 54.) The body of this species is an elongate, compressed mass attached by one edge to its basis. It is a compact of a fine, firm texture; the surface is smooth with inconspicuous oscula. Spicules in this form are needles with heads. Length up to 15 cm; width and height 2 to 8 cm. Color in life, bright yellow. Surface with mottlings that appear as depressions in dried sponge. Found frequently in shallow water on sandy bottoms and on the shells of hermit crabs. Maine to Virginia.

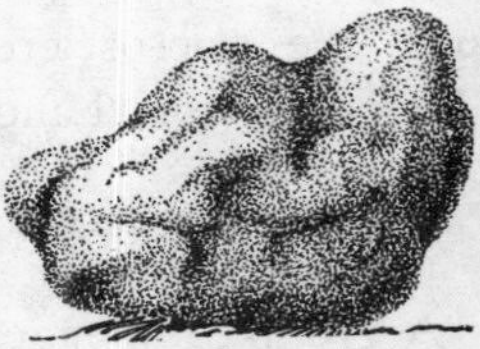

Fig. 54—S. compacta.

S. undulatus. (Fig. 55.) A firm spheroidal sponge made up of a basal undivided portion with closely set ascending lobes. The surface, when examined with a lens, is seen to be minutely roughened and abundantly covered with projecting spicules which in this form are needles with heads. The interior region is considerably denser than the outer portions, but the whole is compressible and easily torn. Height 60 mm; transverse diameters 75 to 90 mm. Color light gray. Fairly common on old shells in muddy tide pools. Cape Hatteras region.

FAMILY POLYMASTIIDAE

GENUS *Polymastia*

P. robusta. (Fig. 56.) A species of irregular form. When young, it forms incrustations on stones and shells; but, as growth proceeds,

Fig. 55—S. undulatus. Fig. 56—P. robusta.

Fig. 57—C. celata.

the incrustations give rise to relatively long and slender fingerlike branches. There are small needle- and pin-shaped spicules as well as long needles that extend radially from the center. The color is also variable, being either red, yellow, yellowish-white, or gray. Diameter up to 30 cm, with fingerlike projections 4 to 10 cm long. Common in offshore waters. Occurs from Maine to North Carolina.

FAMILY CLIONIDAE

GENUS *Cliona*

C. celata. (Fig. 57 and Plate XIV.) The boring sponge. Sometimes known as the sulphur sponge because of its bright sulphur-yellow color. Very common in old clam (Venus) and oyster shells. It will also attack limestone rock and other calcareous objects—even the shells of living oysters. Molluscan shells attacked by this species exhibit numerous round holes about 1.5 mm in diameter scattered over the outside and communicating with a labyrinth of passages in the interior of the shell. Low, retractile, wartlike prominences of the sponge itself, some of which constitute the pores and others the oscula (there are two kinds), may be seen projecting from these holes. (Plate XIV.) The fact that this and related species bore only into shells or other calcareous material indicates that the boring is achieved not by mechanical means but by the action of solvents secreted by the animal. When the shell has been entirely absorbed, the sponge may grow into an individual of considerable size, often forming a mass 18 cm or more in diameter with the surface covered with small elevations. In fact it has been shown that the well-known Neptune's Cup sponge (*Poterion patera*),—a Pacific form that attains to a height of several feet,—may really be the massive or full-grown form of a species of *Cliona*. Our species, *C. celata,* seems to be identical with the European form *C. sulphurea,* and it is often described under the latter name in reference works. It is an abundant sponge and is found from the tide marks to the deeper offshore waters. All ranges.

Suborder HALICHONDRINA

This suborder is a division of the *Monactinellida* that includes those sponges having compact but usually fibrous bodies with a hard or crusty exterior. Well-developed spongin is usually present.

Family CHALINIDAE

Genus *Chalina*

C. oculata. (Fig. 58.) The finger sponge. It is easily recognized by the orange-red or red branches, thick, more or less flattened, and fingerlike, arising from a stout stem of irregular form and breadth. Over the smooth velvety surface, fairly conspicuous oscula are distributed. The general texture is firm, yet delicate, the spongin forming a fine network. Grows to a height of about 12 cm. Very common in shallow and offshore waters. Labrador to Rhode Island.

C. arbuscula. (Fig. 59.) Dead man's fingers. The body consists of a cluster of profuse, forked branches growing from close to the base; the branches are somewhat slender, and the cluster is 10 to 20 cm high with a breadth of about the same measurement. The color of the living animal is white or gray, from which it doubtless receives its popular name; the dried framework, however, is a yellowish-white. It may be distinguished from *C. oculata* by its finer and more delicate texture. The spongin forms a regular network. Common in shallow water. All ranges.

  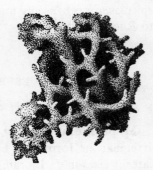

Fig. 58—C. oculata.   Fig. 59—C. ar-   Fig. 60—R. tubifera.
busbuscula.

Genus *Reniera*

R. tubifera. (Fig. 60.) A sponge with a very irregular shape. The body consists of a network of anastomosing cylindrical branches varying in diameter from 3 to 8 mm, very fragile and easily pulverized. Rising vertically from the branches are a number of tubular processes, each measuring from 2 to 10 mm in height and from 1 to 3 mm in diameter, and bearing oscula at the tips. The main skeleton consists of straight needlelike spicules joined at their ends and arranged to form a network. Also there are present spongin and free spicules. Size of sponge, up to 3 cm in height and 13 or more cm in breadth. Common in protected shallow water. Cape Hatteras southward.

Genus *Halicondria*

H. panicea. (Fig. 61.) The breadcrumb sponge. As its popular name indicates, in general appearance this species is not unlike a bread-crumb mass. The form is variable; likewise the color, which may be either gray, yellowish or even orange. The size is not constant, although specimens several inches in breadth are not un-common. It grows on the fronds of algæ, inhabit-ing the bottom at depths from four to eight

Fig. 61—H. panicea.

fathoms, often in its growth entirely spreading over and destroying the plant. Although an offshore species, this sponge is very commonly found washed ashore. It contains little spongin, and its needlelike spicules are confusedly massed together. Occurs from the Arctic Ocean to Rhode Island.

Family **ESPERELLIDAE**

Genus *Microciona*

M. prolifera. (Fig. 62.) An incrusting sponge, irregular in form and bright red when young; later it rises in dark orange-red, fingerlike masses, sometimes reaching a height of 15 cm. In full growth, its branches are profuse, forked, more or less flattened, and sometimes palmate at the ends. It contains spongin fibers and uniaxial spicules. Breadth about 12 cm. Common from low-water mark to ten fathoms. Cape Hatteras southward.

Fig. 62—M. pro-lifera.

Genus *Stylotella*

S. heliophila. (Plate XIV.) Although the body of this sponge is occasionally simply massive in form, it generally rises in protrusive processes from an encrusting base. It lives near the low-water level,

PLATE XIII

1—Sponge section showing foraminifera enclosed in skeleton; 40x.

2—*Leucosolenia botryoides*; 3x.   3—*Grantia ciliata*; 3x.

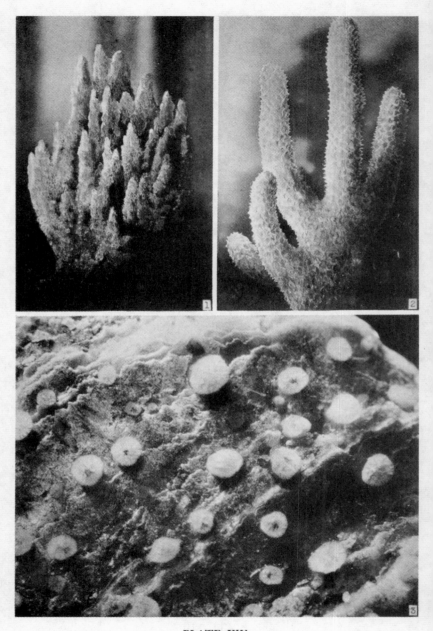

PLATE XIV

1—*Stylotella heliophila*; ¾x.   2—*Hircina ectifibrosa*; ½x.

3—*Cliona celata*; 5x.

and sometimes is so situated that the low spring tides expose certain individuals to the air for several hours; thus, the massive types are usually characteristic of colonies that are from time to time exposed, while the protrusive forms are those that are almost exclusively limited to such as are never exposed by the tides. The texture is soft. Little spongin is present; and the needlelike spicules are in bundles. Size about 5 cm in height by 5 to 10 cm in breadth; that is to say, about as large as a double fist. Color orange, sometimes with a greenish cast. Found very frequently on old shells, stones and piling. Cape Hatteras southward.

Genus *Esperiopsis*

E. obliqua. (Fig. 63.) A bright red branching sponge, with branches cylindrical or sometimes compressed; branches may be either smooth or markedly knotted. Arising from a base, the main branches themselves give rise to branches from 4 to 6 mm in diameter. Fusion occasionally occurs between contiguous branches. The sponge may grow vertically, sometimes to a height of 20 cm, or the branches may extend out variously from the base covering a breadth of 5 cm. Both spicules and spongin are present. Found in offshore waters, but specimens are occasionally washed up on the beach. Cape Hatteras southward.

Fig. 63—E. obliqua.

ORDER **CERAOSPONGIAE**

The sponges comprised in this order are those in which the skeletal framework is without proper spicules and consists of a close network of spongin, such as characterizes the commercial sponges.

Family **SPONGIIDAE**

Genus *Euspongia*

E. officinalis. (Fig. 64.) A commercial variety of the Levant sponges of Europe and Asia Minor. The American form is called the "glove sponge," and is the least valuable of the commercial sponges. The shape is more or less globose, though very variable, frequently being lobed, cuplike, lamelliform or tubular. It has slender elastic fibers with very small meshes. The animal grows to a height of about 3 cm. In life, it is deep dark brown with the sides and under sides lighter in color. The surface is generally even but covered with small tufts. At the top are one or more excurrent oscula. Found on shallow reefs and rocky bottoms. Florida; Bahama; West Indies.

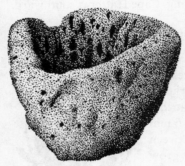
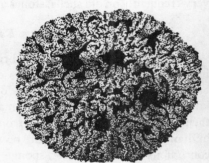

Fig. 64—E. officinalis.   Fig. 65—H. gossypina.

Also classed with this species are the American varieties *adriatica* and *rotunda*.

GENUS *Hippospongia*

H. gossypina. (Fig. 65.) The sheep's-wool sponge, so-called because of its shaggy surface. It has numerous projections and is covered with tufts of fibers, between which are the large oscula. In life, the animal is black. This is the common large bath sponge, and it is the most valuable American species, the fishery being worth more than a quarter of a million dollars annually. Grows in shallow water. Florida; Bahamas.

H. equina. (Fig. 66.) The horse sponge. A commercial sponge, but of less value than *H. gossypina*. The body is massive and cake-shaped with the surface network of fibers produced into numerous tufted conical extensions. This sponge is also characterized by its coarse spongin fibers and extensive canal system with numerous large and closely approximated oscula. It is black when living, but when the animal matter has been removed it is yellow. There are numerous forms of this sponge, some of which have been given both technical and popular varietal names; viz., *meandriformis,* the velvet sponge, with surface marked with convolutions and widening channels; *cerebriformis,* one of the "grass sponges," a massive cup-shaped form with a rough surface; *agaricina,* the yellow sponge, a broadened, flattened, cup-shaped form, but very variable. The horse sponge and its varieties are found in the same localities as is the sheep's-wool sponge, in depths of two to twenty-five feet. Florida; West Indies.

Fig. 66—H. equina.

H. canaliculata, variety *flabellum.* (Also listed as *Spongia graminea* in numerous reference works.) (Fig. 67.) The grass sponge. A commercial variety with the body like a truncated cone having the broad end uppermost. The top may be either flat or funnellike, bearing the oscula on the upper surface and the pores or incurrent apertures between the rows of deep ridges that line the sides. The color in life is black. This form is perhaps the least variable of all the commercial sponges; and it is this persistency of form, the lateral ridges, and the situation of the different kinds of orifices, that make up the characteristic features of the species. The size, however, varies according to the depth in which the animal may be living; those in very shallow water being 12 to 15 cm high, while the ones inhabiting deep water are twice as large. It is an inferior grade of sponge, the fibers being coarse and usually containing considerably foreign material. Found on reefs and hard bottoms in depths of three feet or more. Florida.

Fig. 67—H. cana-liculata.

Genus *Hircina*

H. ectofibrosa. (Plate XIV.) A variable form, but usually occurs with a simple base from which rise a few somewhat cylindrical branches or lobes. The exterior of the sponge is set with numerous sharp conical projections, which divide the surface into polygonal or rounded depressions. Spongin is present, but no mineral spicules. Color when living, black or deep purple. Height, up to 13 cm or more; thickness of branches about 35 mm. Found in offshore waters, but is occasionally washed ashore. Cape Hatteras southward.

H. campana. (Fig. 68.) The form of this species is variable, but normally it is vaselike and may be either ribbed or smooth externally,

excepting for the small conical spines which, like those of *H. ectofibrosa,* project from the surface. In some varieties the vase is open on one side; in others the cup is surrounded with a hedge of branches; in still others the vase is merely a mass with a tendency to erect branches around the top. In all forms the larger apertures are within, and the smaller ones outside, the vasiform body. Color in life, black. Height about 10 cm. Occurs in

Fig. 68—H. campana.

shallow and offshore waters on hard bottoms. Cape Hatteras southward.

H. acuta. (Fig. 69.) One of the so-called loggerhead sponges. It occurs in several varieties. This form may be distinguished from others of the genus by the larger size of the conical spines and the greater spaces between the projections. The excurrent orifices are at the top, and are usually gathered together in one or more sievelike areas. Numerous large-sized canals which are not visible are contained within the sponge, rendering it very light in weight. Color dark gray externally; black internally. Grows on reefs and hard bottoms in about eight feet of water. Florida; West Indies.

Fig. 69—H. acuta.

ORDER **MYXOSPONGIAE**

The slime sponges. The bodies of these sponges are devoid of a skeleton of any kind.

FAMILY **HALISARCIDAE**

GENUS *Halisarca*

H. dujardini. (Fig. 70.) This form may be recognized by the small, soft, irregular, incrusting mass of its body. The oscula are rather large and elevated. The pale-yellow, gelatinous body of the sponge may be found adhering to the fronds of red seaweeds growing on offshore bottoms, from which regions they may occasionally be washed ashore after violent storms. Off Rhode Island.

Fig. 70—H. dujardini.

VI

COELENTERATA

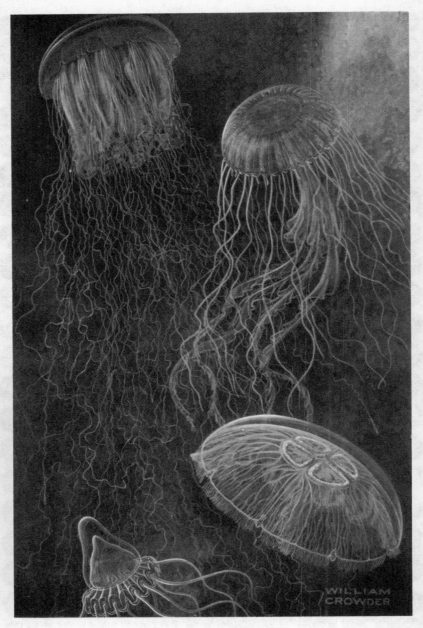

PLATE XV

LARGE JELLYFISHES

Cyanea capillata, variety *arctica* (upper left); *Dactylometra quinquecirrha* (upper right); *Aurelia aurita*, variety *flavidula* (lower right); *Periphylla hyacinthina* (lower left).

CHAPTER VI

PHYLUM COELENTERATA

(POLYPS)

STRUCTURALLY, the group of animals composing the *Coelenterata* are alike in having bodies distinctly radiate; that is to say, the parts of the body are arranged around a common center like the parts of a wheel or of an umbrella; and, also, they are alike in having a single internal cavity which, unlike that of higher animals, is the common digestive space, containing no separate alimentary tract. Some, however, are symmetrically biradiate. Ordinarily there is but a single opening, the mouth, which serves for both ingesting and excreting. The body wall, like that of the sponges, consists of three layers: the outer ectoderm, the inner endoderm, and the middle jellylike layer lying between them, called the mesoglœa. The mesoglœa is merely a supporting layer; it is therefore skeletal in function.

Usually microscopic nettle cells, or stinging organs (nematocysts), are present either on tentacles or on other parts of the body, wherewith the animals are enabled to paralyze and capture their living food. The nettle organs are somewhat egg-shaped (Fig. 71), having thickened walls of an elastic character; and each cell (cnidoblast) contains a coiled, threadlike tube, one end of which is attached to the cell. The basal end of this tubular filament bears several barbs; the free end is pointed, but in certain instances it is barbed in addition. The interior of the cells containing these minute darts is filled with a highly irritant poison (probably formic acid). In some cases the thread tubes act as lassos, capturing animals and food material by winding around the smaller appendages or projections instead of by puncturing. In addition to the parts just mentioned, there is

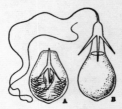

Fig. 71—Nettle cells of a polyp. A, cell containing coiled thread; B, cell with discharged thread.

usually present on the outer surface of each cell a small spine (cnidocil) functioning as a tactile organ or trigger which upon the merest contact causes the poison-tube to be ejected with a force considerable enough to penetrate the skin of the polyp's prey. All cœlenterates are predacious, but they will eat non-living animal matter as well; some are even cannibalistic.

Reproduction and development are often complex. It is in this group that we find numerous conspicuous examples of that developmental phase known as "alternation of generations," stages in the life-history of certain species wherein the offspring resembles not the parents but the grandparents. Thus, in some groups two distinct forms of the same species are met with. (The nature of these distinctions and the groups in which they occur will be specified in the classificatory text.) It is for this reason that the class names of those divisions including such types are double names, viz., *Hydrozoa: Hydromedusae.*

Coelenterates are very low animals; nevertheless, there seems to be little doubt that they stand either with or above the sponges—which latter are by some writers placed within the group—but below the echinoderms. There are more than four thousand known species; these may be grouped in four classes according to the following system:

PHYLUM COELENTERATA

CLASS **Hydrozoa: Hydromedusæ**

ORDERS ....

- **Hydrariae** (Mostly fresh water forms; no species listed in this work.)
- **Hydrocorallinae**
- **Gymnoblastea: Anthomedusae**
- **Calyptoblastea: Leptomedusae**
- **Trachomedusae**
- **Narcomedusae**
- **Siphonophora**

CLASS **Scyphozoa: Scyphomedusae**	ORDERS ...	Stauromedusae Coronatae Cubomedusae Semæostomeae Rhizostomae
CLASS **Anthozoa**	ORDERS ...	Alcyonaria Zoantharia
CLASS **Ctenophora** SUBCLASS **Tentaculata**	ORDERS ...	Cydippida Lobata Cestida
SUBCLASS **Nuda**	(No orders) FAMILY **Beroidae**	

CLASS **HYDROZOA: HYDROMEDUSAE**

For the proper understanding of the creatures that are comprised within this class,—and this applies also to the class next in order, though to a lesser degree,—the beginner must bear constantly in mind that a considerable number of the species occur in two principal forms: the one form, in the present case, being the hydroid, which is called the *trophosome,* and the other form being the medusoid, which is called the *gonosome.* Again, some spe-

cies occur only in the hydroid form; and other species may occur only in the medusoid form. (Fig. 72.)

The hydroid stage is characterized by plantlike, usually colonial, forms that produce by budding, free-swimming, individual medusae. As a rule, hydroids are small, measuring only a few millimeters in

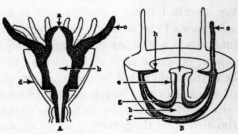

Fig. 72—Diagram of a hydroid (A) and a medusoid (B). a, mouth; b, stomach cavity; c, tentacle; d, hydrotheca; e, manubrium; f, exumbrella; g, subumbrella; h, velum.

length; but there do exist some tubularian representatives in the deep sea measuring a meter or more. Also, the commoner solitary tubularians are exceptions to the rule. Two kinds of individuals commonly make up the hydroid colony; the feeding polyp, or *hydranth,* and the medusa-producing individual called the *gonophore.* The stalks on which these various individuals are borne are termed the *hydrocaulus;* and the root-like holdfast by which the hydroid is attached to the substratum is the *hydrorhiza,* or *stolon.* The individual hydranths of a colony contain a flowerlike cluster of tentacles, invested with nettle organs, and a central mouth leading to a stomach cavity that is continuous through the stalk and is in communication with the mouths of all the other feeding polyps in the colony. Thus it will be seen that the foregoing type of hydroids is composed of animals that are in physical association with one another, thereby living a communal life. The gonophores, or reproductive polyps, are variously shaped, but those that are vasiform and clublike are the commonest. Each consists, at maturity, of a pile of minute, saucerlike individuals attached to one another by the center of the convex side; and these are the growing medusæ, which escape one by one as they approach maturity.

The medusoid stage, or gonosome, of the type of hydroid just described is characterized by free-swimming forms known as *craspedote* medusae. (Fig. 72.) In appearance these are somewhat like tiny umbrellas with very short handles. This handlelike region, the manubrium, of the medusoid has an opening at its free end which forms the mouth. The mouth opening leads to a central cavity in the umbrella-top from which region extend radial canals to the outer margin, where they join the so-called "ring canal" which makes a circuit within the body-substance of the umbrella rim. Projecting inward from the rim of the umbrella is a circular diaphragm or shelf of thin tissue known as the *velum,* or, more commonly, as the "veil." And this latter fact should be particularly noted by the amateur collector; for it is by the presence of the velum that nearly all hydro-medusans can be distinguished at once from the jellyfishes belonging to the *Scyphomedusae.* The umbrella rim also bears the tentacles, which may number one or many, and these are copiously supplied with stinging cells. Swimming is performed by spasmodic contractions of the umbrella. The sexes are usually separate in medusoids; the gonads or sex organs being suspended from the sub-umbrella in the region of the radial canals. The sperm of the male is shed into the water, where it is transported by the currents to the female, fertilizing her eggs, which later develop into planula larvæ. These larvæ are ciliated, and they swim about for a time; after which they settle to the bottom or attach themselves to some object where they grow into hydroid forms; thereupon the life-history cycle begins anew.

But not all hydrozoans are of the foregoing type. In addition to what has already been observed regarding the fact that some representatives have no free-swimming medusoid form, and that others have no known hydroid stage, it may be stated that certain other hydroids give rise to medusoids, by budding, but these latter are never set free and therefore they remain attached to the parents, tugging vainly away until after their eggs have been freed, whereupon they atrophy and disappear. Then, there are free-swimming medusoids that produce other medusoids directly, by budding from the margin of the umbrella or from the manubrium. And there are yet other cases where different species of hydroids produce identical forms of medusoids.

By reason of their smallness, hydroids are not conspicuous, but they are among the most numerous and therefore the commonest animals of the seashore. They are found between the tide marks and in shallow water attached to stones, old and living shells, pilings, and other solid objects. The beginning collector will soon learn to recognize them. And if he takes the trouble to study the living animals under a low-power microscope, he has a revelation awaiting him; for there are few of the smaller forms in nature that are more exquisite and attractive.

The medusoids are pelagic and not so readily found as are the hydroids—unless one essays to hatch them in the tank, which is a comparatively easy matter; for a healthy specimen of hydroids producing free-swimming medusoids will within a few days, or hours, even, liberate a numerous crowd. To keep the medusoids alive, however, is difficult if not impossible, unless one has running sea water plus considerable experience in the matter. They may also be obtained on occasion by the use of the tow-net; by this method specimens of large size are frequently captured.

ORDER HYDROCORALLINAE

The members of this order resemble corals in that the hydroid colony secretes a calcareous covering. But here the resemblance ceases. In the true corals the animals are not connected by a common gastrovascular space with one another as are the hydroids; nor do the different members of a coral colony perform different functions as do hydroids; and, whereas in the individual coral animal the calcareous material forms partitions within the interior of the body, in the hydrocorallines the outside covering only is formed of mineral matter. There are two kinds of polyps in the hydrocoralline colony: the nutritive polyps (gastrozooids) which often have the mouth ends provided with tentacles, and the defensive polyps (dactylozooids) which are without a mouth and are provided with stinging cells.

FAMILY MILLEPORIDAE

GENUS *Millepora*

M. alcicornis. (Fig. 73.) Pepper coral. There is but one species on the Atlantic Coast, but it is very abundant, composing considerable reefs in the waters of Florida. The colonies are treelike, rising erect from the bottom, and the corallike mass is porous, being composed of a network of tubular canals imbedded in the calcareous structure. There are numerous pores on the surface, leading into hollow, cuplike chambers containing the polyps. The pores and chambers are of different sizes, the larger containing the nutritive polyps, and the smaller, which are ranged irregularly around the larger, containing the defensive polyps. On the nutritive polyps, the five or six tentacles which are present are knobbed; it is said that the tentacles of the defensive polyps have unusual stinging powers. As the individual polyp dies, it is succeeded by another which secretes a horizontal platform for a new cup across the old chamber; and it is by these additions that the main mass increases in size. The colony occupies only the canals and chambers in the outer region of the calcareous structure. There are free medusoids, but these are degenerate

Fig. 73—M. alcicornis.

forms without tentacles or a veil or ring canal; they are feeble swimmers and live only five or six hours, long enough for the females to discharge their fertilized eggs. Occurs on the Florida coast.

ORDER **GYMNOBLASTEA: ANTHOMEDUSAE**

In this order are included hydrozoans that have naked hydranths and gonophores; that is, polyps without a protective cup (*hydrotheca*), and gonophores without a hard covering (*gonotheca*) such as distinguish the members of the next succeeding order. However, the stems of these hydroids may have a rigid cuticular covering, the *perisarc*. The hydroids are mostly colonial, and produce either free medusoids or fixed medusoid buds. The medusoids are more or less bell-shaped and bear the sex organs on the manubrium; thus, these forms can be easily told from those of the next order, which in contrast have these sex organs attached to the radial canals. Minute red or blue spots (*ocelli*) usually occur at the base of the medusoid tentacles; and, although the function of the spots is not proved, there is evidence that they serve as light-perceiving organs. The medusoids of this order are known as *anthomedusans*. The hydroids are generally known as *tubularians*. Many anthodusans have never been traced to the hydroids from which they came.

FAMILY **CLAVIDAE**

(No medusoid forms)

GENUS *Clava*

C. leptostyla. (Fig. 74.) This species is a simple unbranched form, that rises from a threadlike hydrorhiza, the latter being protected by a perisarc that extends part way up the pedicel, or stalk. The hydranths are reddish in color, with about twenty to thirty tentacles grouped more regularly than irregularly around a conical proboscis (hypostome) on the tip of which is the mouth. Just below the tentacles are the reproductive organs in the form of buds (sporosacs) arranged in clusters, those of the male being pink, those of the female being purple. Height about 2 cm. Found on *Fucus,* pilings, stones, and other solid objects in sheltered waters where it often forms colonies several feet in area. Very common. Labrador to Long Island Sound.

Genus *Rhizogeton*

R. fusiformis. (Fig. 75.) Although not greatly unlike the much commoner *Clava* in general appearance, this species differs in that it is smaller, being only about 8 mm high, and in having fewer tentacles on the hydranth, these being twelve in number. The reproductive buds are not at the base of the tentacles, but are borne on separate pedicels arising from the hydrorhiza; moreover, these sporosacs are shorter than those of *Clava,* and are covered with the perisarc. Found forming soft fuzzy coatings on rocks in tide pools. Occurs in Massachusetts Bay.

Fig. 74—C. leptostyla.

Fig. 75—R. fusiformis.

Family **CORYNIDAE**

(Hydroid and medusoid forms)

Genus *Syncoryne*

S. mirabilis. (Fig. 76.) The hydroid colony of this animal is slightly and irregularly branched, and has a well-defined perisarc that ends at the base of the hydranth. The hydranth is elongate, cylindrical, large and very stout for its length. Numerous stout tentacles with knobbed ends are irregularly placed around and along the hydranth body. Color rose-red. Height about 20 mm. Found attached to seaweeds, stones, and shells in shallow water. All ranges.

The medusoid form is of two varieties; the attached and the free-swimming form. The former is merely a sporosac without ocelli or a mouth, and with rudimentary tentacles. In the latter form, development is complete. This medusoid is the common *Sarsia mirabilis* of the older reference literature. Its somewhat hemispherical umbrella is provided with four very long marginal tentacles; while from the center there extends considerably beyond the velum the pendulous manubrium. Diameter about 4 mm; height about 7 mm. Found very frequently in tow-net hauls during Spring and Summer. All ranges.

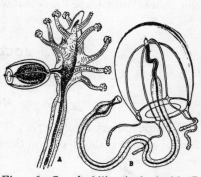

Fig. 76—S. mirabilis. A, hydroid; B, medusoid.

Genus *Gemmaria*

G. gemmosa. (Fig. 77.) The hydroid of this species is similar to *Syncoryne mirabilis,* except that the colony consists of a single elongated hydranth unsupported by a stem; consequently it is devoid of a perisarc

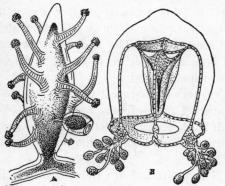

Fig. 77—G. gemmosa. A, hydroid; B, medusoid.

except on the hydrorhiza. It is usually found on mussel shells. Vineyard Sound southward.

The medusoid when young bears two rudimentary and two well-developed tentacles. At this period also, the umbrella walls are very thin and the manubrium is a simple spindlelike tube. Later the animal acquires four long tentacles, each with a large hollow bulb at its base of attachment, and the tentacles are crowded along one side with short branches bearing at their ends large spear-shaped bodies containing nematocysts and having a bristled armature. No ocelli are present. Mouth is without marginal lobes. Diameter about 6 mm. Cape Cod southward.

Family **BOUGAINVILLIIDAE**

(Hydroid and medusoid forms)

Genus *Bougainvillia*

B. superciliaris. (Fig. 78.) The hydroid colony is treelike, and a dense perisarc is developed on the branches as well as on the main stem.

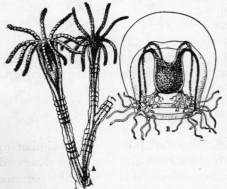

The mouth region, or *hypostome,* is conical, but rather inconspicuous, and carries a circlet of 15–20 tentacles. Each tentacle is ringed at regular intervals with a battery of nematocyst-cells. The medusa buds occur singly on the sides of the stem near the bases of the polyps. Each bud is enclosed in a continuation of the chitinous covering of the stem; but where this chitinous perisarc passes on to the hydranths, it thins out and stops just below the tentacles. Color reddish. The hydroid grows in clusters about 50 mm high, and is usually attached to rocks, shells or rockweed. Common. Greenland to Rhode Island.

Fig. 78—B. superciliaris. A, hydroid; B, medusoid.

The medusoid form has four clusters of marginal tentacles situated at the bases of the four radial canals. Each cluster is composed of 10–15 tentacles. Color yellowish with black ocelli. Diameter 10 mm. Common throughout the Summer north of Cape Cod; common only during early Spring south of Cape Cod.

B. carolinensis. (Fig. 79.) The hydroid form of this species is profusely branched and it tapers gradually from base to summit. Numerous ringed pedicels terminating in polyps grow from the main stem and also from the branches. The hydranths are fusiform with a conical proboscis, the latter being encircled by a row of about 15 long, slender tentacles. At various places along the sides of the stems, medusoid buds occur either singly or in groups. The color of the stem is greenish, while that of the polyps is generally a delicate pink. Height about 25 cm in exceptional cases; commonly about 10 cm. Common on *Fucus vesiculosus* and pilings. Cape Cod southward.

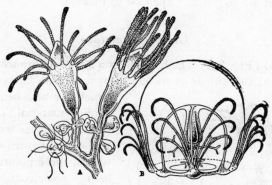

Fig. 79—B. carolinensis. A, hydroid; B, medusoid.

The medusoids are dome-shaped, with 4 radially arranged clusters of tentacles arising from basal swellings on the margin of the umbrella. Each cluster has 7–9 tentacles about as long as the umbrella is high. The manubrium extends about half the distance from the top of the umbrella cavity to the velar opening, and it possesses 4 tentacles, which are arranged radially around the mouth opening, each tentacle branching twice dichotomously, thus making 4 tentacle tips for each tentacle. Color of manubrium brick-red. Body diameter 4 mm. Common Cape Cod to Florida.

B. rugosa. (Fig. 80.) The hydroid is irregularly branched,

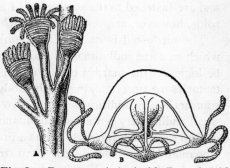

Fig. 80—B. rugosa. A, hydroid; B, medusoid.

with the perisarc extending well up on the hydranth body. That portion of the perisarc which covers the polyp is marked by a number of rings. The proboscis is usually conical, but it may occasionally be much flattened. The tentacles are short and number from 8 to 10. Height about 75 mm. Occurs on pilings near low-water mark. Cape Cod to Florida.

The medusoid is somewhat pear-shaped. There are 4 groups of 3 marginal tentacles, and 4 oral tentacles. The manubrium is short and thick. Diameter 2 to 3 mm. Cape Cod to Florida.

Genus *Stomotoca*

S. rugosa. (Fig. 81.) The hydroid is a simple unbranched stem covered for about two thirds of its length by a delicate film of perisarc to

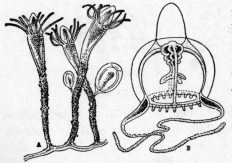

which are attached particles of foreign material. The hydranth has 10 tentacles which point alternately forwards and backwards, those pointing forwards being somewhat the longer. These colonies are very small, being only about .2 mm in height, and are seldom found except by experienced naturalists. They grow on the under side of horseshoe crab (*Limulus*) shells

Fig. 81—S. rugosa. A, hydroid; B, medusoid.

and are fastened to the sand tubes of a worm (*Sabellaria*). Their medusoids, however, are very common. Rhode Island southward.

The medusoid is more or less conical with a solid apical projection which in some individuals is short and blunt, whereas in others it may be long. At the margin of the umbrella are two diametrically opposed tentacles 10 times as long as the body height. The tentacles are highly contractile. In addition to these two long tentacles there are 14 very small rudimenary tentacles. The manubrium is flask-shaped with a cruciform mouth-opening which has 4 prominent, recurved, crenulated lips. The manubrium is brick-red, often streaked with sooty-brown or black. Body height 5 mm; diameter 3 mm. Cape Cod to Florida.

Genus *Rathkea*

R. octopunctata. (Fig. 82.) In this species, the hydroid form is unknown. The medusoid is pear-shaped with a solid apical projection, and 8 clusters of marginal tentacles, 4 clusters being situated radially, and 4 interradially. The radial clusters contain 4 or 5 tentacles; the interradial contain not more than 3. The tentacle-bulbs on the umbrella margin are large and swollen and filled with brownish-green, almost black, pigment granules. The mouth-opening at the end of the manubrium has four prominent lips, each terminating in a pair of tentacles that end in a knob-shaped cluster of nematocysts. Color of manubrium brown-green to black. Body height 5 mm; diameter 4.5 mm. Common during early Spring from Greenland to Rhode Island.

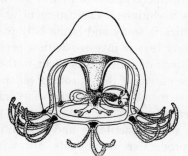

Fig. 82—R. octopunctata.

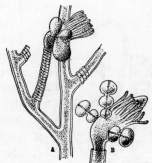

Fig. 83—E. ramosum. A, female hydroid; B, male hydroid.

Family **EUDENDRIIDAE**

(No free-swimming medusoids)

Genus *Eudendrium*

E. ramosum. (Fig. 83.) In this species, the colony is profusely branched. At the base of the branches and of the pedicels supporting the polyps, and sometimes at the base of the internodes, annulations occur. The gonophores, though few in number (usually there are 3, with 3 chambers or only 2), are conspicuous, standing out nearly at right angles to the axis of the hydranth. The male gonophores and hydranths are pink or vermilion; the female gonophores are bright orange-red. Height up to 12 cm. Occurs on piles, large sea squirts, shells, and other firm objects below the low-water mark. Labrador to North Carolina.

Family HYDRACTINIIDAE

(No free-swimming medusoids)

Genus *Hydractinia*

H. echinata. (Fig. 84.) This form, sometimes listed in technical literature as *H. polyclina,* grows in colonies wherein the individuals arise

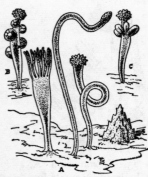

Fig. 84—H. echinata. A, nutritive and defensive hyrdoids; and basal spine. B, female hydroid; C, male hydroid.

from a basal body mass or hydrorhiza which overlies a chitinous, encrusting plate. The hydrorhiza is spiny and is the only part of the colony that is invested with a perisarc. Three kinds of individuals occur in the colony: the nutritive, the reproductive, and the defensive members. The nutritive hydranths have tentacles and are endowed with great powers of contractibility and extensibility; the reproductive individuals are without tentacles, but instead have extensive batteries of nematocysts; the defensive members, also without tentacles, are usually longer than the others, are very mobile, and have a general resemblance to the reproductive individuals without the sporosacs. The different colonies each contain members of only a single sex, the male colonies being white or pinkish with pink gonophores, and the female colonies being darker because of their orange-red or bright red gonophores. Height about 10 mm. This common species is perhaps found oftenest on the shells of the little hermit crab *Pagurus longicarpus,* where it forms the soft fuzzy coating so frequently present; however, it occurs also on shells, stones, wood, and other objects in shallow water and tide pools. All ranges.

FAMILY **PODOCORYNIDAE**

(With free or sessile medusoids)

GENUS *Podocoryne*

P. carnea. (Fig. 85.) The hydroid form of this species might easily be mistaken for *Hydractinia,* except that in the present species the re-

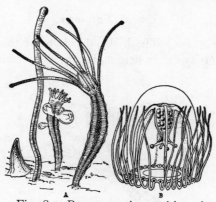

Fig. 85—P. carnea. A, nutritive, defensive, and reproductive hydroids. B, medusoid.

productive individuals have tentacles. Moreover, the feeding polyp pedicels are spindle-shaped, instead of tapering towards the basal end only. The medusoid buds arise from a zone that is slightly below the circlet of tentacles which surround the conical proboscis. Height 5 to 15 mm. Also, like the preceding species, it is common on the shells of hermit crabs and other firm objects—even on the shells of horseshoe crabs. Maine to southern New England.

The mature free medusoid has an ellipsoidal umbrella composed of a thin, but quite tough and rigid, gelatinous substance making it very transparent. From the margin there extend about 24 to 32 tentacles which are as long as the bell height. The tentacles are usually carried curled upward, owing to their lack of flexibility. A tentacle-bulb, filled with pigment granules, is present at the base of each tentacle. The manubrium is flask-shaped and has a mouth-opening surrounded by

4 short tentacles, each of which terminates in a knob-shaped cluster of nematocysts. Both the manubrium and the tentacle-bulbs are red or brown-red in color. Body height 3 mm. Common. Greenland to southern New England.

FAMILY PENNARIIDAE

(With either free or sessile medusoids)

GENUS *Pennaria*

P. tiarella. (Fig. 86.) The hydroid colony of this species is large, sometimes reaching 15 cm in height. It branches featherlike, and a few

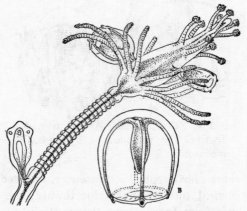

Fig. 86—P. tiarella. A, hydroid; B. medusoid.

annulations occur on the main stem above the beginning of the branch and on the side branches above their origin. The hydranths are large with 10 or 12 filiform tentacles at the base, while around the proboscis occur a varying number of shorter tentacles knobbed at the ends. In this species, the medusoid buds appear growing directly from the side of the hydranth. Color of colony bright pink. Grows on eelgrass, stones, *Fucus,* gorgonians, wharves and other wood structures below low tide. Abundant and very common. All ranges.

The free medusoid form is ellipsoidal, being higher than it is broad. There are no tentacles, but there are 4 rudimentary tentacle-bulbs at the base of each radial canal. The manubrium is rose-pink in color, and there are a number of deep pink patches in the endoderm of each radial canal. Height 2 mm. Common. All ranges.

Family CORYMORPHIDAE

(With free-swimming medusoids)

Genus *Corymorpha*

C. pendula. (Fig. 87.) The hydroid form of this species always grows singly and is never branched. The stem is thick along the mid-region, but it narrows considerably towards the base and towards the upper free end, which is long, slender, and pendulous. Rootlike, tubular, fleshy extensions form a holdfast for the animal. The outer surface is striated and a perisarc is present, but it exists only as a very delicate film. At the base of the large hydranth is a circlet of long, hollow tentacles. The proboscis is flask-shaped and is covered by a couple of rows of numerous tentacles smaller than those at the base of the polyp head. Medusoid buds occur on the sides of the hydranth. Color bright pink. This is one of the largest tubularians. Height 8 to 12 cm; diameter of widest part 6 mm. Not uncommon. Labrador to Vineyard Sound.

Fig. 87—C. pendula. A, hydroid; B, medusoid.

The medusoid form is somewhat pear-shaped with a slight conical extension at the top of the umbrella. This medusoid has large tentacle, and may have from 1 to 3 rudimentary tentacles. The well-developed tentacle is about 3 times the umbrella height, and its surface is surrounded with swollen rings of nematocysts. Pink pigment granules are present in the tentacle-bulbs. The manubrium extends a short distance beyond the veil; and the lips of the mouth-opening are covered with nematocysts. Color of manubrium pink and lilac. Height about 5 mm. Fairly common. Labrador to Vineyard Sound.

Genus *Hybocodon*

H. prolifer. (Fig. 88.) The hydroid occurs singly or in clusters. It is unbranched and the stem is covered with a delicate perisarc which widens and exhibits several conspicuous annulations as it approaches the hy-

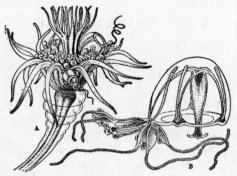

Fig. 88—H. prolifer. A, hydroid; B, medusoid.

dranth. The hydranth is flask-shaped with a broad base surrounded by about 25 long, tapering tentacles, while the cylindrical neck, which is extensible, has at the extremity two circlets of oral tentacles, each circlet composed of about 16 tentacles, those near the mouth being the shorter. The breeding season, which is from January to March, brings out great numbers of medusoids that arise from buds developed from the sides of the hydranth just above the ring of basal tentacles. Color orange. Height about 4 cm. Fairly common. Massachusetts Bay to Vineyard Sound.

The medusoid in the adult form is asymmetrical, being longer on one side than on the other. This longer side bears the single long tentacle for which this creature is distinguished in its immature form. Other tentacles (from 1 to 3) are present at maturity; and at their basal bulbs medusoid buds in various stages of development are often seen, the medusoid buds themselves having a single long tentacle before they are freed. Even more, when ready to be liberated, these young sometimes resemble the parents in that the former may be seen to develop medusoid buds of a third generation on their tentacle bulbs. The color of the tentacle bulbs is an intense orange. The manubrium is a simple tube ex-

tending from the inner apex of the umbrella cavity to about two thirds
of the distance to the velar opening. Height 4 cm. Common. Massachusetts Bay to Vineyard Sound.

FAMILY **DENDROCLAVIDAE**

(Hydroid and medusoid forms)

GENUS *Turritopsis*

T. nutricula. (Fig. 89.) The hydroid form of this species is slightly
branched, each branch bearing a single hydranth. Investing the branches
is a thick perisarc which ends abruptly below
the hydranth. Numerous somewhat shortened
tentacles are arranged in a series of rather
regular rows around the elongate body of the
hydranth. At the base of the hydranth are the
gonophores, which are supported on short
pedicels. The medusoid buds are invested
with a perisarc. Each gonophore gives rise
to a single medusoid. Color of colony a pale
yellowish-red. Height 8 to 10 mm. Found on
piling and submerged wood. Cape Hatteras
southward.

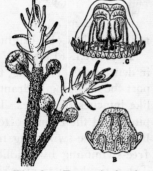

Fig. 89—T. nutricula. A,
hydroid; B, young medu-
soid; C, mature medusoid.

The medusoid form at the time of libera-
tion has 8 tentacles, but the number is con-
siderably increased later, sometimes reaching a total of 70. They are
attached to the margin of the belllike body, which contains 4 simple
radial canals and a narrow ring canal. A well-developed veil is present.
Turritopsis is unusual in the extraordinary development of its vacuolated
endodermal cells that line the courses of the radial canals. Diameter
about 3 mm. Cape Cod to Florida.

FAMILY **TUBULARIIDAE**

(With medusoids attached to hydroid)

GENUS *Tubularia*

T. couthouyi. (Fig. 90.) The individuals of this species are unbranched stems with a chitinous perisarc and with hydranths of large

size, frequently expanding the basal circlet of tentacles 25 mm or more. The total height of the individual animal may be as much as 15 cm.

The hydroids occur in groups of from 5 to 10 rising from a hydrorhiza that forms a creeping, tangled stem. Thirty to forty tentacles compose the basal whorl, while numerous smaller tentacles cover the proboscis, growing successively shorter as they occur near the mouth. From a position immediately above the basal tentacles spring pendent clusters of medusoid buds. No free-swimming medusoids are liberated, but the attached medusoids have distinct radial canals. Color bright pink. Not uncommon. Cape Cod northward.

Fig. 90—T. couthouyi.

T. crocea. (Plate XVI.) This hydroid occurs in colonies growing in dense tufts. The long, tangled stems are sparingly branched and support the drooping hydranths, from which the sporosacs hang in clusters like bunches of grapes. The stems are light yellow; the polyp-head is pink; and the colony often attains to a height of 10 mm. It does not liberate free-swimming medusoids, but the female sporosacs produce free-swimming hydroidlike forms known as actinulæ. (Plate XVI.) The actinulæ later give rise to the hydroid form. The apical processes of the female sporosacs are flattened. There are from 20 to 24 long tentacles in the circlet at the base of the hydranth, while a number of smaller ones are set close to the mouth. Common on wharf piles and submerged wood in shallow water. All ranges.

ORDER **CALYPTOBLASTEA: LEPTOMEDUSAE**

The members of this order are colonial forms having two principal kinds of polyps: the hydranths, or nutritive polyps, and the *blastostyles,* or reproductive polyps. Unlike the forms in the preceding order, these hydroids possess a perisarc that covers the hydranth, forming a protective cup or case (the *hydrotheca*) for the polyp, and a somewhat similar covering or enclosing capsule (the *gonotheca*) for the blastostyles. The rim of the hydrotheca in certain species may possess projections or valves (the *opercula*) which act as lids or doors for closing the chamber.

Not more than one circlet of tentacles surrounds the hydranth, and in the majority of instances the polyp can be extended from or retracted within the hydrotheca. The blastostyle, also, in some forms protrudes from the gonothecal opening and composes a brood chamber (the *acrocyst*) in which the eggs develop. As a rule, however, the blastostyle can not be extended beyond the gonotheca; it produces within this capsule the gonophores which constitute the medusoids, and these may either be liberated as free-swimming forms or they may be kept within the gonotheca, producing sexual products therein that later escape as free larvæ. The medusoids are commonly known as *Leptomedusæ*. Most leptomedusans differ from the anthomedusans of the preceding order in having lithocysts instead of ocelli as sense organs; the lithocysts, however, are organs presumably for the maintenance of equilibrium, and not for the perception of light. Also they are unlike the anthomedusans in that their gonads are borne on the subumbrella beneath the radial canals, instead of on the manubrium.

FAMILY SERTULARIIDAE

The sertularians. Hydrothecæ are sessile and more or less adnate to the hydrocaulus (never stalked), and arranged on *both sides* of the stem and branches. No free-swimming medusoids are produced; the gonophores give rise to sporosacs.

GENUS *Sertularia*

S. pumila. (Plate XVI.) The colonies of this species are more or less branched and rise from a creeping hydrorhiza. Both stem and branches are divided into short regular internodes, each internode bearing a pair of opposed hydrothecæ with opercula composed of 2 flaps. The gonothecæ are sessile but not adnate, and are urn-shaped, the male gonophores being somewhat more slender than those of the females. Often the gonothecæ bear acrocysts. Height from 1 to 5 cm. Found on *Fucus,* stone and old shells. Common. New England and southward.

GENUS *Sertularella*

S. polyzonias. (Fig. 91.) Colonies of this species are irregularly growing, irregularly branched with irregularly spaced internodes. Their hydrothecæ are smooth and somewhat stout, and have square apertures with 4 low teeth on the margin with 4 opercular flaps. The gonothecæ are ovate and deeply ringed throughout; and around their margins are 4 conspicuous hornlike projections. Maturation of the sex-elements causes the gonothecæ to become surmounted by globular acrocysts bearing eggs that develop into planula larvæ. Height about 12 cm. Common. New England.

Genus *Thuiaria*

T. argentea. (Fig. 92.) The silvery sertularian. This species occurs in big bushy colonies having silvery branches on dark stems, and specimens 25 cm high are not uncommon. The branches are spirally placed

Fig. 91—S. poly-
zonias. A, portion of
colony; B, gono-
thecæ.

Fig. 92—T. argentea.
A, portion of colony;
B, gonothecæ.

on the stems, and have irregular nodes. The hydrothecæ are somewhat cylindrical and are nearly adnate to the hydrocaulus; they may be arranged either sub-opposite to each other in pairs or alternately. Circular apertures surmount the gonothecæ, and usually 2 spines are present on the gonothecal opening; the operculum is 2-valved; and the gonophores are broad and taper toward the base. Acrocysts are borne on the mature gonophores. This one of the commonest as well as one of the most beautiful of the sertularians. It is found from low-tide mark to depths of 100 fathoms. Arctic Ocean to New Jersey.

T. cupressina. (Fig. 93.) The sea cypress. Somewhat similar to *T. argentea,* this species branches alternately and dichotomously. It differs, however, in being more slender and elongated with arched or drooping branches instead of straight. No internodes are present on the main stem; those on the branches are irregularly spaced. The hydrothecæ are sub-opposite to alternate, and have large 2-lipped apertures with 2-flapped opercula.

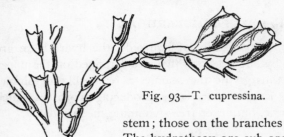

Fig. 93—T. cupressina.

Rather stout gonothecæ, each with 2 very strong, pointed, lateral spines projecting upward, are borne on the upper sides of the branchlets where they occur in rows. Acrocysts are present at the maturity of the gonophores. Height from 15 to 25 cm. Common, though not so common as *T. argentea.* Low-tide mark to 150 fathoms. New England.

Genus *Hydrallmania*

H. falcata. (Fig. 94.) In the colonies of this species, the main stems are somewhat twisted spirally and bear branches on each of which occur regularly arranged and feather-like secondary branches. The whole structure has a superficial resemblance to a plumularian. The hydrothecæ are arranged in groups on the front of the branches and branchlets, the top of one reaching the middle of its close neighbor overhead, and they bend alternately to the right and left when seen from above. A single flap composes the operculum. The ovate gonothecæ are usually borne on the distal parts of the branches and the proximal parts of the branchlets. They have several indistinct longitudinal striations, and end in a short tubular neck with a round aperture. Height 30 cm. Fairly common and abundant

Fig. 94—A, gonotheca; B, portion of colony.

on old shells, rocks and stones from low-tide mark to deep water. Occurs on the New England coast and in Long Island Sound.

FAMILY **PLUMULARIIDAE**

The plumularians. In this family of hydroids, the hydrothecæ are sessile, usually adnate by one side, and they are arranged on the upper side of the branches, thus giving the colonies a featherlike or plume-like appearance. The gonophores are large, producing sporosacs which give rise to planula larvæ, but never free-swimming medusoids.

GENUS *Antennularia*

A. antennina. (Fig. 95.) The colony of this hydroid consists of a dense cluster of upright stems, each bearing at the internodes a whorl of fine branches on which are the hydranths and special pedicelled stinging organs, the nematophores. The gonothecæ are ovate and are borne in the axils of the branches and the main stem. Height from 16 to 20 cm. Found in offshore waters. Bay of Fundy to Vineyard Sound.

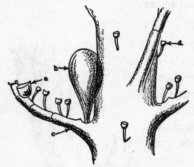

Fig. 95—A. antennina. a, hydranth; b, gonotheca; c, hydrocladium; d, nematophore.

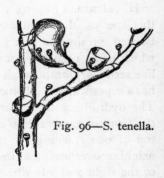

Fig. 96—S. tenella.

GENUS *Schizotricha*

S. tenella. (Fig. 96.) Colonies of this species grow in clusters. They are branched dichotomously, and each stem is divided into alternate long and short internodes, the short internodes each bearing a hydranth

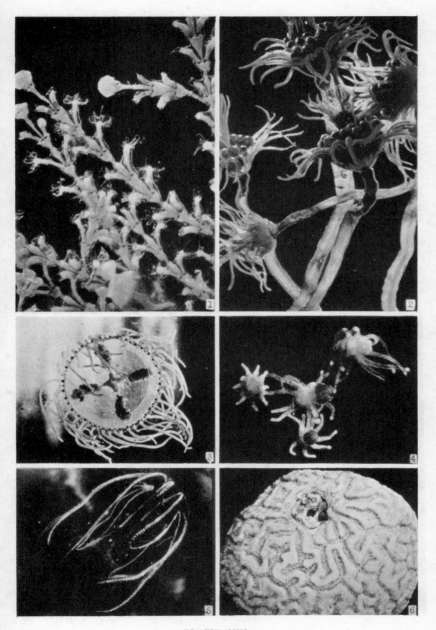

PLATE XVI

1—*Sertularia pumilla*; 15x.   2—*Tubularia crocea*; 8x.
    3—*Gonionemus murbachii*; 2¾x.   4—*T. crocea*; actinulae; 10x.
5—*Mnemiopsis leidyi*; ¾x.   6—*Meandrina sinuosa*; ½x.

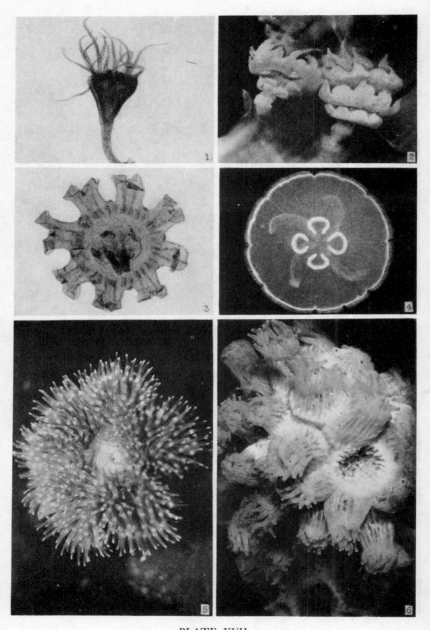

PLATE XVII

1—Scyphistoma of *Aurelia*; 18x.　　　2—Strobila of *Aurelia*; 15x.
3—Ephyra of *Aurelia*; 15x.　4—Adult form of *Aurelia*; about 3/16x.
5—*Metridium dianthus*; 1x.　6—*Astrangia danæ*; 3x.

and a slender side branch which in turn is often branched. On each internode of the stem and branches are 1 or 2 special stinging organs, the nematophores; and on each alternate internode of the branches occurs a hydranth equipped with lateral nematophores. The gonothecæ are curved like cornucopiæ, and carry from 3 to 4 nematophores at their bases. Height 5 cm. Abundant on wharf piles. Cape Cod southward.

GENUS *Aglaophenia*

A. acacia. (Fig. 97.) In the colony of this plumularian, no branches appear toward the base of the stem, but distally they are given off quite regularly, occurring in opposite or sub-opposite pairs. Also, these branches may again branch in pairs, thus forming a twice-pinnate arrangement. The internodes of the main stem are indistinct, but those of the branches are very marked and divided regularly, each carrying a hydranth supplied with stinging organs, the nematophores, 1 of which is mesial and situated about halfway up the front of the hydrotheca, and 2 nematophores occurring in pairs at the top of the hydrothecal margin near the nodal joint. The margins of the hydrothecæ are notched, bearing 9 rather sharp teeth. The gonophores are protected by peculiar gonothecal coverings known as the *corbulæ,* which consists of 6 paired plates, or leaves, called the *costæ.* Height 12 cm. Found occasionally in offshore waters. Cape Hatteras southward.

Fig. 97—A. acacia. A, portion of colony; B, gonotheca.

FAMILY **CAMPANULARIIDAE**

The campanularians. The hydroids of this family are distinguished by their bell-shaped, usually stalked, hydrothecæ. The proboscis, or hypostome, is trumpetlike. Large gonophores are the rule, and these produce planula larvæ, never free-swimming medusoids.

Genus *Campanularia*

C. poterium. (Fig. 98.) This hydroid occurs in unbranched colonies. The hydranths are borne at the end of stalks that are ringed throughout their lengths, and which rise from tortuous creeping hydrorhiza. No teeth appear on the bell margin. A circlet of 24 tentacles surrounds the base of the proboscis. The gonothecæ are somewhat slender and ovately cylindrical and are supported on short pedicels. Height from 6 to 10 mm. Found in shallow and offshore waters· on seaweeds, other hydroids, spider crabs, etc.; seldom on rocks or other solid objects. Common. Labrador to Vineyard Sound.

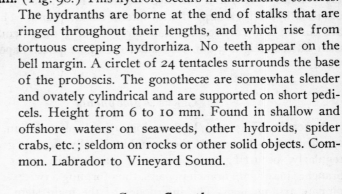

Fig. 98 —C. poterium.

Genus *Gonothyrea*

G. loveni. (Fig. 99.) The trophosome of this hydroid consists of a clump of irregularly branched stems with a series of 3 to 6 annulations immediately above each pedicel base. It is a colonial form and the animals often grow to a conspicuous size, sometimes attaining to a height of 4 cm or more. The hydrothecæ are gracefully campanulate, or belllike, in form, and the margins are each ornamented with 10 to 12 teeth the ends of which are squared off, thus giving to the bells a castellated appearance. Supporting the hydranths and gonophores are short pedicels ringed nearly throughout their lengths and occurring alternately on the stems and branches. The gonothecæ are borne near the junction of the pedicels and the hydrocaulus, and often occur in pairs. When mature they project from 3 to 5 fixed medusiform sporosacs that have radial canals and tentacles, but these are never liberated as free-swimming medusoids. These fixed medusoids do, however, have eggs that develop into planula larvæ which escape and lead a free life. Not uncommon.

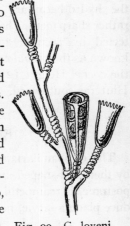

Fig. 99—G. loveni.

Found in shallow and offshore waters on stones and shells. Maine to Long Island Sound.

FAMILY **EUCOPIDAE**

Like the campanularians, the hydroid members of this family are also distinguished by their bell-shaped hydrothecæ, but they differ from the hydroids of the preceding family in their life-history. They invariably produce free-swimming medusoids. These are with 4 radial canals, each canal bearing a gonad on the subumbrella.

Although there are upward of 34 genera in this family, in only 2 are the hydroid forms well known, *Obelia* and *Clytea;* the remaining genera being represented only by medusoids. In the older literature, some members of this family were grouped with the *Campanulariidae.*

GENUS *Obelia*

O. commissuralis. (Fig. 100.) In this species, the hydroid form occurs in treelike colonies in which there is a long central trunk giving

off side branches at right angles somewhat spirally arranged. The trunklike stem intervals are slightly curved, and the base of each branch and pedicel bears 4 or 5 annulations. Several annulations also are present at the bases of the hydranths, and often the pedicels are annulated throughout their entire length. The hydrothecæ have no marginal teeth; their sides are often slightly incurved; and they are 12-sided around the marginal region. The gonophores are elongate and are on short pedicels, occupying the angles of the branches. Height

Fig. 100—O. commissuralis. A, hydroid; B, medusoid with inverted umbrella, in which form it often appears.

about 20 cm. Common on wharf piles, *Fucus,* etc. All ranges.

The medusoid form when first set free has 16 tentacles and no developed gonads; later, however, the gonads achieve the characteristics of the family. The shape of the medusoid is a somewhat flattened hemisphere bearing on the rim the slender tentacles which are about as long as the diameter of the umbrella, which latter is about 3 mm. All ranges.

Genus *Clytea*

C. bicophora. (Fig. 101.) The hydroid form of this species is colonial, occurring either sparingly branched or unbranched, the pedicels rising from a creeping hydrorhiza. Annulations occur on the pedicels usually at the distal and the proximal ends, leaving the midway section unmarked. Twelve to fourteen rounded teeth compose the marginal rims of the hydrothecæ. In this region the hydrothecal walls are thin, and when the polyps are retracted, a number of longitudinal folds appear according to the number of teeth. The gonopthecæ are oblong ovoid in shape and are ordinarily borne on the hydrorhizæ, but they may occasionally occur on the pedicels of the hydranths, particularly when these are branched. Height about 5 mm. Fairly frequent. Found in shallow water on seaweeds and the stems of other hydroids. All ranges.

Fig. 101—
C. bicophora.
A, hydroid;
B, young medusoid.

The mature medusoids are somewhat hemispherical, flattened and with flaring sides. The tentacles number 16 and are with well-developed tentacle bulbs. Spindle-shaped gonads about one fourth as long as the radial canals are present. The manubrium is short, and the mouth has 4 slightly recurved lips. Brownish pigment occurs in the tentacle-bulbs, gonads and manubrium. Diameter about 5 mm. Sometimes numerous. All ranges.

Genus *Oceania*

O. languida. (Fig. 102.) The hydroid form of this species is obscure. The medusoid is a somewhat flattened hemisphere with exceedingly flexible walls. There are 32 or more short, slender tentacles between each two of which, on the rim of the umbrella, are usually 2 lithocysts. The manubrium is small and tubular, and the mouth opening is provided with 4 slightly recurved lips. It is inconstant as to color: the manubrium and tentacle-bulbs may be either brilliant green, pink, or pale yellow. Diameter about 20 mm. Abundant during the Summer months. All ranges.

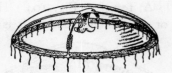

Fig. 102—O. languida.

Genus *Eutima*

E. mira. (Fig. 103.) In this species the hydroid form is not well known; young animals, only, having been observed, which were ob-

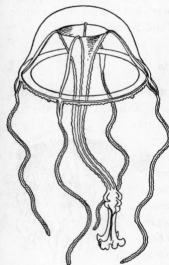

Fig. 103—E. mira.

tained from the eggs of the medusoid. The medusoid is typically umbrella-shaped, having a broad flattened body with a long, slender, manubrial appendage that in some specimens is fully 4 times as long as the umbrella height. Four tentacles are present, each of which is about 3 times as long as the body diameter; and there are also about 100 rudimentary nodules present on the umbrella rim. The gonads are located at two different points on the radial canals: on the subumbrella and on the long peduncle that carries the manubrium. Some specimens have been pigment granules in the tentacle-bulbs. Diameter from 15 to 30 mm. Very abundant some years and rather rare in others. All ranges.

Family **AEQUOREIDAE**

The members of this family may be recognized chiefly by their numerous radial canals, which may number from 8 to 100, and by their reproductive organs, which are usually ribbonlike. In most instances, the hydroid form is unknown.

Genus *Aequorea*

A. tenuis. (Fig. 104.) Hydroid form obscure. The medusoid is 3 or 4 times as broad as it is high, with 48 to 90 tentacles around the

rim of the umbrella. The tentacles are usually 2 or 3 times as numerous as are the radial canals, and they are long and slender and have hollow bases. In addition to these long tentacles there are present about 200 to 250 small rudimentary tentacles. Forty-eight to ninety

Fig. 104—A. tenuis.

lithocysts. The flat stomach is funnel-shaped and has a small mouth-opening surrounded by minute lappets. Diameter 8 to 10 cm. Common in late Summer; but commoner in some years than in others. Vineyard and Long Island Sounds.

Genus *Zygodactyla*

Z. groenlandica. (Fig. 105.) Hydroid form obscure. The medusoid body is much flattened, approaching the form of a disk. Usually the top surface of the disk is flat or slightly concave in the center. Only a rudimentary vellum is present. About 100 long tentacles with hollow, tapering bulbs depend from the rim of the disk, these being only slightly more numerous than are the radial canals. Between each two pairs of tentacles are 8 to 12 very small lithocysts. A row of 6 to 13 wart-like prominences is situated on the under surface of the subumbrella between each two successive pairs of radial canals. The stomach is broad and somewhat funnellike, terminating below in a

Fig. 105—Z. grœn-landica.

cylindrical throat-tube. Tapering tentacles, equal in number to the radial canals, surround the mouth-opening. This is the largest American hydro-medusan, often measuring more than 12 cm in diameter. It occurs in all ranges; those in the northern and central ranges being colorless; the endoderm of the radial canals and the circular canal, and of the gonads and the tentacles in the southern varieties is a delicate madder-pink.

ORDER **TRACHOMEDUSAE**

In most of the marine forms belonging to this order, the hydroid stages are unknown or where known are apparently degenerate and of minute size. Many trachomedusans are independent of a fixed hydroid stage, the medusoids developing directly from the eggs of other medusoids and not being produced by a hydroid. As these animals are not associated with the shores by a fixed hydroid stage, they are essentially creatures of the open sea. They are, however, often carried close to the shore and into harbors and bays by storms and other agencies; therefore, their study is well within the province of the seashore naturalist. A distinguishing characteristic of the trachomedusans is an uncleft umbrella margin. Five families comprising 80 species are contained in the order. Two representatives are described here.

FAMILY **PETASIDAE**

GENUS *Gonionemus*

G. murbachi. (Fig. 106 and Plate XVI.) The hydroid form is a small saclike creature with 4 tentacles around a proboscoid mouth having a cruciform oral opening. Height about 1 mm. It apparently attaches itself to solid objects on the bottom in water of various depths; but it is very rarely collected. Southern Coast of New England.

The medusoid is a slightly flattened hemisphere with 60 to 80 stiff tentacles about three fourths as long as the body diameter. The tentacles project from the sides of the umbrella a slight distance above the rim: and each tentacle has near the outer end a small adhesive pad by which the animal is enabled to attach itself to objects on the sea bottom. At the point where the adhesive organ occurs, the tentacle makes an abrupt right-angled turn.

Fig. 106— Hydroid stage of G. murbachi.

There are 4 radial canals, upon which the gonads are developed; the latter being ribbonlike and deflecting into series of folds. The manubrium is short and has four recurved crenated lips. Pigment spots of grass-green are at the roots of the tentacles near the circular canal; the gonads and manubrium are brown. Diameter about 20 mm. Common. Vineyard and Long Island Sounds.

FAMILY GERYONIIDAE

GENUS *Liriope*

L. exigua. (Fig. 107.) There is no hydroid stage in this species, development being direct. The hydromedusan is somewhat hemispheri-

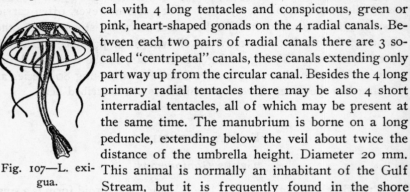

cal with 4 long tentacles and conspicuous, green or pink, heart-shaped gonads on the 4 radial canals. Between each two pairs of radial canals there are 3 so-called "centripetal" canals, these canals extending only part way up from the circular canal. Besides the 4 long primary radial tentacles there may be also 4 short interradial tentacles, all of which may be present at the same time. The manubrium is borne on a long peduncle, extending below the veil about twice the distance of the umbrella height. Diameter 20 mm. This animal is normally an inhabitant of the Gulf Stream, but it is frequently found in the shore waters of all ranges.

Fig. 107—L. exigua.

ORDER NARCOMEDUSAE

No hydroid forms are known to occur in the order, development being direct. The hydromedusans may be distinguished from those of the preceding order by their marginal clefts and lappets. Two families with about 50 species are contained in this order, 1 species being described here.

FAMILY AEGINIDAE

GENUS *Cunoctantha*

C. octonaria. (Fig. 108.) In this species the umbrella is a much flattened hemisphere. Eight tentacles project from the sides about midway from the rim and the top of the umbrella, the tentacles being rigid and incapable of much movement and are curved downwards. The tentacles are a little longer than the body radius. The short cone-shaped manubrium does not extend below the velum in the full-grown animal. This velum is quite complex, consisting not only of a simple membranous ring, but also of an upward extension in 8 webs between the 8 nerve-cords that bind the rim of the umbrella. No radial canals or ring canals are present in this species. Except for a delicate green in the mouth region, and occasionally in the tentacles, the creature is colorless. Diameter 7 mm. The larva of this species leads a parasitic life within the umbrella of *Turritopsis nutricula* medusoids. Common. Cape Hatteras southward.

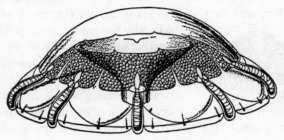

Fig. 108—C. octonaria.

ORDER **SIPHONOPHORA**

In this order of hydromedusans are included free colonial forms, in which the individuals that compose the colony are of several different types, varying according to their separate special functions in bringing about a division of labor. Although the various members of the colony are unlike in appearance, and seemingly heterogeneously associated, there is a common gastrovascular connection uniting them all. Two general types of colonial structure prevail in this order: in the one, the form is a large float in which the individual members bud off from the under side; in the other, a long axial tube is present from which the various individuals are budded off. In most instances, this axial tube is expended at the upper end to form a float. This float, in both general types of colony, is called the *pneumatophore*. The order is composed of 4 suborders which contain about 250 species, all the members being pelagic or deep-water forms; notwithstanding, many of them find their way to the shallow waters of the shores.

Suborder DISCONECTAE

Floating siphonophores, having no individual swimming members of the colony. The pneumatophore is disklike and usually carries a central stem beneath, which bears the principal mouth and stomach.

Family VELELLIDAE

Genus *Velella*

V. mutica. (Fig. 109.) The disk of this siphonophore is elliptical and very flat, and carries diagonally across the top a triangular sail

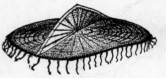

whereby the colony is propelled as it floats on the surface of the water. The discoid pneumatophore is of a complex structure, being divided into a number of concentric, communicating air-chambers. On the under side, besides the single mouth and the stomach, are small reproductive individuals bearing gonads, and surrounding those near the rim of the disk are a number of longer tentacle-like individuals bearing stinging organs. Color bright blue. Length of disk about 4 cm; breadth about 2 cm. Occasionally occurs in all ranges; common in southern waters.

Fig. 109—V. mutica.

Suborder CYSTONECTAE

Floating siphonophores, without swimming individuals; from the under side of the large pneumatophore depend numerous nutritive individuals. No central stem.

Family PHYSALIIDAE

Genus *Physalia*

P. pelagica. (Plate XII.) Portuguese man-of-war. In this siphonophore, the conspicuous pneumatophore is somewhat pear-shaped and bears on its upper side a chambered crest which helps to act as a sail when the colony floats at the surface. The pneumatophore on such occasions is filled with air or gas. But the colony can sink and float below the surface when it elects to do so, by compressing the pneumatophore and thus expelling the contents through a pore on the upper side. When it attempts to rise to the surface again, it generates a buoyant gas to fill the float. Depending from the under side of the pneumatophore are numerous very long and powerful stinging tentacles, sometimes attaining to a length of more than 15 meters. These tentacles are clustered at one side near the broader end of the float, and they seem to serve the purpose of sea-anchors as well as that of killing and capturing prey. In addition to these longer tentacles are alternately situated clusters of similar, but shorter, tentacular organs that aid in directing the course of the colony. The reproductive individuals, which resemble bunches of grapes, and the nutritive individuals are adjacent to the tentacles and hang in associated clusters from the under side of the pneumatophore. Color of pneumatophore iridescent peacock-blue and sometimes orange; color of gonads bright blue. Length of pneumatophore about 12 cm. Common in the Gulf Stream, which is its native habitat; occasionally in shore waters of all ranges.

Suborder **PHYSONECTAE**

Swimming siphonophores. With swimming bells, or *nectophores,* budding off from the central axis immediately below the pneumatophore.

Family **AGALMIDAE**

Genus *Cupulita*

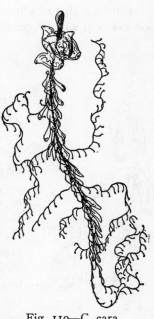

C. cara. (Fig. 110.) In this siphonophore, the members of the colony are grouped around a long stem, or axis, looking somewhat like the flowers on the raceme of the lily-of-the-valley. At the uppermost end of the stem is the pneumatophore, a small, ovoid, bubblelike float containing air. Just below the float are a number of closely set swimming bells without a manubrium or mouth, but with radial canals and

Fig. 110—C. cara.

a velum. It is these individuals that propel the colony through the water; which feat they accomplish by contracting rhythmically, thus pumping water into their cavities and immediately expelling it. Below this group of swimming individuals, there immediately comes another of a totally different appearance and containing several different sorts; these are arranged in assemblages occurring one after the other at regular intervals, thus dividing the stem into regions like the nodes and internodes of certain hydroids or plants. From some of the nodal points there arise polyps bearing no circlet of tentacles but a single long branched tentacle endowed with stinging properties. Other nodal points give rise to mouthless polyps, each bearing an unbranched tentacle that is used as a sensory organ or "feeler." The gonophores are at the bases of the various polyps. From the internodes project delicate, leaflike structures, the *bracts,* which partly cover the gonophores. Color of colony rose-pink. Length about 11 cm. Not uncommon. New England.

Suborder CALYCONECTAE

Swimming siphonophores, with large swimming individuals, or nectophores, and without pneumatophores or palps, or feelers.

Family DIPHYIDAE

Genus *Diphyes*

Fig. 111—D. bipartita.

D. bipartita. (Fig. 111.) In this species, the 2 large conical swimming bells each contain a groove into which the remainder of the colony can be retracted. The remaining groups of individuals are attached to a common stem depending from the nectophores, and they are somewhat widely separated. From time to time, the stem separates at the internodes, detaching the oldest individuals; and each of these independent groups, called a *Eudoxia,* swims about on its own and becomes sexually mature. Colorless and transparent. Length of colony variable, but about 30 mm; length of nectophores 10 mm. Fairly frequent. All ranges.

CLASS **SCYPHOZOA: SCYPHOMEDUSAE**

The large jellyfishes. In this class of creatures are found many re-semblances to the *Hydromedusae.* In some only the hydroid stage is passed through; in others, the medusoid alone exists, there being no hy-droid generation; in still others, both the hydroid and the medusoid forms are present. Scyphozoan jellyfishes, however, differ from most hydromedusans in having a very rudimentary velum or, what is most usually the case, in having this structure wanting entirely. This absence of the veil in the scyphozoans has characterized them as *acraspedote,* in contradistinction to the craspedote *Hydromedusae,* which are commonly distinguished by its presence. Also they differ in their sense organs, which are small tentacular structures bearing statoliths at their tips. Tech-nically, these sense organs are known as *tentaculocysts* or *rhopalia.* A protective hood, or fold of the umbrella rim, commonly covers the sense organs. In some species, one or more minute eye spots are present. Then, too, the stomach cavity in the scyphozoan jellyfishes is more complex than that of the hydromedusans, extending out toward the rim of the umbrella in four radial pouches and forming a large space in the center of the animal. In those species in which a ring canal is present, the stom-ach may often be continuous with it. The group is carnivorous in its habits. The sexes are separate, and the gonads are usually conspicuous and brightly colored. Many species are luminescent and these contribute largely to the nightly phosphorescence of sea. The class includes 5 orders with more than 175 species.

ORDER **STAUROMEDUSAE**

Hydroidal jellyfishes. Sedentary but not permanently fixed forms. No typical medusoid stage.

FAMILY **LUCERNARIIDAE**
GENUS *Haliclystus*

H. auricula. (Fig. 112.) The body of this form consists of a stemlike region, which in cross-section is cruciform, one end of which is usually attached to a base, and bearing at the free end an umbrella with 8 groups of tentacles on extensions of the rim. The tentacles are short and terminate in small knobs covered with nematocysts; the tentacles numbering

Fig. 112—H. auricula.

about 100 in each cluster. Between each two prolongations of the umbrella rim, and situated perradially and interradially on the margin, is a large adhesive organ, or anchor as it is commonly called. This organ is bean-shaped and mounted on a short cylindrical base. The 4-sided mouth is in the center of the umbrella. Eight gonads are present; each is wide and triangular, and begins just above the junction of the stem and the umbrella and extends to the base of a cluster of tentacles. This attractive creature is iridescent, transmitting various shades of blue, green, yellow, or orange; however, individuals colored with actual purple or brownish substances are not uncommon. Height about 25 mm; diameter of umbrella about 25 mm. Although seemingly fixed when collected on eelgrass and other objects to which it attaches itself, this animal can move about at will, either floating free or crawling from place to place with the aid of its anchors and adhesive stem base. Also, its body is quite flexible, and with its introversible umbrella it can assume a number of protean attitudes. Not uncommon. Greenland to Cape Cod.

ORDER **CORONATAE**

Jellyfishes with a constriction usually around the middle of the umbrella. Free-swimming forms with a hydroidal stage in life-history.

FAMILY **PERIPHYLLIDAE**

GENUS *Periphylla*

P. hyacinthina. (Plate XV.) This jellyfish is recognized by its high, narrow umbrella, having somewhat the shape of a foolscap. Below the deep, broad constriction girdling the bell, the margin is divided into 8 pairs of well-developed lappets, between each two pairs of which is a deep longitudinal furrow. Twelve tentacles a trifle longer than the body height are inserted, 3 in a group, between the folds of every four lappets in series. Four rhopalia, or sense organs, are present: 1 in each lappet-cleft between the tentacle groups. Colorless except for the lappets, which are densely pigmented with dark purple-brown. Height 8 cm; diameter 4 cm. This is a deep-sea animal, but occasionally it is found at the surface and very close to shore. All ranges.

FAMILY **EPHYROPSIDAE**

GENUS *Linuche*

L. unguiculata. (Fig. 113.) The thimble jelly. This species, sometimes appearing under the name of *Linerges mercurius,* is a thimblelike form with a constriction around the top of the umbrella. Like the object from which it derives its popular name, it is small and cylindrical, but the sides of the umbrella bear 16 longitudinal grooves. The margin is adorned with 16 blunt lappets, 8 very small tentacles and 8 sense organs. The manubrium is 4-sided and urnshaped with 4 recurved lips at the mouth-opening. Color pale blue. Height 13 mm; diameter 16 mm. Inhabitant of the Gulf Stream, but occasionally very numerous near the shore.

Fig. 113—L. unguiculata.

ORDER **CUBOMEDUSAE**

Free-swimming jellyfishes with cuboidal umbrellas. A false velum, or *velarium,* is present, which may lead the beginning collector to mistake these forms for craspedote medusoids. The life-history contains a hydroidal generation.

FAMILY **CHARYBDEIDAE**

GENUS *Tamoya*

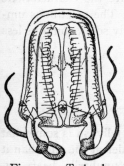

Fig. 114—T. haplonema.

T. haplonema. (Fig. 114.) Body somewhat rectangular with 4 long tentacles. The outside, or exumbrella, is covered with wartlike clusters of stinging cells. The tentacles are situated 1 at each corner and are expanded at the bases, forming flattened structures known as *pedalia.* Four sen-

sory organs are present. Velarium well developed, but is unlike the
velum of a craspedote medusoid in that it is not structurally a part
of the exumbrella. Gonads curtainlike with frilled edges, and situated
in the 4 interradial regions of the body. Body colorless with bluish or
milky-white tentacles and warts. Height 9 cm; diameter 5 cm. Sometimes common. Southern New England to Florida and the West
Indies.

ORDER SEMAEOSTOMEAE

Mostly large free jellyfishes having 4-sided mouths, the lips of which
are extended into lobes sometimes much folded and frilled and of considerable length. With one exception (*Pelagia*) the members of this order pass through a hydroidal stage in their development. This fixed form
is a small hydroidal animal of less than a centimeter in height, usually
occurring solitary, and is known as the *scyphistoma* (Plate XVII). The
scyphistoma is asexual; it produces medusoidal jellyfishes by a process
of terminal budding. As the scyphistoma grows in height, a number of
constrictions appear along its length dividing it into a graduated series
of saucerlike individuals, each tiny saucer attached to its neighbor by
the center of the disk, the topmost and first-formed disk being the largest
and farthest developed. In this stage the now compound animal is called
the *strobila* (Plate XVII); each saucerlike individual is termed an
ephyra, or *ephyrula* larva, and is a young medusa, or jellyfish (Plate
XVII). As the ephyrula larvæ become fully developed, they successively
liberate themselves; and after a few weeks acquire the characteristic
appearance of the adult and sexual jellyfish form. The gelatinous umbrella of most jellyfishes, though firm and seemingly solid, contains less
than 5 per cent. of solid matter; the rest is water.

FAMILY PELAGIIDAE

GENUS *Pelagia*

P. cyanella. (Plate XV.) The umbrella of this jellyfish is somewhat
fuller than a hemisphere; that is, it is a trifle broader at the rim than it
is a short distance above. The outer surface, or exumbrella, bears numerous reddish, wartlike bodies, nematocyst capsules, arranged in radi-

ating lines extending from the top. Sixteen lappets adorn the rim, bearing alternately between each two pairs either a sense organ, or a very long, contractile tentacle; thus, making 8 rhopalia and 8 tentacles. The manubrium is very long and branches into 4 flexible mouth-arms with crenulated, veillike edges. Four much folded horseshoe-shaped gonads are present. Color variable; sometimes bluish, sometimes yellowish. Height 4 cm; diameter 5 cm. This species is not represented by a hydroidal or strobila generation. The eggs give rise to planula larvæ which in turn develop directly into free jellyfishes. Common in southern waters; occasionally in the north.

GENUS *Dactylometra*

D. quinquecirrha. (Plate XV.) The speckled jellyfish. Umbrella nearly hemispherical and with numerous wartlike groups of nematocysts on the exumbrella. Forty-eight marginal lappets, 40 long tentacles and 8 sense-organs adorn the rim. Mouth-opening surrounded by 4 very long mouth-arms having their free edges produced into a complexly crenulated veillike structure. Four gonads are present, their position being marked by 4 deep subgenital depressions. Color variable; sometimes yellowish, sometimes bluish, sometimes pinkish. Height about 4 times the umbrella diameter, the latter being up to 25 cm. This species produces eggs that develop into free-swimming planula larvæ which become attached to the bottom and pass through the scyphistoma stage. Occasionally frequent. Southern New England to Carribean Sea.

FAMILY **CYANEIDAE**

GENUS *Cyanea*

C. capillata, variety *arctica*. (Plate XV.) The sea blubber. The umbrella of this jellyfish is quite flattened, being shaped like a watch crystal. Eight deep clefts divide the rim into 8 principal lappets, which in turn are each divided by a minor median cleft. Within the major clefts are situated the 8 sense-organs. From the floor of the inner cavity, or subumbrella, depend about 800 long tentacles in 8 clusters. The tentacles are highly contractile, and when expanded to their full length are about 25 times the diameter of the umbrella. A 4-cornered mouth is located at the center of the subumbrella, and is provided with 4 long mouth-arms.

These latter structures have the margins much folded, and hang curtain-like from the oral floor. The gonads are contained in 4 folded pouches which project from the subumbrella floor in the region of the 4 inter-radial sides of the stomach. Color variable, but commonly reddish-brown or yellow. This is the largest known jellyfish, specimens having been found that measured more than 2 meters in diameter, with tentacles 40 meters long. Individuals of this size, however, are rare; more commonly they measure from 10 to 50 cm in diameter. Very common. Greenland to North Carolina.

The foregoing described form is a variety of *C. capillata,* a European species found in the English Channel and the North Sea. Other American forms as follows:

Variety *fulva.* Similar to *C. arctica* except for its characteristic yellowish color and smaller size, rarely measuring more than 20 cm in diameter. Common. Cape Cod southward.

Variety *versicolor.* Similar to *C. fulva* except for its color and geographic range. It may be bluish-white or pink. Common. Cape Hatteras southward.

Family ULMARIDAE

Genus *Aurelia*

A. aurita. (Plates XV and XVII.) The moon jelly. In this species, sometimes known as *A. flavidula,* the umbrella is a somewhat flattened hemisphere composed of tough gelatinous substance which is thick in the center and thin at the rim. Eight broad lappets extend around the margin, between each two pairs of which is a sense-organ covered by a hoodlike structure. A slight distance above the lappet edge, numerous short tentacles arise which alternate with an equal number of very small lappet-like projections forming a fringe. The central mouth is 4-sided and is produced into 4 mouth-arms, each of which is almost as long as the radius of the umbrella, and each of which has the free edges much convoluted and provided with a row of small tentacles. The mouth-arms are broader at the bases than at the tips, and are folded at their free edges in such a manner that a trough or gutter is formed on the under side along their entire lengths. The gonads alternate with the mouth-arms, and are centrally situated; they are curious horseshoe-shaped structures, those of the males being pink while those of the females are yellow. Diameter from 14 to 25 cm. Very common. All ranges.

ORDER **RHIZOSTOMAE**

The root-mouth jellyfishes. Distinguished usually by the absence of marginal tentacles; and by having a large oral structure consisting of 8 lobes produced into fleshy, branched appendages with numerous pores. The pores are each bordered by minute, constantly moving tentacles, and often entirely take the place of the single large mouth characteristic of other jellyfishes. So far as is known, the root-mouth jellyfishes pass through both the scyphistoma and the strobila stages. Nearly 65 species are contained in the order, the great majority of which are tropical.

FAMILY **STOMOLOPHIDAE**

GENUS *Stomolophus*

S. meleagris. (See the frontispiece.) Umbrella thick and rigid; half-eggshaped and higher than a hemisphere. No marginal tentacles. Marginal lappets number about 128 in groups of 16 between the 8 sense-organs around the rim. A manubrium, forming a thick-walled mouth-tube with branching, flaring end processes, projects downward from the central region of the subumbrella beyond the level of the rim. In addition to a central mouth opening, numerous slitlike lateral orifices are present on 16 leaflike extensions of the upper part of the manubrium. Small tentacles are present on free edges of lateral mouths and mouth grooves at free end of manubrium. Color of exumbrella yellowish with brown recticulation, and with a broad, dense band of brown pigment near the margin. Diameter about 18 cm. Occasionally in northern waters; frequent in temperate waters; common in southern waters.

CLASS **ANTHOZOA (ACTINOZOA)**

Sea fans, sea pens, sea anemones, corals, etc. The coelenterates in this class are typically barrel-shaped. One end of the cylindrical body is either permanently or temporarily attached, and is known as the pedal disk, or foot; the other end is usually flattened and bears a central, slitlike mouth surrounded by hollow tentacles, and is known as the oral disk. The animals may occur singly or in disconnected groups, such as the sea anemones; or they may form associated colonies as in the sea fans and corals. Unlike the hydroids, in which the mouth opens directly into the stomach, a gullet is present between the mouth and the gastrovascular space. The stomach cavity is divided into a number of communicating chambers by six or more vertical partitions, or *mesentaries,* which rise from the body wall and extend towards the center, some mesentaries in the upper part joining the gullet, thus forming chambers continuous with the hollow tentacles and leaving the mesentaries between them, and the mesentaries in the lower part, with their edges free. (Fig. 115.) Along the free edges of the mesentaries the gonads, and also nematocysts, are borne. Most anthozoans are unisexual. The eggs and sperm find their way to the stomach cavity and are then carried out through the mouth by the way of a groove called the *siphonoglyph,* often situated at one or both ends of the slitlike opening. In some cases development of the eggs takes place within the stomach cavity; in others the young may be carried for a period in the body walls. Nonsexual reproduction by budding is common. The group is divided into two orders: the *Alcyonaria* and the *Zoantharia.* The fully developed members of the first-named order can be recognized by their possession of only eight tentacles and eight mesentaries. Members of the last-named order are known by their having tentacles and mesentaries that number six or some multiple of six —never eight.

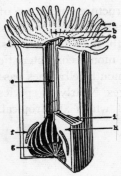

Fig. 115—Diagram of a sea anemone. a, tentacle; b, disk; c, siphonoglyph (one at each end of the mouth); d, mouth; e, gullet; f, gonads; g, mesenteric filament; h, primary mesentery; i, secondary mesentery.

ORDER **ALCYONARIA**

Sea pens, sea feathers, sea pansies, and gorgonian corals. Polyps are of two kinds: large 8-tentacled and 8-partitioned nutritive polyps, called *autozooids,* and small polyps without tentacles and incomplete partitions, called *siphonozooids*. Tentacles, when present, are fringed. Nearly all are colonial, the individuals generally united to a common stalk. The skeleton consists either of calcium carbonate or of horny (*ceratine*) spicules, or may be wanting entirely. The spicules are often fused together in the center of the colony so as to constitute a compact and rigid axial framework. Because of this characteristic stiffening of the mesoglœa in alcyonarians by the spicules, the mesoglœa and the outer ectoderm have been given the name *coenenchym*. Small depressions in the cœnenchym contain the polyps, and these individuals are in communication with one another by means of endodermal canals. More than 600 species of alcyonarians are known.

Suborder **ALCYONACEA**

Fixed colonial forms without a central axis.

Family **ALCYONIIDAE**

Genus *Alcyonium*

A. carneum. (Fig. 116.) Colony lobed or tree-like. Polyps usually buried up to their tentacle-bases in the cœnenchym. Color of colony yellowish or reddish. Height about 8 cm. Frequently found at low-water mark, but normally an inhabitant of deeper water. Gulf of St. Lawrence to Long Island Sound.

Fig. 116—A. carneum.

Suborder **GORGONACEA**

Family **GORGONIIDAE**

Genus *Gorgonia*

G. flabellum. (Plate XXXIV.) The sea fan. Colony a fan-shaped network. The meshes being 2 to 6 m wide. Color yellowish or reddish. Height up to 50 cm; width 50 cm. Key West and West Indies.

Suborder **PENNATULACEA**

Colonial forms; not fixed, but capable of independent movement. A central calcareous or hornlike axis is usually present.

Family **PENNATULIDAE**

Genus *Pennatula*

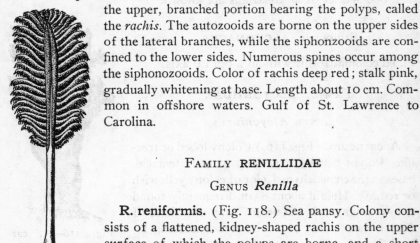

P. aculeata. (Fig. 117.) Sea feather. In this form the colony consists of two parts, the stalk which is imbedded in the sand or mud, and the upper, branched portion bearing the polyps, called the *rachis*. The autozooids are borne on the upper sides of the lateral branches, while the siphonzooids are confined to the lower sides. Numerous spines occur among the siphonozooids. Color of rachis deep red; stalk pink, gradually whitening at base. Length about 10 cm. Common in offshore waters. Gulf of St. Lawrence to Carolina.

Family **RENILLIDAE**

Genus *Renilla*

R. reniformis. (Fig. 118.) Sea pansy. Colony consists of a flattened, kidney-shaped rachis on the upper surface of which the polyps are borne, and a short smooth stem without an axis. Color of polyps white; rachis violet or amethyst; tip of peduncle white. Length 7 cm when extended, but can contract to half this length. Frequent in shallow water. Occurs on the Carolina coast and in the West Indies.

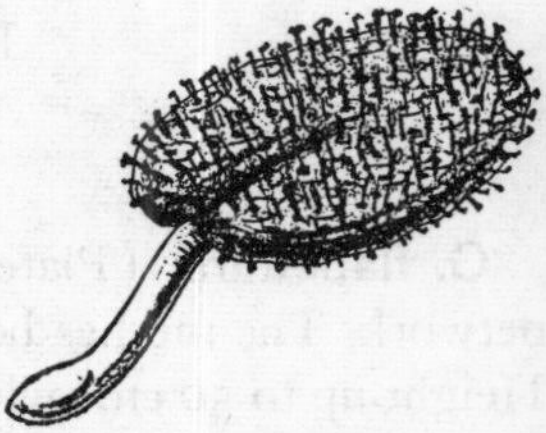

Fig. 117—P. aculeata.

ORDER **ZOOANTHARIA**

Sea anemones and stony corals. Aside from the fact that the members of the present order differ from the alcyonarians in having tentacles and mesentaries that occur in sextets or some

Fig. 118—R. reniformis.

multiple of six, the zooantharians differ also in that the tentacles are usually simple; whereas, in the *Alcyonaria,* the tentacles are commonly pinnately branched. In the *Zooantharia,* the tentacles may number from six to several hundred; and in addition to the six primary mesentaries there are usually present numerous secondary mesentaries. In some forms, such as the anemones, nettle organs known as the *acontia* occur. These organs are usually long threadlike structures armed with nematocysts, which can be extruded through the mouth or pores of the body when the animal is irritated. They seem to be used solely for defense. About 1500 species are contained in the order.

SUBORDER **ACTINARIA**

The sea anemones. Animals usually solitary and often of large size. Commonly attached to firm objects by the broad, adhesive foot, or pedal-disk; but capable of moving about at will.

DIVISION **EDWARDSIAE**

GENUS *Edwardsiella*

E. sipunculoides. (Fig. 119.) A burrowing species. It imbeds itself in sand or mud leaving only the flowerlike disk of expanded tentacles exposed above the surface of the bottom. There are about 36 tentacles, 8 of which are pointed outward and the remainder directed upward. The body is long and cylindrical and its outer surface bears 8 longitudinal ridges. Color transparent brown. Length about 12 cm; diameter 4 mm. Cape Cod northward.

DIVISION **HEXACTINIAE**

GENUS *Eloactis*

E. producta. (Fig. 120.) This elongate, wormlike anemone burrows in the sand or mud or under stones, leaving only the tentacles exposed. The tentacles number about 20. On the sides of the yellowish-gray body are about 20 longitudinal rows of whitish wartlike structures. Length, extended, 25 cm; diameter 18 cm. Fairly frequent in shallow water. Cape Cod to South Carolina.

Fig. 119—
E. sipunculoides.

Fig. 120—E. producta.

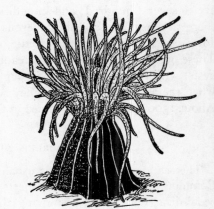

Fig. 121—S. luciæ.

Genus *Sagartia*

S. luciae. (Fig. 121.) The orange-striped anemone. The body of this species is olive-green with about 12 thin orange longitudinal stripes.

Fig. 122—S. leucolena.

The tentacles are numerous, sometimes reaching a total of 84, and are arranged in 4 rows around the mouth. Height about 8 mm; diameter 6 mm. Very common on stones and shells in tide pools. Gulf of Maine to Long Island Sound.

S. leucolena. (Fig. 122.) (See Plate I.) The white anemone. Notwithstanding its popular name, the semi-translucent body of this anemone is a delicate amber or flesh-color; although the color of the body might easily be mistaken for white in contrast to the dark situations in which it is sometimes found. The tentacles, however, are pure white, and number 96

arranged in 4 rows. Height, extended, 6 cm; diameter 10 mm.
Found in shady situations and under stones; often half buried in
sand or gravel. Shallow water. Very common. Cape Cod to North
Carolina.

GENUS *Metridium*

M. dianthus. (Plate XVII.) The brown anemone. This species,
also occasionally listed as *M. marginatum,* is the largest as well as the
commonest in the temperate waters of the Atlantic Coast. Its color is
variable, often being yellowish, occasionally white or pink or olive, but
more usually it is brownish. Tentacles numerous and rather short. Height
up to 10 cm; diameter about three fourths of body-length. Found on
stones, rocks, shells, wharf-piles, etc., from high-water mark to deep
and offshore waters. Very abundant. Labrador to New Jersey.

GENUS *Adamsia*

A. tricolor. (Fig. 123.) This species is a commensal animal; that
is, it is a form that lives in close company with another species, partaking

of its food. In this case the anemone attaches itself to the shell of some hermit
crab, and in return for the food it derives
from its companion, it is said to offer
protection to the other by its armature
of acontia and other stinging cells. A
band of wartlike prominences containing pores for the extrusion of the
acontia surrounds the base of the animal.
The tentacles in large individuals number about 500. Height about 7 cm; diameter about 4 cm. Not uncommon. Shallow water. Cape Hatteras to Florida.

Fig. 123—A. tricolor.

SUBORDER **MADREPORARIA**

The stony corals. Coral animals are virtually anemones that secrete
a hard calcareous skeleton. The individual animals of a coral colony

are each contained in a skeletal structure, commonly cuplike in form, which is composed of an outer wall, or *theca,* from which radial partitions, or *septa,* extend into the body of the polyp, and which is called the *corallite.* An assemblage of corallites is known as the *corallum.* A continuation of the body-wall of each polyp extends over the theca, forming a basal membrane that covers the entire corallum connecting the animal with its neighbors. Sometimes a central column, called the *columella,* is formed in the corallite. The growing coral animals increase the size of the corallites by dividing or by budding new individuals, and by withdrawing from the lower part of the corallites and building platforms, or *tabulæ,* over which they add extensions to the corallites. In certain species, the corallite is solid; in others it is perforated or forms a network through which the body substance, or cœnosarc, of the various members of a colony forms a continuous unit. The shape of the corallum is usually determined by the manner in which the polyps bud or divide. Thus, the brain-corals and other hemispherical types of corallum are produced by fission of the individual animals; while such

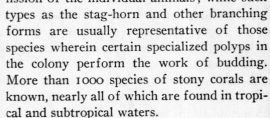

types as the stag-horn and other branching forms are usually representative of those species wherein certain specialized polyps in the colony perform the work of budding. More than 1000 species of stony corals are known, nearly all of which are found in tropical and subtropical waters.

Family **ACROPORIDAE**

Genus *Acropora*

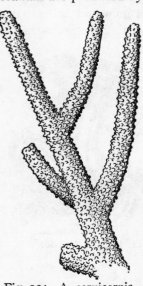

Fig. 124—A. cervicornis.

A. cervicornis. (Fig. 124.) Stag-horn coral. This form, which is a variety of the species *A. muricata,* grows in large branched colonies. The calcareous portion is porous and contains canals connecting the polyps, the latter being small in comparison with the corallum. At the end of each branch is a larger terminal polyp with 6 tentacles. This individual is contained in a cup with 6 septa, and is the original animal that started the branch.

The smaller lateral polyps, in corallites of 12 septa, bear 12 tentacles, and are the offspring budded from the base of the terminal polyp during its constant additions to its corallite structure. Considerable reefs are built by this species, sometimes extending for miles. The individual colonies attain to dimensions of 1 m or more in height by about 50 cm in breadth. Florida and the West Indies.

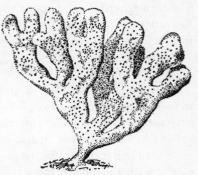

Fig. 125—P. porites.

Family **PORITIDAE**

Genus *Porites*

P. porites. (Fig. 125.) A massive reef-building form with thick lobular branches. Corallite with about 12 short septa and a poorly developed columella. Coral structure porous and consisting of a system of trabeculæ and cross-bars. Polyps small, but numerous and set close together. Florida and the West Indies.

Family **ASTRAEIDAE**

Genus *Astrangia*

A. danæ. (Plate XVII.) Star coral. This species forms small colonies of usually not more than 25 or 30 individuals. It is an encrusting species with starlike corallites having 6 principal septa, 6 secondary septa and 6 incomplete tertiary septa. The coral structure is solid. The polyps are whitish and nearly transparent, and have from 18 to 24 tentacles, which are knobbed at the ends and are stippled with white wartlike nematocysts. Colony increases in size by buds arising from the bases of the polyps. Size of corallum up to 10 cm in breadth and 5 cm in height. This species is found farther northward than any other stony coral. It occurs in shallow and offshore waters on stones, old shells and other hard objects. Fairly common. Cape Cod to Florida.

GENUS *Meandrina*

M. sinuosa. (Plate XVI.) Brain coral. Colony encrusting, massive and hemispherical in shape with corallites arranged in serpentine configurations. Corallum solid. Breadth of corallum about 25 cm. Occurs in Florida and the West Indies.

CLASS **CTENOPHORA**

The comb jellies. This is an aberrant group of animals; although in certain respects they have a superficial resemblance to medusoids and scyphozoans, there is considerable doubt that they are even remotely related to them. Indeed, it has been held that there is little evidence to warrant their inclusion in the phylum *Coelenterata.* Thus, their classification with the cœlenterates is largely a matter of convenience rather than one of proved relationship. Ctenophores are biradially symmetrical animals having at one pole a slitlike mouth, and at the other pole a sense-organ containing lithocysts which are presumably composed of phosphate of lime. Their bodies are nearly transparent, and on the outside are eight meridional lines that are made up of short transverse rows of cilia. It is from the comblike appearance of these meridional lines that the ctenophores derive both their scientific and popular names,—*Ctenophora* meaning "comb-bearer." The cilia are the locomotor organs of the animals. The mouth leads into a laterally compressed chamber, or gullet, which extends about two thirds through the length of the body and opens into another chamber called the funnel or infundibulum. In those ctenophores with tentacles, the funnel connects with these organs by a canal running to each basal bulb; a canal is also sent from the funnel upwards to the sense-organ. From the funnel-cavity two other vessels, the paragastric canals, arise and extend downward and close to the gullet. In most species the canal system is complex. There are eight meridional canals, which in certain forms end blindly near the mouth; in other and higher forms these canals and the two paragastric vessels fuse in various ways in the oral region, and form more or less complete circuits. All comb jellies are hermaphroditic. The gonads lie near the outer wall of the paragastric vessels. There is no alternation of generations. Only 21 species of the hundred-odd found throughout the world are known to occur along the Atlantic Coast; but the rarity of species is amply compensated for by the numbers of individuals; for these delicate, rainbow-tinted, phosphorescent and superlatively beautiful creatures often occur in extensive swarms containing billions.

Subclass **TENTACULATA**

Ctenophores that possess a pair of long tentacles during either the larval or the adult stage. When long tentacles are absent in the adult stage, these members of the group may be distinguished by the presence of oral lobes.

ORDER **CYDIPPIDA**

These ctenophores have very long retractile tentacles on opposite sides of a spherical or cylindrical body. The body usually is compressed in a plane transverse to the tentacular axis.

Family **PLEUROBRACHIIDAE**

Genus *Pleurobrachia*

P. pileus. (Plate IV.) The body of this species may be spherical or egg-shaped but is slightly compressed on two sides. From a point near the eye-spot, or sense-organ, at the aboral pole, the eight ciliary combs extend down the sides of the body for nearly the whole length. The 2 tentacles arise from deep clefts, or tentacle-sheaths, in the upper part of the body, into which the entire contractile portion can be withdrawn. The tentacles are fringed with lateral filaments giving them a delicate, featherlike appearance, and when fully extended are from 15 to 20 times the length of the body. The color of the exceedingly transparent body is delicate rose-pink; while the vibrating cilia and tentacular filaments reflect all the subtle tints of the spectrum. Body about 20 mm long and 18 mm wide. *Pleurobrachia* is a predatory animal and subsists on other small pelagic creatures that it encounters. Common. Greenland to New York.

ORDER **LOBATA**

In the ctenophores of this order the body is ovate and compressed in a plane transverse to that of the stomach. Two large oral lobes, one on each side of the wide mouth, are at the lower extremity of the body. A pair of flexible, ribbonlike organs, called *auricles,* bearing cilia along their narrow edges, project at the base of each lobe at the lower ends of the shorter combs. (See Fig. 126.) In the adult stage ordinary long tentacles are not present, but a fringe of tentacular filaments may surround the mouth opening. In the larval stage these animals possess a pair of long tentacles, thus indicating that the *Lobatæ* may have come from cydippiform ancestors. In fact, in one genus (*Leucothea*) these larvæ become sexually mature.

Family **BOLINOPSIDAE**

Genus *Bolinopsis*

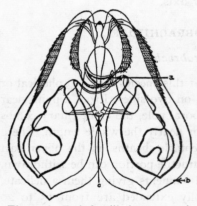

Fig. 126—B. infundibulum. a, auricle; b, oral lobe; c, mouth.

B. infundibulum. (Fig. 126.) Body very transparent. Sense organ sunk in a pit; combs not extended on sides into the oral lobes. In this species the auricles are comparatively short; and this characteristic, together with the fact that the oral lobes are of medium size, will enable the collector to distinguish it from the next described species. Body length up to 15 cm. Phosphorescent and exceedingly delicate. Predatory on pelagic forms, and occasionally cannibalistic. Very common. Labrador to Vineyard Sound.

Family **MNEMIIDAE**

Genus *Mnemiopsis*

M. leidyi. (Plate XVI.) This species, except for its long auricles and larger oral lobes, resembles *Bolinopsis* in form. It usually travels in vast swarms. Very common. All ranges.

ORDER **CESTIDA**

Ctenophores with bodies flattened and extended ribbonlike laterally. Deep tentacles sheaths with tentacles more or less rudimentary. Four of the 8 combs are rudimentary; the other 4 are well developed.

GENUS *Cestus*

C. veneris. (Plate XII.) Venus' girdle. This beautiful species, although occasionally seen along the shores in temperate waters, is a tropical form. When young it is transparent, but when full grown it may become violet with a greenish blue or ultramarine fluorescence. In the larval stage it is cydippiform, thus showing its evident derivation from pleurobrachian types. *Cestum* attains to a length of 1 m; height 8 cm. Like other comb jellies, it is predatory.

SUBCLASS **NUDA**

Ctenophores without tentacles.

FAMILY **BEROIDAE**

GENUS *Beroë*

B. cucumis. (Plate I.) Body somewhat ovate and compressed to a flattened shape so that it assumes the appearance of a miter. The 8 ciliary combs extend about three fourths of the distance from the apical sense-organ to the very wide mouth. A network of peripheral canals is present, but unlike the canals in other species of the same family, they do not anastomose. This species is identical with *Idyia cynthina* and *B. roseala* of older textbooks. Color rose-pink. Body 10 cm long, 9 cm wide, and 6 cm thick. Predatory and cannibalistic. Very numerous and common. Labrador to Long Island.

VII

VERMES

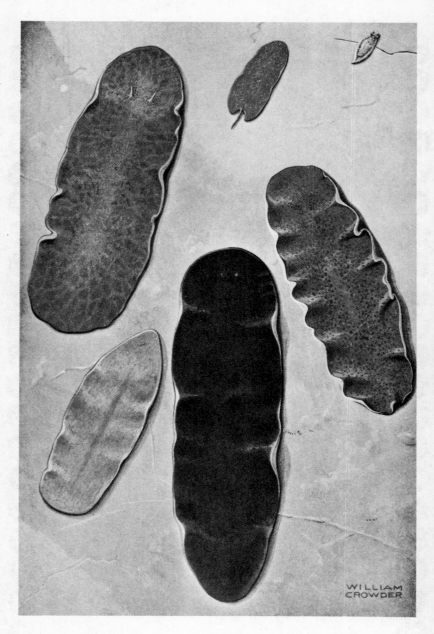

PLATE XVIII

FLATWORMS

Stylochus ellipticus (upper left); *Polychoerus caudatus* (upper middle); *Plagiostomum wilsoni* (upper right); *Bdelloura candida* (lower left); *Phagocata gracilis* (lower middle); *Leptoplana variabilis* (middle right).

CHAPTER VII

VERMES

Phylum **VERMES**

(THE LOWER WORMS)

THIS is the lowest group of animals in which the individuals are bilaterally symmetrical. Although the worms of this group are very numerous in species and individuals, because of their secretiveness and specialized habits and modes of living, they do not often come under the observation of others than professed naturalists. Yet the casual collector can readily find most of the salt-water forms if he knows where to look. Their hiding places are numerous. Virtually no part of the tidal zone is uninhabited by one or several species. Many are found in the water living a free life. They are found under stones, in rock crevices and on seaweeds; they burrow in the sand and mud, and are concealed in old shells; some species occupy the homes also tenanted by other creatures, and some are parasitic. Often they are objects of considerable beauty; always their habits are interesting.

The lower marine worms are of primitive structure and can usually be easily distinguished from the higher worms, or annelids, which have segmented bodies bearing locomotor appendages. In most cases the lower worms are small and sluggish of movement. There are an anterior and a posterior region, but commonly no distinct head, such as in the higher worms, is present.

The group may be subdivided according to the following classification:

PHYLUM VERMES

SUBPHYLUM **Platyhelminthes**

CLASSES ....
- **Turbellaria**
- **Trematoda** (Mostly parasitic; not listed here.)
- **Cestoda** (Mostly parasitic; not listed here.)
- **Nemertea**

SUBPHYLUM **Nemathelminthes**   (The round worms, thread worms, etc. These are chiefly parasitic forms; although some are free and are found under stones and on algæ. Of interest to few but specialists; no species described in this work.)

SUBPHYLUM **Trochhelminthes**

CLASSES .... {
Rotifera
Gastrotricha
Kinorhyncha
}

SUBPHYLUM **Chaetognatha**

GENERA .... {
Sagitta
Eukrohnia
Pterosagitta
}

SUBPHYLUM PLATYHELMINTHES

The flatworms. As their name would indicate, these worms usually have a flattened body; in some species, however, the body is quite cylindrical. The body may be thin and leafllike, or long and ribbonlike, or vary considerably from these typical shapes. In most cases, a body cavity is wanting, the spaces between the internal organs and the outer walls being filled with a connective tissue known as the *parenchyma*. Sense organs; when present, may consist of simple eyes, tentacles, or *statocysts*—the latter probably functioning as organs of equilibrium, or orientation. With the exception of the nemerteans, the flatworms are generally hermaphroditic. Propagation is usually by means of eggs, but in certain species asexual reproduction may take place by budding or fission.

CLASS TURBELLARIA

This division includes the free-living, soft-bodied flatworms that have a ciliated skin. The presence of a ciliated epidermis can often be determined superficially by the fact that the constant motion of the cilia causes a noticeable blur in the water immediately around the living animal; usually, however, a microscope is necessary. The body generally is flat and leaflike, sometimes nearly circular; and, although usually less than 25 mm long, may in certain cases attain to twice this length. One specimen recorded from off Ceylon was found to measure more than 12 cm in length and 8 cm in breadth. Turbellarians are usually predatory, probably all are chiefly carnivorous; they seize their food by thrusting out their pharynx (when this organ is present) through the mouth and then withdrawing it within the body. No anus is present, so the fecal matter must be ejected through the mouth; however, in some forms the intestinal branches have external pores. There is an excretory system consisting of one or more canals which connect with excreting cells, or so-called "flame-cells," by tiny capillaries. The genital opening may be either single or double, and is located ventrally back of the mouth—the latter opening being located usually at or near the middle of the under side. Although the turbellarians, with rare exceptions, are hermaphroditic, and in certain circumstances are capable of self-fertilization, cross-fertilization is usually perforce the rule because the male organs mature before the female organs do; thus, it is only during the periods of maturation when the two elements overlap, that self-fertilization can be effected. That this comparatively neglected group opens a considerable field to the researches of the collector is attested by the fact that more than a thousand species are already known.

ORDER **ACOELA**

Small marine turbellarians in which no intestine is present. Ingested food enters directly into the parenchymous tissue.

Family **CONVOLUTIDAE**

Genus *Polychoerus*

P. caudatus. (Plate XVIII.) This species has a somewhat elongate body with the lateral margins in parallel lines. At the hinder end is a deep indentation from the middle of which arise from 1 to 3 caudal filaments. Color red. Length 4 mm; breadth 1.5 mm. Occasionally frequent. Found in tidal zone among stones. New England and southward.

ORDER **RHABDOCOELIDA**

In this order the intestine is a straight, unbranched tube or a saclike organ only slightly lobed. (Fig. 127.)

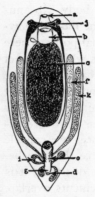

Fig. 127— Diagram of a rhabdocoele worm. a, mouth; b, pharynx; c, intestine; d, genital pore; e, penis; f, testes; g, vagina; i, uterus; j, brain; k, yolk sac.

FAMILY **PLAGIOSTOMIDAE**

GENUS *Plagiostomum*

P. wilsoni. (Plate XVIII.) Body elongate, blunt anteriorly, tapered to a point posteriorly. Intestine a simple sac without lateral extensions. Mouth at forward end of body; genital pore at hinder end. Eyes present on forward dorsal surface. This is a small, semitransparent form about 1.5 mm long, and the internal organs can be seen with more or less distinctness. Very common on *Ulva*. Cape Cod southward.

ORDER **TRICLADIDA**

The intestine in the species of this order consists of 3 principal trunks with numerous lateral extensions; from the pharynx, which connects with the mouth opening in or behind the middle of the body, one intestinal trunk extends forwards and two extend backwards, one on each side. (Fig. 128.)

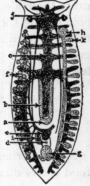

Fig. 128— Diagram of a triclad worm. a, mouth; b, pharynx; c, intestine; d, genital pore; e, penis; f, testes; g, vagina; h, ovary; i, uterus; j, brain; k, yolk sac.

FAMILY **PLANARIIDAE**

GENUS *Phagocata*

P. gracilis. (Plate XVIII.) A planarian worm with body elongate, flat and having a rounded anterior end and semi-blunt tail. At the junction of the three principal intestinal trunks is a large pharynx. Color black. Length 30 mm; width about 4.5 mm. Often abundant in brackish water.

FAMILY BDELLOURIDAE

GENUS *Bdelloura*

B. candida. (Plate XVIII.) A planarian worm. Body elongate and flat with anterior end tapering, and posterior end wide and blunt. Two eyes present. At the hinder end is an adhesive disk for attachment. Color gray. Length 15 mm; width 4 mm. This species is often common on the gills and outer surface of the horseshoe crab (*Limulus polyphemus*) where it probably lives commensally rather than parasitically. All ranges.

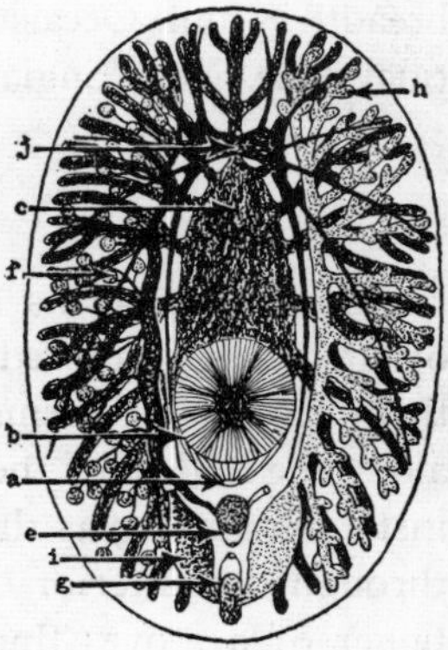

ORDER POLYCLADIDA

The intestine in this order has numerous ramifying branches that extend to all parts of the body. Two genital pores are present. No reproduction by budding or fission takes place. (Fig. 129.)

Fig. 129—Diagram of a polyclad worm. a, mouth; b, pharynx; c, intestine; d, genital pore; e, penis; f, testes; g, vagina; h, ovary; i, uterus; j, brain.

FAMILY PLANOCERIDAE

GENUS *Stylochus*

S. ellipticus. (Plate XVIII.) A flat, elliptical-bodied planarian worm with undulating margins. Mouth centrally located. Towards the forward dorsal end, 2 short, white tentacles are present, each with a group of ocelli. In the forward region, near the brain, are from 8 to 12 ocelli; while numerous other ocelli are borne along the dorsal margin. Color yellowish-brown forming an irregular radially arranged pattern. Length 20 mm; breadth 6 mm. Common under stones, in old shells and on piling in shallow water. New England and southward.

Family LEPTOPLANIDAE

Genus *Leptoplana*

L. variabilis. (Plate XVIII.) This planarian is an elliptical, leaf-like form in which the body margin is undulating. The mouth is centrally situated. No tentacles or marginal eyes are present, but numerous conspicuous ocelli occur in 4 clusters near the anterior end; in one pair each cluster contains about 30 ocelli; in the other pair each cluster consists of about 15 ocelli. Color yellowish-brown. Length 18 mm; breadth 8 mm. Occasionally common among rocks and algæ below the tidal zone. New England and southward.

CLASS NEMERTEA

Nemertean worms are long, ciliated, usually narrow and flat, and often brightly colored. No parasitic forms are known. Nearly all of them are marine forms. They are chiefly characterized by the presence, along the middle of the dorsal region, of a long tubular cavity (in some instances as long as the body itself) that can be nearly entirely everted through the anterior opening, much like the turning inside out of the finger of a glove; thus forming a proboscis. In some forms the proboscis is thrust through the mouth-opening. This organ serves both tactile and defensive purposes, some species have a sharp spine, or stylet, at the end, while in other species it is invested with stinging cells. It is endowed with great flexibility and can be rapidly extended or withdrawn. Occasionally, however, it breaks off, exhibiting a considerable degree and duration of vitality in its reflexes; but such accidents are not serious, for a new proboscis is soon regenerated. Regeneration of lost or mutilated portions of the body is common among these worms; some species can be cut into several pieces and each piece will grow into a new worm; often, however, it is only the anterior portion of the body that will regenerate an entire individual, the posterior part in most species eventually dying.

Nemerteans are carnivorous and frequently cannibalistic; but their bodies are so soft that oftentimes the bristles of chætopod worms which they usually devour work their way through the body walls to the outside. The mouth is at the forward ventral end; the anus is at the hinder end. There is no body cavity, the spaces between the organs being filled with a parenchymous tissue of gelatinous nature. The digestive canal consists of an œsophagus, a stomach, an intestine and a rectum, the intestine composing considerably the greater part of the length of the canal. A system of blood vessels is present, as is also a nervous system. An excretory system, somewhat similar to that of the turbellarians, is usually present; the minute terminal branches of the system do not end in single flame-cells, but are provided with *ciliary flames,* each branch surrounded by a group of cells.

The majority of nemerteans are unisexual; hermaphroditism occurs only in a few species. In some nemerteans development is direct from the egg; in others the young go through a metamorphosis; in still a few others the individuals are *viviparous,* or live-bearing.

ORDER **PALEONEMERTEA**

Species of this order have the mouth usually situated **far back.** Proboscis without spines. Brain usually absent.

FAMILY **CEPHALOTRICHIDAE**

GENUS *Cephalothrix*

C. linearis. (Fig. 130.) Individuals long, slender and cylindrical, tapering to a point at the forward end. Without excretory canals. Proboscis very long and threadlike. Eyes and brain wanting. Mouth located very far back. Color variable, from whitish to yellowish, or flesh-color. Length up to 15 cm; by 1 mm thick. Not uncommon. Found under stones and in sand in tidal zone. Nova Scotia to Long Island Sound.

Fig. 130—C. linearis.

ORDER **HETERONEMERTEA**

Nemerteans with mouth behind the brain. Brain always present. No stylets on proboscis.

FAMILY **LINEIDAE**

GENUS *Cerebratulus*

C. lacteus. (Fig. 131.) The ribbon worm, tape worm, etc. This worm, formerly listed as *Meckelia ingens,* is the best known of all the nemerteans, being commonly and extensively used along the Atlantic Coast as fish bait. The body is long and broad with thin edges, and is flattened throughout its length. Its head is pointed and at the hinder end is a caudal filament. There are no eyes present. The proboscis is very long and is thrust out through an opening on the dorsal side of its head. Body flesh color; proboscis white. One of the largest of the seashore worms, often attaining to a length of 2 meters or more, with a breadth of 25 mm; but individuals measuring 1 m by 20 mm are far more common. It is a burrowing animal, but is also a good swimmer. Very common in sand near the low-water level. All ranges.

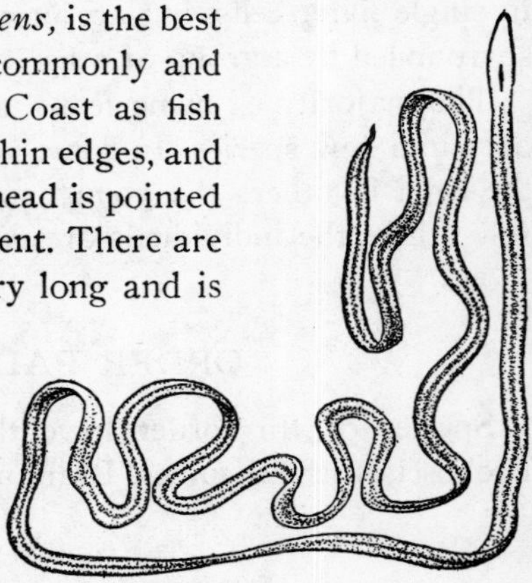

Fig. 131—C. lacteus.

ORDER **HOPLONEMERTEA**

In the nemerteans of this order the mouth is situated in front of the brain. The proboscis is provided with stylets.

FAMILY **AMPHIPORIDAE**

GENUS *Amphiporus*

A. ochraceus. (Fig. 132.) Body thick and somewhat flattened, with head slightly broader than adjacent portion. Eyes numerous at for-

ward end where they form an
angle with the tip directed backwards. Proboscis with a long central spine and several pouches of
accessory stylets. Color yellowish.
Length about 7 cm; width 3 mm.
Common. Found in tidal zone and
shallow water under stones and
creeping on piles or among algæ.
Long Island and Vineyard Sounds.

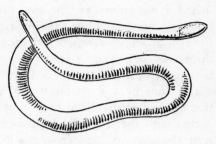

Fig. 132—A. ochraceus.

Subphylum TROCHELMINTHES

The wheel animalcules and others. These are minute forms, generally
microscopic, in which the body is commonly, but never completely,
ciliated, and in which the hinder end is usually forked.

CLASS ROTIFERA

The rotifers, or wheel animalcules, are best generally known through
their fresh-water representatives. There are, however, numerous forms
that make their home in the sea. All are minute; most of them are microscopic. And for the student who wishes to do original work in a comparatively unworked field, it may be observed that the salt-water rotifers
constitute a considerable group and there is much work yet to be done
in classifying and recording the many new species that are bound to
come within the observation of the diligent collector.

Most members of the group are easily identified. They are chiefly
characterized by the ciliated area (usually having the appearance of one
or several disks which look like revolving wheels when the cilia are
in motion) at or near the anterior end of the body. This ciliated area
may serve as a swimming organ or to create a vortex in the water, thus
bringing food to the mouth. The body is without true segments, but is
often marked with external rings making it appear segmented; and it
is composed of three principal parts: the head, the trunk, and the foot.
The head bears the corona, or ciliated wheels, and the mouth. The
mouth which is in the middle of the ciliated region, opens into a
pharynx, or *mastax,* provided with large jaws, known as *trophi,* which

in the living animal can be seen through the transparent tissues continuously working while the cilia are in action. The head also contains the special sense organs, when these are present, consisting of one to three eyes and one to four tentacles. There is a large stomach, which is joined to the pharynx by a narrow œsophagus, and which leads to a dorsal anus by a short intestine. In some forms, however, the intestine is blind at the distal end. A pair of large gastric glands are connected with the stomach. A distinct brain is present situated dorsally near the pharynx, and from the brain extend the trunks of the nervous system. There is an excretory system consisting of two kidney tubules containing flame-cells, the tubules leading to a kidney in the form of a contractile vacuole which in turn connects with the intestine.

In numerous forms the trunk is encased in a *lorica,* or cuticular shell, which in some cases is provided with spines or other projections. The terminal foot is commonly forked or divided into two toes, but it may often end with one or several toes. It is usually retractile, or telescopic, and can be used as a rudder when the animal swims, or as an organ of attachment, for which purpose the toes are provided with glands that secrete a sticky mucus. By means of this adhesive mucus, the animal can anchor itself so firmly that the motion of its food-bringing cilia with their attendant tractor-action does not dislodge the body.

Among rotifers the males are minuter than the females, and are seemingly very rare. In certain species they have never been found. They are most of them degenerate forms and either are entirely without a digestive tract or retain only a vestige of it. Reproduction among the females may take place parthenogenetically or by sexual methods. There is commonly but one ovary; it lies ventral to the intestine in the posterior part of the body. Some species of rotifers are oviparous; others are viviparous. Budding and fission do not take place. No hermaphroditic rotifers are known; the sexes always occurring separately.

ORDER **RHIZOTA**

Free-swimming colonial forms and sessile tube-dwelling forms.

Family MELICERTIDAE

Genus *Conochilus*

C. unicornis. (Fig. 133.) This form, which occurs in colonies or clusters of only a few individuals, is free-swimming. Each animal is in a transparent gelatinous tube, and has a single large antenna located ventrally on the corona. Pelagic. Also found in fresh water. Southern New England waters.

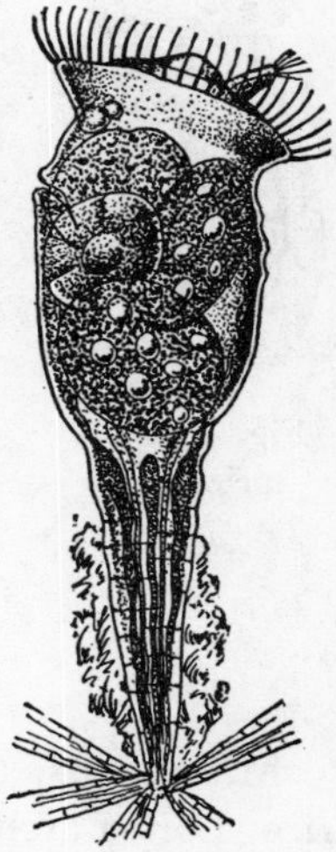

ORDER BDELLOIDA

The rotifers in this order are usually non-tubiculous. They are able to swim with the aid of the corona, and to creep leechlike by attaching alternately the head and the foot.

Family PHILODINIDAE

Genus *Rotifer*

R. vulgaris. (Fig. 134.) The elongate body of this rotifer is covered with a cuticula consisting of segments that can be telescoped one into the other. The head bears 2 ciliary

Fig. 133—C. unicornis.

disks, or wheels, one on each side of the middle mouth. Behind the corona is a dorsal proboscis provided with a pair of eyes. Half of the animal is taken up by the foot, which ends in 3 toes. A short distance from the toes is a pair of spurs. The body is whitish and semitransparent to opaque. Length 0.5 mm. This is a common form found in fresh water as well as in the sea. It occurs among algæ. All ranges.

Fig. 134— R. vulgaris.

ORDER PLOIMA

Non-tubiculous forms. These rotifers can swim and creep, but they creep with the toes only—not leechlike.

Family ASPLANCHNIDAE

Genus *Asplancha*

Fig. 135—A.
priodonta.

A. priodonta. (Fig. 135.) In this form the foot is absent. The body is short and sac-shaped and rather transparent, revealing the inner structure and organs with more or less distinctness. The head is provided with 3 eyes on the dorsal side posterior to the ciliary and near the forward end, which is in place of a true corona. Length 0.5 mm. This rotifer is viviparous, and the embryo can often be seen in clear detail within the gravid female. Not uncommon. Pelagic. Southern New England waters.

Family COLURIDAE

Genus *Metopidia*

M. colurus. (Fig. 136.) Body with a lorica. Lorica arched and laterally compressed. The head bears an arched shield, which in a side view shows a heavy, hooklike, forward projecting spine. Foot and toes not exceptionally long. Length 1 mm. Sometimes common among algæ. Northern and southern New England waters.

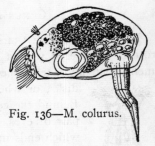

Fig. 136—M. colurus.

CLASS GASTROTRICHA

The worms in this class are among the commonest of microscopic forms occurring along the seashore. Wherever vegetation is present, or organic débris in which rotifers and protozoans abound, one or more species of these minute, interesting and gracefully moving worms are almost invariably present. They are very abundant and widely distributed; yet they are limited in variety of species; of the 75-odd known to occur throughout the world, not more than one third of this number have been found in this country—and this includes both fresh-water and salt-water forms. The truth is, the *Gastrotricha* have been but little studied in America; but there is little doubt from the evidence already at hand that intensive investigation will reveal the occurrence of varieties that are not inferior in numbers to those in other parts of the world. Although strictly microscopic, being from 0.54 mm long to as small as one eighth of this length, they are easily reared and their habits are easily observed; the only thing necessary for this purpose being a cell of sea water containing a fragment of seaweed.

They can be recognized by their elongated bodies,—most usually having a forked posterior,—provided with cilia on the ventral surface. The dorsal surface may be bare or it may have scales or spines or both. At the head end are the mouth and paired groups of sensory bristles, and very often a pair of eyes. There is a large brain; excretory organs are present in the form of lateral coiled tubes with flame-cells near the center of the body; paired ovaries and testes occur together in the posterior part of the animal. The individuals are hermaphroditic. Their eggs are very large and are deposited on algæ and firm objects, such as shells or pebbles, and are usually provided with projections that appear to anchor them in place.

Family **CHAETONOTIDAE**

Genus *Chaetonotus*

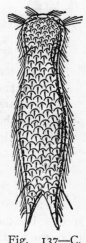

Fig. 137—C. brevispinosus.

C. brevispinosus. (Fig. 137.) Body with rounded head and short forked tail. Dorsal scales present, each of which bears a short recurved spine, both scales and spines being larger toward the posterior end than toward the anterior end. On the head are 4 eyes and 4 tufts of bristles arranged in pairs. Paired single bristles occur dorsally at the head and neck and near the hinder forked end. Length about .12 mm. This is a common freshwater form as well as an inhabitant of the sea. It is frequently recorded in technical literature as *C. larus*. Found among shore algæ and débris. Frequent in indoor salt-water aquaria. Southern New England.

CLASS KINORHYNCHA

These minute worms are strictly marine, no fresh-water forms being known. They are less than 0.5 mm long and have a slight resemblance to the gastrotrichans; also like the latter they have the hinder end usually forked. They differ radically from the gastrotrichans, however, in having the body composed of a series of rings, and in being without cilia on the outer surface. Although the body is marked with external rings, these do not represent true body segments, the body cavity having no transverse structural divisions. In most cases, spines or bristles or both are present. The mouth is surrounded by a ring of hooks and is provided with a number of locomotor spines. The sexes are separate; the reproductive organs are paired, and their external openings are near the hinder end. The group is somewhat imperfectly known; but this circumstance is not owing to lack of material; rather it is because relatively few students have applied themselves to the study of these forms. Fewer than 50 species are known, most of which have been recorded from European waters.

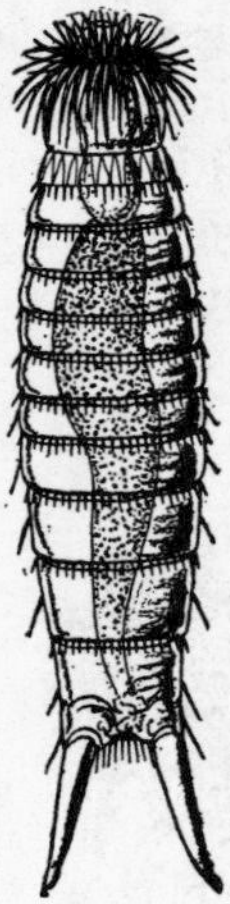

FAMILY ECHINODERIDAE

GENUS *Echinoderes*

Fig. 138
—E. du-
jardini.

E. dujardini. (Fig. 138.) Body elongate; arched dorsally; concave ventrally, and consists of 12 rings. Head retractile and bearing a ring of long oral bristles and paired eyes. The annulations of the trunk are each provided with a fringe of short spines. Color reddish; red eyes. Found in mud and on algæ. Southern New England.

SUBPHYLUM CHAETOGNATHA

The arrow-worms. A small group in the number of known genera and species, but the individuals are multitudinous and are found in all seas from the surface to the greatest depths. The body is long and slender; and its distinguishing features are the long, paired groups of sickle-shaped bristles surrounding the mouth, and the presence of a tail fin and paired horizontal lateral fins. The bristles that surround the mouth are used in seizing prey. Within the hooded oral opening are from one to four rows of short spines serving as teeth. The body is divided into a head, a trunk, and a tail, the head being usually marked by its rounded form or terminal mouth, while the ventral anus generally indicates the junction of the trunk and tail. A straight digestive tract runs from the mouth to the anus. There are no special respiratory, circulatory, or excretory organs; but, on the other hand, there is a well-developed brain and nervous system. Also there are two eyes, one on each side of the dorsal surface of the head; and there is a ringlike structure or collarette, just behind the head, supposed to be an olfactory organ.

Chætognaths are hermaphroditic. The ovaries are in the trunk, and the testes are in the tail. Internal impregnation is the rule, the fertilized eggs being strewn in the water where they float at or near the surface while undergoing further development. Except in size, the newly hatched young are similar to the adults. About 35 species have been described. Nearly all of them are pelagic, being swift and agile swimmers.

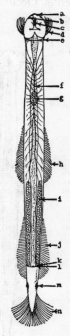

Fig. 139—S. elegans. a, anterior teeth; b, posterior teeth; c, mouth; d, seizing jaws; e, collarette; f, intestine; g, ventral ganglion; h, anterior fin; i, ovary; j, posterior fin; k, female genital pore; l, anus; m, male genital pore; n, tail fin.

GENUS *Sagitta*

S. elegans. (Fig. 139.) The glass worm. Body slender and having as its chief characters 2 pairs of lateral fins besides the tail fin, 9 to 12 oral hooks, 4 to 8 anterior teeth, and 6 to 9 posterior teeth. The teeth are in two paired rows. A collarette is present, but it does not extend more than halfway to the ganglion. Length up to 3 mm. Very abundant. Northern Atlantic seas.

S. hexaptera. (Fig. 140.) This form is of the same general proportion as the preceding species, but smaller in size, measuring about 6 mm. The paired hooks at the mouth number from 5 to 10; anterior teeth from 3 to 4;

posterior teeth from 3 to 5. The teeth are in 2 paired rows. It has 2 pairs of lateral fins besides the caudal fin, this being a characteristic of the genus. Occurs seasonally in prodigious numbers. Vineyard Sound southward.

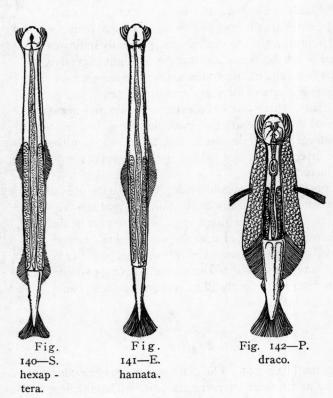

Fig. 140—S. hexaptera.

Fig. 141—E. hamata.

Fig. 142—P. draco.

Genus *Eukrohnia*

E. hamata. (Fig. 141.) In common with other members of the genus *Eukrohnia,* this species possesses but 1 pair of lateral fins near the middle of the body. Body slender with somewhat transparent integument revealing reddish ovaries. There are 8 to 10 oral hooks, and a single row of 15 to 28 teeth. Length 4 mm. Seasonally abundant. Vineyard Sound southward.

Genus *Pterosagitta*

P. draco. (Fig. 142.) With the generic characters in which the body is broad and has 1 pair of lateral fins near the tail. On each side of the body, extending from each lateral fin to the head, is an expansion of the cuticular covering. Oral hooks number from 4 to 10; anterior teeth from 6 to 9; posterior teeth from 12 to 18. Teeth arranged in 2 paired rows. An odd characteristic of this species is the paired tufts of long bristles that project from sides of the trunk. Often numerous. Southern New England waters.

VIII

MOLLUSCOIDEA

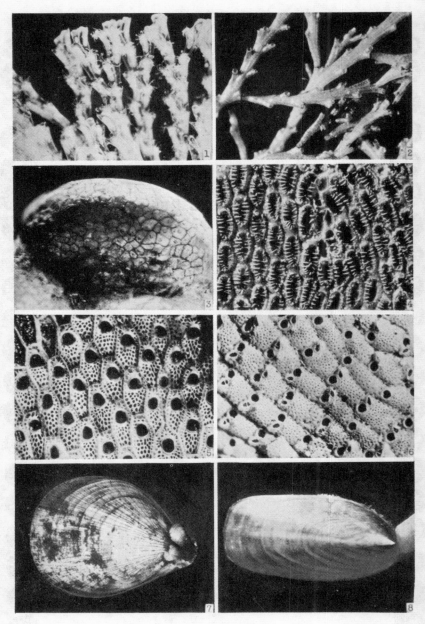

PLATE XIX

1—*Bugula turrita*; 18x.   2—*Crisia eburnea*; 18x.   3—*Alcyonidium mytili*; 11x.

4—*Membranipora pilosa*; 18x.   5—*Lepralia pallasiana*; 18x.

6—*Schizoporella unicornis*; 18x.   7—*Terebratulina septentrionalis*; 2¼x.

8—*Glottidia audebarti*; 1½x.

CHAPTER VIII

PHYLUM **MOLLUSCOIDEA**

THREE divergent groups of animals are usually included in this phylum: the *Bryozoa,* or moss animalcules; the *Phoronida,* a small group of tube-dwelling, wormlike forms; and the *Brachiopoda,* or lamp shells. In some respects, the known larval forms in these three groups have certain similarities to one another (and to a moderate degree to the larvæ of the lower worms on the one hand, and to those of the annelids on the other); furthermore, the adults agree in possessing a structure known as the *lophophore,* which is a tentacle-bearing ridge containing a special chamber of the body-cavity, and often having an *epistome,* a sensitive, liplike process, or projection, overhanging the mouth. Also, they are alike in having a true body cavity within which is an alimentary canal consisting of an œsophagus, a stomach, and an intestine. The oral opening and the anal opening are commonly situated close together.

But it should be added that despite these similarities, their relationships are uncertain; indeed, they are held by some naturalists to be of no relation whatever to one another. Nevertheless, in the present state of our knowledge, convenience, if not the rules of custom, demands their inclusion in one phylum. Therefore, they are accordingly grouped herewith.

PHYLUM MOLLUSCOIDEA

CLASS **Bryozoa** SUBCLASSES........ { **Entoprocta** **Ectoprocta**

CLASS **Phoronida** (No large subdivisions)

CLASS **Brachiopoda** ORDERS......... { **Ecardines** **Testicardines**

CLASS **BRYOZOA**

The moss animalcules, known in this country and continental Europe as the *Bryoza,* but in England as the *Polyzoa,* are among the most attractive-appearing of the minuter animals of the sea, not deferring in delicacy and exquisiteness of structure even to the most beautiful of the corals, which many bryozoans superficially resemble. In fact, it is because of this resemblance that the calcareous members of the group have often been termed "corallines"; but no less frequently—perhaps because of the thin frondlike expansions of the colonies formed by other members—the name "sea mats" has been given; but the commonest general term in use is "moss animals" or "moss animalcules," a term derived from their mosslike appearance in certain situations along the seacoast. It is this descriptive term which is used in its latinized form as the name of the class.

The great majority of the *Bryozoa* are marine. The individuals are small, but most of them form colonies of appreciable size; consequently the detection and collection of them are comparatively easy. Some chitinous forms are like fixed hydroids in general appearance and might easily be mistaken for them until the hand lens or microscope is used to examine their structure; the greater number, however, are calcareous and encrusting forms, the identity of which can not readily be mistaken for that of any other type of marine animal.

Bryozoans form colonies, called *zoaria,* consisting of hundreds or thousands of individuals that have multiplied in number by a process of budding; but the first individual from which the colony arises, known as the *ancestrula,* develops from a free-swimming larva. Each individual unit, or zooid, of the colony is composed of a double-walled chitinous or calcareous compartment, the *zooecium,* within which is the visceral mass, or the soft animal individual known as the *polypide.* The mouth and anus being situated close together causes the freely suspended alimentary canal to assume the characteristic U-shaped tract so common in the animals of the phylum. In the marine forms, the lophophore consists of a circlet of hollow, slender, ciliated tentacles surmounting a ring-shaped (sometimes oval) ridge. In certain fresh-water forms, this ridge is horseshoe-shaped. The tentacles have a respiratory function and are used to gather diatoms and other microscopic food organisms which are swept into the central mouth by ciliary action. A few bryozoans have both the mouth and the anus situated within the lophophore in a depression called the "vestibule"; but in the vast majority of species, only the mouth is within the lophophore, the anus being situated just outside of it. The reader will observe that although this type of internal structure is quite simple, it is considerably in advance of that characterizing the fixed hydroids, wherein a true alimentary canal is wanting, the digestive cavities of the different hydranths of a colony being continuous with one another. Some bryozoans have connecting body cavities, but their alimentary tracts are separate.

A heart and a vascular system are wanting, but numerous white blood corpuscles occur floating in the general body cavity. The nervous system is simple, consisting of a ganglion located between the oral and anal openings, and from which extend delicate filaments to the tentacles and the œsophagus.

Both sexes are often combined in the same zooid. The reproductive organs are located in various parts of the body cavity, though the spermatozoa are more usually produced in the lower region, while the eggs are generally formed in the upper part. In some species the eggs are passed directly into the water, but more frequently they develop in a special structure or compartment known as the *ovicell* or *ooecium*. The ovicell may be modified zooecium or it may be an inflation of the surface.

A strictly natural classification of the *Bryozoa* is impossible at the present time owing to our want of knowledge concerning the early development of the greater part of the group. Each family is characterized essentially by its larva, but it is in only comparatively few families that the larval forms have ever been seen. Consequently, the present system must rest entirely on a structural or physical basis. Thus, the two principal subdivisions of the group, the *Entoprocta* and the *Ectoprocta,* are based on the position of the anal opening with reference to the tentacles; yet these subclasses differ widely in many other respects, so much so, in fact, that here again some authors have maintained that they are not even distantly related. However, as the vast majority of the *Bryozoa* are contained in the subclass *Ectoprocta,* and the greater part of these are comprised in the single order *Chilostomata,* the learner will soon familiarize himself with the commonly accepted characteristics of the class.

In all there are about 2000 known species of bryozoans, and of these all but about 50 are marine.

SUBCLASS **ENTOPROCTA**

In this division the forms are distinguished principally by the fact that the anal opening is within the circle of the lophophore. Another feature that sets them apart from the other members of the Bryozoa is that the tentacles of the individual polypides cannot be retracted into the body but roll up instead. Although widely distributed, being found in all seas, this group contains not more than a couple of dozen species comprising both fresh-water and marine forms.

ORDER **PEDICELLINEAE**

FAMILY **LOXOSOMIDAE**

GENUS *Loxosoma*

L. davenporti. (Fig. 143.) The body of this form is slightly vase-shaped and is supported on a cylindrical contractile stalk, or pedicel,

which is about equal in length to the rest of the animal. The lophophore is in an oblique position and is provided with from 18 to 30 tentacles. Within the tentacular ring arises a conelike projection containing the anal opening at the tip. From 1 to 6 pairs of buds are usually present attached to the body near the lower side of the stomach. A singular feature of this animal occurs in the method by which the young are developed. In the floor of the vestibule is a mammary organ to which the young embryos cling and from which they obtain nourishment, a process

Fig. 143—L. davenporti.

of nurturing not essentially remote from that of far higher animals. Length about 2 mm. A solitary form found in worm tubes in sand. Sometimes common. Southern New England.

FAMILY **PEDICELLINIDAE**

GENUS *Pedicellina*

P. cernua. (Fig. 144.) A colonial entoproctan with a cup-shaped zooecium supported on a stout contractile stalk that rises from a creeping stolon and tapers gradually toward the upper end. The body is marked off from the stalk by a diaphragm. The lophophore is transverse and there are from 12 to 24 tentacles present. The cup and the stalk are often provided with a number of spines. The stolon is branched and flexuous. The color of the stalk is yellowish-red; stolon more or less transparent. Length about 1.5 mm. Occasionally abundant. Occurs on pilings and submerged objects in company with other creeping forms. All ranges.

Fig. 144—
P. cernua.

SUBCLASS **ECTOPROCTA**

Bryozoans in which the anal opening is situated without the lophophore are all included in this group. In these animals the tentacles can be retracted into the zooecium. All of them form large colonies.

ORDER **GYMNOLAEMATA**

In this single order are contained nearly all the known species of marine *Bryozoa*. It is, however, divided into three suborders, termed and characterized as follows: suborder *Cyclostomata,* in which the zooecium opening is circular and not closed by an operculum; suborder *Chilostomata,* in which a movable, chitinous operculum is present; suborder *Ctenostomata,* in which an operculum composed of a fringe of setæ is present.

SUBORDER **CYCLOSTOMATA**

Zooecia tubular and well calcified. The rounded orifice is wide. No opercular or appendicular structures (avicularia and vibracula, described on page 183) are present. There is no external brood chamber, but the embryos develop in ovicells consisting of a modified zooecium.

FAMILY CRISIIDAE

GENUS *Crisia*

C. eburnea. (Plate XIX.) Grows in dense, bushy tufts and is usually attached by a single stem. The branches are characteristically curved inward and are jointed, the internodes being short, somewhat flattened, and usually composed of 5 or 7 zooecia. The zooecia are in two rows and alternate, each zooecium having the free upper portion bearing the aperture bent forward nearly at a right angle to the stem. Sometimes a pointed process occurs on the outer angle of the aperture. A large bulbous ovicell generally occurs in the position normally occupied by the second (sometimes the third) zooecium of an internode. The ovicell aperture is conspicuous and is elevated on a short tube. Because of the hornlike joints, the colony is more or less flexible. The joints are yellow, otherwise the colony is colorless, except that the older parts may attain to a dark brown. Reaches to a height of from 8 to 25 mm. Found on piles, stones and shells, often attached to hydroid and other stems; and occurs in all depths from the low-water level to 80 fathoms. Abundant. New England and southward.

FAMILY TUBULIPORIDAE

GENUS *Tubulipora*

T. atlantica. (Fig. 145.) In this attractive species the zoarium rises erect and spreading from a small base. It is irregularly branched dichotomously, the branches being mostly in the same plane. The branches are triangular in section and have the back side marked with minute stipples and striations. The tubular zooecia are arranged in rows of from 1 to 4 or 5, the innermost individuals being the longest. They are confluent for a considerable part of their length, having the apertures directed somewhat outward, often leaving a free space in the middle of the front side of the stem, in which is developed an irregularly shaped, elongate ovicell spreading laterally around the bases of the adjacent zooecia. The ovicell aperture is on a long tube similar in size to that of a zooecium. Attached to stones and shells in offshore waters. All ranges.

T. liliacea. (Fig. 146.) This species may either form flat adnate colonies or grow high and erect with the branches usually in one plane, the adnate types being seemingly characteristic of the more temperate

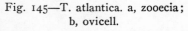

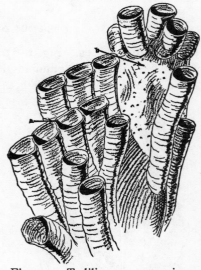

Fig. 145—T. atlantica. a, zooecia; b, ovicell.

Fig. 146—T. liliacea. a, zooecium; b, ovicell.

waters. Young colonies are uniformly flabellate; older colonies are irregularly lobate. The tubular zooecia grow in series, more or less alternate, and are confluent. Numerous minute pits cover the outer surface. The ovicells are flattened and spreading, assuming a very irregular shape and involving the bases of a number of zooecia. Ovicell aperture opens sidewise at the end of a short tube. Color white or purple. Height about 9 mm. Grows on eelgrass, piles, stones, shells, hydroids and like objects on which it can secure attachment. Not uncommon. Cape Cod region.

T. flabellaris. (Fig. 147.) The zoarium of this species is usually adnate, flabellate when young, but more or less discoidal and lobate when old. Minute pits cover the outer surface of the zooecia. The zooecia are oftentimes confluent and oftentimes free, arranged radially in series. Numerous, well-developed, irregularly shaped ovicells involving the bases of the zooecia are usually present in the older colonies. Ovicells coarsely punctate. The orifice of the ovicell is compressed from side to side and is at the end of a tubular process inferior in size to that of the zooecia. Color pale purple. Diameter of colony 12 mm. Found in shallow water on algæ, stones and shells. Not uncommon. Cape Cod region.

Genus *Stomatopora*

S. diastoporoides. (Fig. 148). A species in which the colony forms a thin, irregular crust commonly having a lobed or sinuate outline. The zooecia are embedded for the greater part of their length, the free open

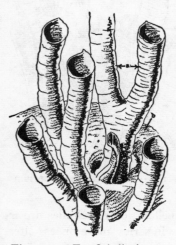

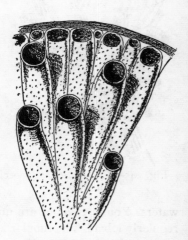

Fig. 147—T. flabellaris. a, zooecia; b, ovicell.

Fig. 148—S. diastoporoides.

ends turning outward at right angles and being well apart from their neighbors. Outer surface rather coarsely punctate, often striated transversely. When fresh, the colonies are milk-white or semitransparent. Diameter about 18 mm. Found on pebbles, etc., in offshore waters. Cape Cod southward.

Family LICHENOPORIDAE

Genus *Lichenopora*

L. verrucaria. (Fig. 149.) In this form the zoarium is a calcareous, dislike structure, usually adherent, but sometimes attached by a greater or lesser part of the under side or with only the edges raised. The disk is often modified in shape, not infrequently being raised into a dome. The zooecia occupy a region well out toward the periphery of the disk, leaving the central area free, except for the presence of the ovicells, which occur as inflations of the surface and which have apertures on short oval or trumpetlike elevations. This central area, as well as the spaces between the zooecia, is coarsely punctate. The zooecia are comparatively large and are not confluent, all except those at the margin being raised; each zooecium usually has a pronounced rib on the side facing the central area, which extends into a pointed projection above the large ánd oblique aperture. This is a not unhandsome species when viewed under the binocular microscope, and it well repays study. Although common, it is small, measuring about 3 mm across the disk, and is apt to be overlooked by the ordinary collector. It occurs on other bryozoans, such as *Bugula,* as well as on hydroids, stones and shells in shallow and offshore waters. Northern New England to Long Island Sound.

Fig. 149—L. verrucaria. A, colony; B, ovicell; C, C, zooecia.

Suborder CHILOSTOMATA

In this large suborder, many variations occur; but nearly all the species agree in possessing a movable door, or operculum, for closing the principal aperture when the polypide is retracted. In the *Chilostomata* are to be found the most singular of external appendages among the *Bryozoa.* These are the *avicularium,* so-called from its resemblance in its most highly specialized form to the head of a bird; and the *vibraculum,* a whiplike appendage occurring with less frequency than the avicularium, but like that organ giving considerable aid to the systematic study of the group. (Fig. 150.) The avicularium commonly has a fixed upper and a movable lower beak, which latter is worked by muscles and is in constant snapping motion. This structure is presumed to have a defensive function, as it is usually located near the principal aperture and the pore of the "compensation-sac," a delicate organ that by hydrostatic means assists in the extension and retraction of the tentacles. The

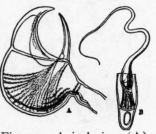

function of the vibraculum is not precisely known, but evidence seems to indicate that it is solely sensory, notwithstanding that this appendage, also, is generally located in particular relation to the orifices just mentioned. Some of the chilostomes are without avicularia or vibracula; although on the other hand, in a certain genus (*Scrupocellaria*) both organs coexist.

Fig. 150—Avicularium (A) and vibraculum (B).

In the *Chilostomata,* many of the forms are encrusting, and frequently the frontal wall is composed of calcite assuming a lacelike pattern that in delicacy and exquisiteness of design is seldom surpassed by that of other natural objects. Again, other forms are erect and plantlike, reminding one of the hydroids, their closest rivals for the palm of general structural beauty.

FAMILY AETEIDAE

GENUS *Aetea*

A. anguina. (Fig. 151.) A delicate form in which the colony is made up of creeping branches, or stolons, from which arise upright, more or less bent, club-shaped tubes. The terminal portion of the tube is somewhat spoon-shaped with the back finely punctate, and having across the front a membranous wall at the upper end of which is the operculum. The general texture is white and glossy. The ovicell is situated close to the end on the back, opposite to the membranous area, and is a round, transparent, saclike structure in which the eggs are clearly visible. There are no appendages. The ovicells in this

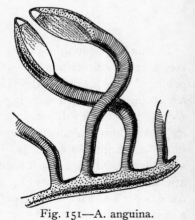

Fig. 151—A. anguina.

species, however, are rarely present; in fact, for a long time it was supposed that they did not occur, and older descriptive accounts were made to confirm this supposition. This bryozoan, on account of its small size—1 mm high—is inconspicuous, though it is fairly abundant. Found in shallow and deep water on eelgrass, algæ, stones, shells and stems of various hydroids. Occurs in Long Island Sound and northward.

Family EUCRATEIDAE

Genus *Eucratea*

E. chelata. (Fig. 152.) A branching colonial form rising from a creeping stolon; each branch consisting of zooecia arranged end-to-end in a single row. The zoarium usually is straggling and more or less erect, but recumbent specimens are not infrequent. In the living polypide, the tentacular sheath terminates above in a ring of bristlelike projections. The zooecia are narrow at the lower end with a gradual enlargement upward to the base of the aperture. The aperture is oval, with a thin, raised margin, and slants toward the top of the cell. No operculum is present in this species; nor do any appendages occur. When present, the ovicells are borne on rudimentary zooecia situated

Fig. 152—E. chelata.

immediately below the front of the apertures. Occasionally numerous on shells and stones, or spreading over algæ, hydroids and other bryozoans. Occurs in tidal zone and shallow water. Vineyard Sound northward.

Genus *Gemellaria*

G. loricata. (Fig. 153.) The zoarium is erect, forming bushy colonies in which the zooecia are joined back-to-back in regularly formed double rows. A large, slightly oblique aperture with a thin raised margin occupies more or less of half of the front of the zooecium. No ovicells or appendages are present. Height about 20 cm. Comparatively common in offshore waters. Vineyard Sound northward.

Family CELLULARIIDAE

Genus *Caberea*

C. ellisi. (Fig. 154.) A branching species, which forms fan-shaped colonies; branches stout, widening upward. The zooecia are short and quadrangular and are arranged in 2 to 4 rows. Occupying nearly the

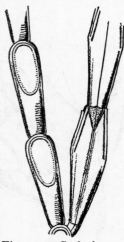

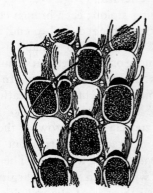

Fig. 153—G. loricata.   Fig. 154—C. ellisi.

whole of the front of each cell is an elliptical aperture with a broad margin. There are two stout spines on the outer side of the lateral zooecia, and one spine on the inner side; the median cells bear a single spine on each side. Two kinds of avicularia are present in this species: a small lateral one with a rounded mandible and situated slightly below the top of the zooecium; the other appendage is raised and rounded and located below the aperture. Large vibracular cells are present, almost covering the back of the zooecium on which they are situated, and they have long vibraculæ bearing numerous teeth, especially near the tip. The ovicells are flattened with a surface that is either smooth or finely striated with radiating lines. Color yellowish-brown. Occasionally abundant on shells, stones or other objects in offshore waters. Vineyard Sound northward.

Genus *Menipea*

M. ternata. (Fig. 155.) A delicate branching species forming white, bushy tufts. The branches are somewhat straggling and are jointed with internodes consisting usually of 3 zooecia; although 5 or 7 cells are not infrequent. The zooecia are elongated, and taper to a narrow diameter

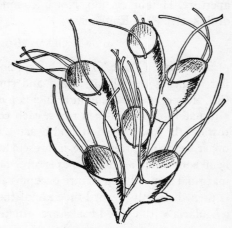

Fig. 155—M. ternata.   Fig. 156—B. ciliata.

at the lower end; and each bears two spines at the top, and one at the outer margin of the aperture. The avicularia are of two kinds: a conspicuous sessile one at the outer upper angle of the zooecium, and a small one just below the aperture, the latter kind often being wanting. Ovicells are present; somewhat elongated with a smooth surface. From the zooecia extend long fibers; those arising from the lower part of the zoarium being simple, and those on the upper branches being tendrillike and enlarged at the ends. Height about 25 mm. Found attached to stones, shells, hydroids and other bryozoans in offshore waters. Sometimes abundant. Cape Cod northward.

Family **BICELLARIIDAE**

Genus *Bicellaria*

B. ciliata. (Fig. 156.) The colonies of this species form white feathery tufts with the branches curved inward at the tips. The zooecia alternate in 2 rows and are somewhat cone-shaped with an elliptical, oblique aperture, and have from 4 to 7 very long spines on the upper margin, and 1 centrally situated spine on the lower margin. On the outer side below the aperture of the zooecia are often present small avicularia with saw-edged beaks. The ovicells are helmet-shaped and are borne on narrow stalks arising from side of the zooecium near the inner side of the aperture. Height 12 mm. Not uncommon. Found on shells, stones, hydroids, etc. New England.

Genus *Bugula*

B. turrita. (Plate XIX.) This species grows in colonies composed of flat branches spirally arranged and curling inward somewhat at their tips, thus giving each main branch a pyramidal outline. The zooecia are alternately arranged in 2 rows, each cell being elongate and narrowed toward the lower portion, and bearing a short, stout spine on the outer upper margin. Often there is present on the inner angle a short spine bent somewhat across the aperture, and usually one just behind the marginal spine. The aperture occupies about two thirds of the front and is turned slightly toward the axis of the branch. Small, somewhat stout avicularia with curved beaks are situated near the margins of the apertures of various zooecia. Large, rounded ovicells arise on the upper margin of one side of the zooecia. Strong and well-developed root fibers are present. Color ranges from pale yellow to bright orange. Height up to 30 cm or more. This is a very common bryozoan, being found almost everywhere attached to various objects; from low-water level to deep water. Casco Bay to North Carolina.

B. flabellata. (Fig. 157.) The zoarium of this species is branching with the branches growing broad and fanlike and somewhat whorled; it is with a very short main stem with the larger branches rising near the base. The zooecia are elongate and with a membranous area over the en-

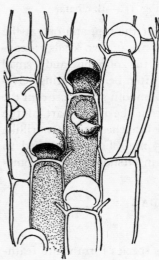

Fig. 157—B. flabellata.

tire front; and the cells are arranged in rows of from 3 to 6 with the zooecia more or less alternating. From 2 to 4 heavy spines are present on the upper margin, those situated anteriorly being the stouter and extending somewhat forward or curving inward. Avicularia are present, but these are located only on the outer rows of zooecia; they are situated about one fourth of the distance below the upper margin of the cell, and are of appreciable size with strongly recurved beaks. Ovicells

occupy positions immediately above the zooecia; they have a hemispherical or hoodlike form with a wide opening at the lower part, and are attached to the zooecia by a short, broad stalk. Flesh-color when living ash-color when dried. Height 25 mm. Common on piles, shells, etc., in shallow water. All ranges.

B. murrayana. (Fig. 158.) In this species, the zoarium branches into broad bushy tufts, often with the branches

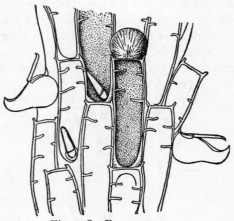

Fig. 158—B. murrayana.

in ribbonlike strips, abruptly ending, or truncate, at the tips. The zooecia are elongate, truncate at the top and somewhat tapering toward the bottom, and are arranged in rows numbering from 4 to 12. Each upper angle has a stout erect spine, and there are from 1 to 5 long slender spines on the lateral margins, which curve over a large aperture reaching nearly to the bottom of the cell. There are two kinds of avicularia: a large type situated on the outer margin of the aperture of the lateral zooecia, only; a type several times smaller situated on the front at the bottom of the zooecia, and having the mandible turned upward. In both types, the beak is strongly hooked. The ovicells are large and somewhat globular, and have the surface marked with radiating striae. Long root-fibers, or holdfasts, are present; these are stout and wrinkled and arise from the marginal cells at the base of the colony. Color light yellow or brownish. Height up to 50 mm. Very common in offshore waters. Arctic Ocean to Vineyard Sound.

B. avicularia. (Fig. 159.) Like nearly all other species of the genus *Bugula,* this one has the aperture occupying nearly the whole of entire front of the elongate zooecium. The zooecia are arranged in 2 rows and bear spines in the following manner: 1 spine on the upper inner angle, 2 spines on the upper outer angle, and 1 spine on the outer lateral margin. Ovicells are present, these being somewhat globular structures occupying a position at the top of the zooecium. A distinguishing characteristic of this species, however, is the large avicularium usually situated about midway on the outer lateral margin of the orifice. This appendage is elongated and has a long slender beak gently curved or sometimes abruptly decurved toward the tip. Not uncommon. In shallow and offshore waters. New England waters.

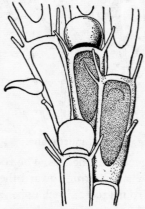
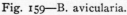
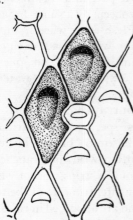

Fig. 159—B. avicularia.   Fig. 160—C. fistulosa.

Family CELLARIIDAE

Genus *Cellaria*

C. fistulosa. (Fig. 160.) The zoarium of this bryozoan is calcareous and consists of slender, erect branches that are jointed at intervals, the internodes being connected by flexible horny tubes. Usually the zooecia are lozenge-shaped, but the shape of the cells is variable. They are ranged in rows around a central axis; each zooecium is surrounded by a raised margin, and the surface of the cell is marked with numerous minute pits. The principal aperture is arched above and somewhat in-

curved below. Just above the orifice is the opening of the concealed ovicell; this opening may be round or oval. Avicularia are present; they are
of simple type, resembling somewhat the zooecia, and are situated above
the aperture in line with the cell. Found on rocks and shells in shallow
and offshore waters. Northern New England.

FAMILY FLUSTRIDAE

GENUS *Flustra*

F. foliacea. (Fig. 161.) In this species the zoarium is hornlike and
flexible, and consists of broad, leaflike branches, attached by a slender

base, in which the zooecia are ranged in
contiguous rows and in 2 layers. It is
said that the fragrant odor of violets
emanates from freshly collected specimens of this bryozoan. The zooecia are
roughly quadrangular in shape, and
rounded above and have raised margins.
They are provided with a pair of spines
near each of the two upper angles. The
aperture is wide and narrowed longitudinally, and is usually arched above and
decurved below. Inconspicuous ovicells
of shallow structure are present; the

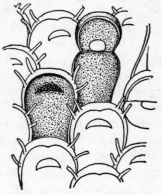

Fig. 161—F. foliacea.

opening forming an arch over the upper margin of the zooecium. The
avicularia have a general resemblance to the zooecia, and are usually
situated in line with them. Color brownish. Height up to 15 cm. Occasionally common. Occurs on stones, shells, etc., in shallow water.
Northern New England.

FAMILY MEMBRANIPORIDAE

GENUS *Membranipora*

M. monostachys. (Fig. 162.) Zoarium calcareous and forming irregular, frequently radiate encrusting colonies. Zooecia comparatively
small. Each zooecium has an oval membranous area of considerable
size at the front. This space is surrounded by a slightly raised margin,

or border, usually provided with 8 or 10 pairs of sharp spines which project about midway across the area. In addition to these spines, there

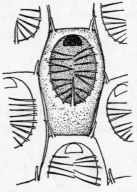

is a much stouter spine occupying a position at the base of the membranous area. Often the marginal spines will be wanting; but the basal spine is invariably present, a consideration which gave the species its name. In some old colonies, a secondary calcareous layer not infrequently occurs partly closing the membranous area; and abortive or completely closed zooecia are not uncommon. No ovicells or avicularia are present in this species. Fairly frequent on stones, shells and algæ in shallow and offshore waters. Arctic Ocean to New Jersey.

Fig. 162—M. mono- stachys.

M. pilosa. (Plate XIX.) This handsome species is a calcareous encrusting form with large ovate zooecia having the basal portion coarsely punctate and the membranous area surrounded by a somewhat high, smooth border from which usually project 4 to 12 stout, curved spines. Immediately below this area a horny spine is present, varying in length from rather short to very long. The region around the aperture is numerously perforated with minute oval pores. There are no ovicells. The zoaricum occurs in irregular patches, often with a silvery sheen, on stones, shells, algæ, etc., in shallow and offshore waters. Very common. Arctic Ocean to Long Island Sound.

M. lineata. (Fig. 163.) This species, although rarer than the two species of *Membranipora* just described, usually occurs in the same situations and might easily be taken for either because of its superficial resemblance to them. The zoarium forms small, rounded patches, however; and the oval or elongate aperture of the zooecium bears 4 to 6 pairs of slender pointed spines on the narrow raised margin. The zooecium, when viewed from the dorsal side, reveals 2 pairs of lateral pore chambers and 1 large anterior one, with spines projecting into the chambers. There are large ovicells with a smooth, glossy surface transversely crossed with a raised rib. Not infrequently there are present at the bases of some zooecia moderately-sized, raised avicularia with rather prominent beaks. All ranges.

M. aurita. (Fig. 164.) Zoarium calcareous and encrusting. Zooecia fairly large and narrowed at the anterior third portion, and with high, broad margins, finely tuberculate on the inside, from which project

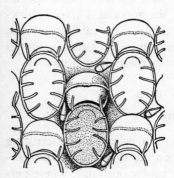

Fig. 163—M. lineata.

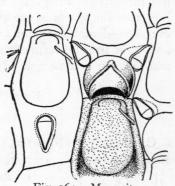

Fig. 164.—M. aurita.

from 1 to 4 more or less erect spines. It should be noted that in the younger stages the colony is entirely membranous; but the older colonies are strongly calcified and have the aperture partly closed by a calcareous lamina. Moreover, in adult colonies, there are usually present but 1 or 2 spines, one of which, in the latter case, is larger than the other. When present, the ovicell is rounded and more or less sunken according to age and degree of calcification, and bears on the front a raised, triangular rib which in older colonies often terminates in a knoblike apex.

Paired avicularia with tips pointing forward occur on zooecia in which the neighboring posterior cell bears an ovicell; when the neighboring posterior zooecium is without an ovicell, only a single median avicularium with a backward-pointing tip is present. Somewhat common on shells and algæ in offshore waters. Southern New England.

M. tenuis. (Fig. 165.) A calcareous, encrusting species in which the zooecia are of moderate size, oval or oblong in shape, and with a high, raised margin

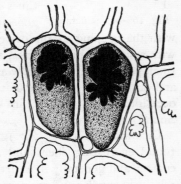

Fig. 165—M. tenuis.

finely tuberculate on the inner side. The inner part of the frontal area is partly closed over by a calcareous lamina having a finely pitted surface and bearing an aperture irregularly bordered with spinulose projections. Frequently there occur rounded knoblike processes rising from the spaces in the angles between the zooecia. There is, however, considerable variation in the area and shape of the calcified lamina, and in the size and form of the knobs. Not uncommon. Occurs on stones and shells in shallow and offshore waters. Southern New England.

Family CRIBRILINIDAE

Genus *Cribrilina*

C. punctata. (Fig. 166.) This species is a calcareous, encrusting form. The zooecia are rather small, somewhat cylindrical, and per-

forated by large irregular openings. These perforations, however, sometimes tend to become almost closed in older colonies. The principal aperture is a semicircular orifice bearing on its lower lip a small, pointed median process, often cleft, which in old cells may become strongly pronounced, sometimes altering or obscuring the shape of the aperture. Usually the aperture bears 4 marginal spines, of which the posterior pair is the larger; this pair in fertile cells is sometimes curved inward across the orifice of the

Fig. 166—C. punctata.   zooecium, while the anterior pair may be fused with the opening of the ovicell. The ovicell is somewhat globe-shaped with a smooth shiny surface, and is perforated by a varying number of small pores. There is usually an avicularium pointing obliquely forward and outward on each side of the principal aperture. Common on shells, occasionally on pebbles, in shallow and offshore waters. Vineyard Sound northward.

C. annulata. (Fig. 167.) Like the preceding species, this form is found encrusting shells and small stones, but the zooecia are much coarser and generally more irregular in form; also the perforations are relatively smaller and are arranged in rows, those in the anterior region being transverse, while those in the posterior part tend to radiate. Often there is present extending from the principal aperture to the posterior end of the cell a median ridge. The orifice is semicircular and usually bears a small tooth on the lower lip; but the shape is often obscured in older individuals, particularly in fertile cells, wherein the lips become thickened and a secondary lip frequently extends over the ovicell. There are commonly 4 short spines projecting forward on the anterior lip, the hindmost pair being the smaller and more divergent. The ovicells are small, hemispherical structures with few perforations, and are situated in line with the zooecium. There are no avicularia. The colonies of this species form rounded reddish or brownish patches of crust. Although commonly

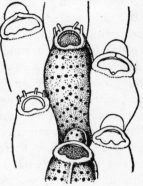

Fig. 167—C. annulata.

growing attached to shells and stones, they are not uncommon on kelp. From shallow to rather deep water. Cape Cod region and northward.

FAMILY MICROPORELLIDAE

GENUS *Microporella*

M. ciliata. (Fig. 168.) Zoarium a delicate, irregularly shaped calcareous crust of a frosty or silvery sheen. The distinguishing feature of this species, and in fact of all other members of the family, is the special median pore that occurs just below the principal aperture. Also, a secondary character helpful in determination is the shape of the principal aperture itself, which is semicircular above with the lower lip usually straight and without prominences or indentations. The principal orifice occurs at the anterior end of the zooecium, the latter being an ovate or sometimes elongated hexagonal cell often punctured with numerous minute

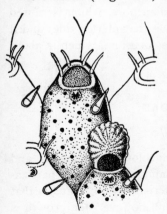

Fig. 168—M. ciliata.

pores. Generally there are from 3 to 7 outward-curving spines present on the aperture, but in certain collections these spines may be wanting more often than not. In the present species the median pore is lunate or more or less semicircular, is fringed with internally-projecting teeth, and has a projection extending from the anterior border into the opening. However, this median pore is not always clearly evident; for in certain conditions of calcification it is apt to become obscured by an umbonate process that frequently occurs just posterior to it. The ovicell is more or less rounded, sometimes bearing perforations, sometimes marked with radiating ridges, and sometimes having an umbonate process at the top. A large avicularium is generally present (sometimes there are two) on each zooecium, and it is situated at one side somewhat behind the principal opening; it has an acute mandible often prolonged into a long slender spine usually directed forward and outward. This is a very variable species, any one or all the characters being subject to considerable modification according to the degree of calcification undergone by the cell; then, too, there are vast differences in the sizes of the zooecia, and in the structural appearance of the apertures and appendages. Somewhat common on seaweeds, shells, etc., from low-tide level to fairly deep water. New England and southward.

M. ciliata, variety *stellata.* This variety of the preceding form usually occurs with the normal *ciliata,* but may be distinguished from it by the nearly circular stellate character of the median pore, which is caused by the absence of any process or projection on the anterior side; and it may also be differentiated by the size of the zooecium, which is about twice that of *ciliata.*

FAMILY MYRIOZOIDAE

GENUS *Schizoporella*

S. hyalina. (Fig. 169.) In this species the colony is calcareous, and encrusting, forming on stones, shells, algæ and other objects, particularly when the zoarium is young, the characteristic hyaline patches that give it its specific name. In the older zoarium, however, the growth is irregular, the cells being piled on one another and the colony frequently assuming a more or less erect form. The zooecia are elongate and somewhat cylindrical, marked with transverse furrows, and arranged irregularly in radiating rows. The aperture is situated anteriorly and is characterized by a distinct indentation in the posterior margin. This, by the way, is a family feature. In some cases, however, this marginal sinus may be considerably obscured from above by an overhanging umbonate process. The ovicells are somewhat globular with coarse perforations, and occur on small zooecia that rise above the other cells. There are no appendages. Very common. Found from the tidal zone to deep water. Arctic Ocean to Long Island Sound.

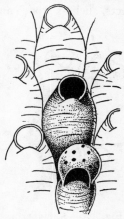

Fig. 169—S. hyalina.

S. unicornis. (Plate XIX.) A calcareous, encrusting form. The colonies are often many layers in thickness and will attach themselves to almost any object that offers a substantial base. In general outline the shape of the zooecia is quite variable, being either ovate, hexagonal, rectangular, square, or even broad. The surface, too, is variable, sometimes being smooth and glossy, but more often tuberculate or rough; and it is punctured with numerous pores. In some specimens the cells are separated by a raised wall; in others only a deep groove is present. The principal aperture may be circular or semicircular, but it is normally characterized by an indentation on the posterior border. Often an umbo occurs just below the orifice. The ovicell is globose and perforated, and is marked with radiating grooves on the sides; occasionally there is a small umbonate process at the top. Avicularia are present, occurring at one or both sides of the opening; often, however, these appendages may be lacking in the entire colony. With the exception of the character of the principal opening, the zooecia of this species may show a very great degree of variation. In life the color ranges from pale orange to dark brick-red, and often silvery and colorless forms occur. Found from the tidal zone to deep water, on all bottoms except those of pure mud or sand where there is nothing solid to afford a holdfast. This is one of the most common and abundant of the calcareous bryozoans. Occurs in all ranges.

S. biaperta. (Fig. 170.) The calcareous zoarium of this species is encrusting or not, according to the nature of its support. On solid objects, such as stones and shells, it forms smooth, flat colonies, but when attached to algæ, hydroids and other such bases, it develops into projecting, shelflike expanses of unusual beauty. The zooecia are ovate or hexagonal. In the younger stages the cells are conspicuously perforated with small pores and are smooth and glossy; but, as calcification increases, the pores tend to become obscured, and the surface becomes dull and rough. Advancing calcification also may obscure the raised border which separates the zooecia. The principal opening has the family character wherein a sinus, or indentation, is present in the lower lip. The ovicells are rounded in outline when viewed directly, but have the upper surface flattened and marked with radiating lines; surrounding the flattened area is a heavy border rising from the base. At the side of or facing the aperture, 1 or 2 avicularia with rounded or oval mandibles may be present situated on a rounded prominence. Living colonies vary in color from translucent or colorless to bright pink or red. Often abundant. Found in shallow and offshore waters. New England and southward.

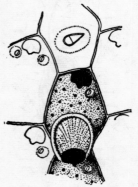

Fig. 170—S. biaperta.

FAMILY ESCHARIDAE

GENUS *Lepralia*

L. pallasiana. (Plate XIX.) A calcareous and encrusting form; the colonies tending to become circular in outline when the nature of the base is favorable. The zooecia are large, often quadrangular or somewhat hexagonal, but very variable, and coarsely punctate. When young, the surface is smooth and glossy, but older specimens are often rough and have thick ridges between the pores. The principal opening is large, longer than it is wide, contracted on each side below the middle, and widened near the posterior end. Ovicells wanting. Occasional specimens will have avicularia present, situated below the orifice. An umbonate process sometimes occurs just below the opening. Very common. Found

on shells, stones, submerged wood, etc., in the tidal zone and shallow
water. Arctic Ocean to New Jersey.

L. americana. (Fig. 171.) Zoarium calcareous and encrusting, forming whitish to reddish colonies sometimes several cells in thickness. The
zooecia are large but not generally so large as
those of *L. pallasiana*. Although rather variable,
the outline of the cells is commonly roughly
quadrangular or hexagonal. In young specimens, the surface is coarsely punctured with
large pores, but, as calcification progresses,
rough ribs appear. The principal orifice is
somewhat quadrangular and slightly longer
than broad, and bears a denticle on either side
near the posterior end. Numerous specimens
occur, however, in which the opening is nearly
rounded. The peristome, or border of the aperture, is thin and slightly raised, and is often

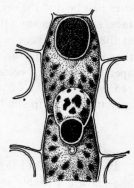

Fig. 171—L. americana.

produced into a kind of projecting lip on either side of the opening.
Sometimes there is a prominent umbo behind the orifice. Large, sub-
globular ovicells are often present, having a few very conspicuous
pores on the upper surface. At the top of the umbo, just below the
opening, there is frequently a rounded avicularium. As this species often
occurs in the same situations as *L. pallasiana* and has a superficial resemblance to that species, the collector can readily distinguish them by noting
that the present species differs in the shape of the aperture, in the possession of ovicells, in the more radiate character of the ribbed surface

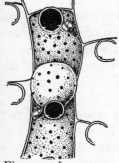

in older zooecia, and in the occasional presence
of a raised border between the cells. Although
frequently inhabiting the same waters and forming the same attachments as the foregoing bryozoan, this one usually occurs on deeper bottoms. Not uncommon. Southern New England
coast.

L. pertusa. (Fig. 172.) Calcareous and encrusting. Zooecia large, of variable shape, separated by a raised border, and perforated with numerous pores. When young, the surface is smooth

Fig. 172—L. pertusa.

and glossy; when older and fully calcified, it is often considerably roughened. The principal aperture is rounded and is provided with a pair of teeth situated laterally behind the middle. The peristome, or oral border, is slightly raised and thickened. A rough umbonate process sometimes occurs just behind the orifice. Large, globose ovicells with a punctured surface, sometimes smooth and sometimes rough with an occasional umbo at the top, are frequently present. Avicularia, however, are rare. When they do occur, they are small oval structures situated at one side of the aperture. Color of colony, silvery white when young; various shades of red when older. Not uncommon. Occurs on stones and shells and occasionally algæ in shallow and offshore waters. All ranges.

Genus *Porella*

P. concinna. (Fig. 173.) Colony calcareous and encrusting. The zooecia are variably ovate or elongate, are with granular surface and

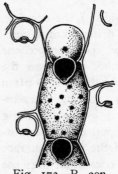

possess a row of conspicuous marginal pores. Often the cells are separated by a raised margin. The primary, or first-formed, aperture is arched above, straight behind, and with a broad tooth, but secondary calcification produces a deep opening with a peristome, or secondary aperture wall, enclosing the primary orifice and a rostrum on the lower lip on which is mounted a round avicularium. Occasionally the lateral sides of the peristome will be produced into a pair of short, blunt spines. In the younger zoaria, the ovicells are globose and prominent, but these become less conspicuous as calcification proceeds; they are

Fig. 173—P. concinna.

situated adjacent and anterior to the orifice, and are usually provided with a single median pore. Not uncommon. Occurs on stones and shells in comparatively shallow and offshore waters. Gulf of St. Lawrence to Cape Cod.

P. proboscidea. (Fig. 174.) In this species the colonies rise erect from an encrusting base, forming calcareous, folded frills composed of two layers of cells. The zooecia are elongated, and around the margin of each cell is a row of areolæ with strong ribs running between them toward the middle, the ribs, however, often becoming obscured in old and much-calcified individuals. The primary, or first-formed, orifice is rounded anteriorly and straight posteriorly with a platform on the lower lip on which is situated the small, round avicularium. A raised peristome, or border, surrounds the opening, which embraces the rostrum and is continuous with the ovicell. As calcification proceeds, this border is raised still higher, thus producing a secondary aperture which is somewhat pear-shaped with the smaller end directed posteriorly. The ovicell is large and prominent in the young stages, but is completely suppressed when the colony is old. Grows on stems, pebbles and shells. Occurs below the tide marks and in offshore waters. Sometimes abundant. Southern New England coast.

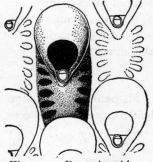

Fig. 174—P. proboscidea.

Genus *Smittia*

S. trispinosa. (Fig. 175.) Zoarium calcareous and encrusting. Zooecia more or less quadrangular to ovate, separated from one another by a raised border, and with surface smooth and shining to more or less granular. Around the border is a row of areolæ, which in older cells are separated by strong, short ribs. The orifice is rounded in front and nearly straight behind. The primary opening bears from 2 to 4 spines; the second orifice produced by increased calcification bears a sinus in front. Ovicells large with several perforations. There are two kinds of avicularia present, one large kind with a pointed mandible, and a smaller kind with a rounded or blunt mandible. The colonies form large whitish or yellowish crusts on stones and shells; at first they are thin and smooth, later they are thick and roughened. Not uncommon. Found in inshore and offshore waters. New England.

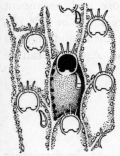

Fig. 175—S. trispinosa.

S. trispinosa, variety *nitida*. This variety of the foregoing described species may be distinguished from it chiefly by the absence of the large pointed avicularium. The cells usually average smaller than those of *S. trispinosa;* and the aperture is also correspondingly smaller. It occurs in the same situations, but oftentimes in shallower water and in greater abundance.

GENUS Mucronella

M. peachii. (Fig. 176.) A calcareous, encrusting form with rhomboidal zooecia.

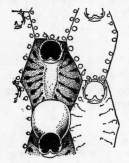

Fig. 176—M. peachii.

The cells, when young, are smooth with large pores around the margin, between which are raised ribs; but increasing calcification tends to obliterate the pores. Near the anterior end of the zooecium is the orifice, longer than wide and rounded in front with the lower lip nearly straight. In the primary, or early-formed, aperture, the lower lip has a short, median, two-pronged tooth. There are a pair of well-developed lateral teeth. A slightly raised peristome with 6 anterior spines and a mucro, or pointed projection, posteriorly is present, but the spines may be wanting in old cells. The ovicells are globose and prominent in young individuals; in older specimens these are more or less obscured. There are no avicularia. Found on stones, shells and algæ in shallow and offshore waters. Sometimes abundant. Arctic Ocean to Long Island Sound.

FAMILY CELLEPORIDAE

GENUS Cellepora

C. americana. (Fig. 177.) Colonies of this species may be encrusting or rising into nodular branches. All are calcareous. The zooecia are somewhat ovate or pear-shaped and, like all others of the genus to which it belongs, the cells stand more or less erect with the principal aperture at the top. The arrangement of the cells is somewhat irregular, these being usually much crowded and piled upon one another, besides being turned in various directions. Their surfaces are smooth and shining, and they are irregularly perforated around the base. The orifice is nearly circular and bears a V-shaped notch in the posterior margin which is overhung by a very prominent pointed rostrum situated just behind and a little to one side of the opening. There are often numerous conspicuous, smooth ovicells, flattened above and bearing a number of perforations. Avicularia are usually present at the bases of the rostra, located laterally and somewhat internally. Found from the low-tide level to offshore waters; grows on algæ, hydroids and other bryozoans. Sometimes abundant. All ranges.

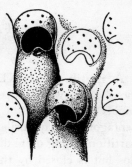

Fig. 177—C. americana.

C. canaliculata. (Fig. 178.) Zoarium calcareous and encrusting or branching. Zooecia somewhat ovate, erect and punctured around the base. Cells irregularly disposed. The orifice is rounded and has a rather broad notch in the posterior margin. This is overhung by a long, stout, curved rostrum. In well-calcified individuals there is a thin peristome, from the sides of which a broad flange rises to the sides of the rostrum, forming a wide spout reaching to the primary aperture, or first-formed, opening at the top of the zooecium. Prominent ovicells are often present, flattened above and bearing a variable number of punctures. In this species the avicularia are rounded and are situated at the tips of the rostra. Found on algæ, hydroids and other bryozoans. Occurs in inshore and offshore waters. Not uncommon. All ranges.

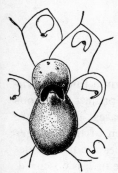

Fig. 178—C. canaliculata.

SUBORDER CTENOSTOMATA

In this suborder the zooecia are fleshy or membranous, or more or less chitinous, never calcareous. The principal aperture of the cell is situated terminally and it is closed usually by an operculum of setæ, or bristles, borne on a membrane. There are no ovicells, avicularia or vibracula present.

FAMILY ALCYONIDIIDAE

GENUS *Alcyonidium*

A. mytili. (Plate XIX.) The zoarium of this species forms a gelatinous crust, which appears covered with small prominences when the zooids are retracted. Usually the zooeci are hexagonal in shape, but pentagonal and quadrilateral forms not infrequently occur, the marginal partitions being quite distinct. The orifice is not provided with external lips but is closed by the retraction of the tentacle sheath. Sometimes abundant, forming irregular reddish or yellowish patches on stones, shells, piles, algæ, and the coverings of living animals, such as barnacles, spider crabs, skate egg-cases, etc., in shallow water. Arctic Ocean to Long Island Sound.

FAMILY FLUSTRELLIDAE

GENUS *Flustrella*

F. hispida. (Fig. 179.) Colonies of this species form a thick brownish incrustation from which project numerous large spines having swollen bases. The cells are large; but, with the exception of the younger individuals in which the spines are not yet developed, the structure is more or less obscured. They are roughly hexagonal in outline and have a flat, smooth surface. The aperture is slightly raised and is provided with a pair of distinct lips capable of opening and closing. Spines are often present around the orifice as well as around the margin of the cell. Occurs on stones and shells, but seems to show a preference for the stems of *Ascophylum* and *Fucus*. Often abundant in very shallow waters; it is eminently littoral in its choice of habitat. New England southward.

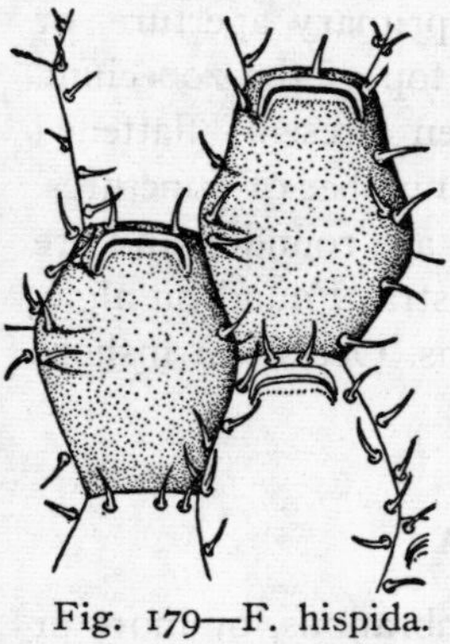

Fig. 179—F. hispida.

Family **VESICULARIIDAE**

Genus *Bowerbankia*

B. gracilis. (Fig. 180.) In this form the colony is branched and creeping, the cylindrical zooecia rising singly or in clusters from the recumbent stolon. The cells are without appendages at the orifice, or what in this instance may be termed the margin; this being closed by the retraction of the zooid. The zooid has 8 ciliated tentacles and an alimentary tract provided with a strong gizzard. Fairly common on stones, piles, hydroid stems, and algæ. Occurs in very shallow and in offshore waters. New England coast.

Fig. 180—B. gracilis.

B. gracilis, variety *caudata*. This variety of the preceding species occurs in great abundance and over a wider range. It does not, however, seem to inhabit waters of so great a depth as does the other, being apparently a strictly inshore form. It has a general resemblance to *B. gracilis,* but may be distinguished from that form partly by the truncated, or squared-off, tops of the zooecia, but chiefly by the short caudate appendage projecting from near the base of the zooecium. This projection may end in a single point or it may be bifid or trifid. All ranges.

Genus *Amathia*

Fig. 181—A. dichotoma.

A. dichotoma. (Fig. 181.) Colonies of this species are erect and plantlike in form, often reaching 5 cm or more in height. The branches are white in color with a short, dark-brown segment at the base of each fork. The zooecia are short and are arranged in double rows winding spirally partly around the branches just below each fork, each cluster containing from 6 to 12 individuals. Fairly frequent; on rocks, oyster shells, algæ, in shallow water. New Jersey northward.

Family **VALKERIIDAE**

Genus *Valkeria*

V. uva. (Fig. 182.) A creeping form in which the colony is composed of delicate jointed tubes with the zooecia grouped at intervals on the stems. The zooecia are small, transparent, cylindrical structures, pointed below and ending obtusely at the top. When the tentacles of the living animal are expanded, two of them are characteristically bent outward. Not uncommon. Found creeping over rocks and hydroids in tide pools and shallow water. New England southward.

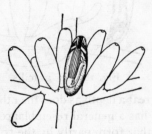

Fig. 182—V. uva.

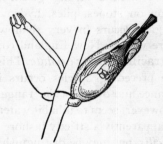

Fig. 183—H. armata.

Family **TRITICELLIDAE**

Genus *Hippuraria*

H. armata. (Fig. 183.) The colony in this form consists of slender, threadlike stems, often anastomosing, to which are attached, usually in pairs, stout oval zooecia, broad at the base and with short pedicels, or stalks. The zooecia have 4 conical processes at the distal end, each of which is provided with a considerable spine; and the cells are further characterized by a flattened membranous area occupying the greater portion of the ventral side. There is a small but quite distinct gizzard present. This species occurs characteristically creeping and adnate, but erect shoots are not uncommon. Found on piles, rocks, and hydroids, in shallow water. Not uncommon. Southern New England.

H. elongata. (Fig. 184.) In this species the zoarium is creeping and consists of a slender, branched and jointed stolon. The branches are paired and arise from later projections on each side at the end of the internodes. The zooecia arise from the same projections that give rise to the branches, and are elongate structures with slender pedicels of variable length. The cells usually arise in pairs, and each has a membranous area of considerable size extending over the ventral side. This is a commensal species; that is to say it derives its nourishment from the food material obtained by another animal. It is found in the gill chamber of the blue crab (*Calinectes sapidus*) and that of the large spider crab (*Libnia sp.*). Not infrequently it occurs extending along the bases of the legs and over the back of the small crabs (*Pinnixia sp.*), occupying the tubes of the marine worm *Chaetopterus*. Sometimes abundant. Cape Cod to South Carolina.

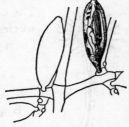

Fig. 184—H. elongata.

CLASS **PHORONIDA**

This is a group of wormlike animals which usually live in colonies, the individuals of which occupy chitinous tubes often in contact and twisted together, but never communicating with one another. At the anterior end of the animal is the horseshoe-shaped lophophore which consists of a double ridge, each part bearing a single row of tentacles, and in most cases terminating in a spiral coil on each side. In the middle of the lophophore are the oral and anal apertures, situated close together but separated by an epistome, a process rising from the body wall. The body cavity is comparatively large. There is a U-shaped digestive tract made up of œsophagus, stomach and intestine. The kidneys are a pair of tubes that lead from the body cavity to the outside through two pores situated near the anus. Both male and female sex organs are contained in the same individual, gonads being near the stomach and their products passing out through the paired kidney orifices. The animals contain two kinds of blood, a colorless fluid circulating in the body cavity, and a red fluid circulating in a system of closed vessels lying along the digestive tract and extending to the tentacles and other organs. During development the young pass through a metamorphosis wherein the larvæ are of the characteristic form known as *actinotrocha* larvæ.

The class is a small one, containing only about a dozen species; of these, but one is represented on the Atlantic Coast of this country.

GENUS *Phoronis*

P. architecta. (Fig. 185.) The tubes of this species are generally straight and covered with sand grains or small shell fragments. Although the animals may occasionally tend to be gregarious, the tubes are more often found isolated and alone. The lophophore is not produced into lateral spiral coils. In color the body is pinkish anteriorly and yellowish or reddish posteriorly. Length of tube 13 cm; width 1 mm. Occasionally common. Found in sand near the low-water level. Cape Hatteras region.

Fig. 185—P. architecta.

CLASS **BRACHIOPODA**

The lamp shells. These animals have a superficial resemblance to the clams, but a study of their internal structure soon reveals them to be totally unrelated to the mollusks. Their affinities, indeed, are obscure; but there is evidence in certain of their characters that they cannot be far removed from the worms, and it is for this reason that they are usually grouped with the *Bryozoa* and the *Phoronida*.

The body of a brachiopod is enclosed in a bivalve shell of calcareous or horny nature. The valves usually differ slightly in shape and are of unequal size; one is dorsal and the other is ventral. Between the valves or through a hole in one of them, a peduncle, which is a stout muscular stalk, extends from the body, wherewith the animal fastens itself to other objects. In certain species (*Crania*), however, the peduncle is wanting, the animal attaching itself by the whole ventral valve. A large, conspicuous lophophore is present; this consists of a pair of coiled ridges projecting from the anterior soft parts of the animal, along the surface of which is a ciliated groove bearing on one side a row of ciliated tentacles. The cilia cause the food-bearing currents to flow toward the mouth which lies at the base of the ridges. Besides their use in the ingestion of microscopic food organisms, the arms of the lophophore have respiratory and sensory functions. There is a tubular digestive tract consisting of œsophagus, stomach and intestine. The nervous system consists of a pair of dorsal ganglia situated near the œsophagus, connected with a ventral pair by commissures, or a ring of nerve tissue, around the œsophagus. A so-called heart forms part of the circulatory system; but it has been demonstrated that circulation may also result from ciliary action. Excretory organs open into the mantle cavity at one end of the nephridial tubes, and are connected at the other end with the body cavity. The sexes are usually separate, and they bear two pairs of genital glands, the products of which find their way out through the nephridial openings. The animals pass through a trochophore larva stage during development. Although the brachiopods were in ancient times a very numerous group, there are but about 125 species living today.

ORDER **ATREMATA**

The distinguishing features of members of this order are the absence of a hinge joining the valves, and the presence of an anus.

FAMILY **LINGULIDAE**

GENUS *Glottidia*

G. audebarti. (Plate XIX.) This species, formerly bearing the name *Lingula pyramidata,* possesses a narrow, white, hornlike shell tapering at the beak, and having the front margin nearly straight. The dorsal valve has 2 internal, sharp-incurving laminæ which diverge from the beak and extend along the length of the shell for about a third of the distance. The ventral valve has a septum extending along the middle, forward from the beak. There is a long, slender, contractile peduncle. The mantle, or soft fleshy folds covering the internal surface of the shell, is provided with long, stiff setæ, or bristles, on the edges. This is a hermaphroditic form. Color of shell white with transverse bands of green. Length of shell 27 mm; breadth 10 mm. Length of peduncle 16 mm. This form lives in the sand in the tidal zone where it occupies a vertical burrow, and usually remains sessile with the opening of the shell even with the surface of the substratum. Cape Hatteras to Florida.

ORDER TELOTREMATA

Members of this order are characterized by having a calcareous shell, the two valves of which are joined by a hinge; the ventral valve is the larger and bears a beak at the hinder end which has a perforation through which the peduncle extends. Projecting from the inner side of the dorsal valve are calcareous loops, or arms, forming a support for the lophophore. No anus is present.

FAMILY TEREBRATULIDAE

GENUS *Terebratulina*

T. septentrionalis. (Plate XIX.) The lamp shell. In this species the shell is somewhat oval with a slightly projecting beak truncated horizontally and with a large peduncular opening. The outer surface is punctate and with 5 radiating ribs crossed by concentric striations. The calcareous loop is short and simple. Shell thin and semitransparent with a whitish or yellowish cast. Length 13 mm; breadth 8 mm. Occasionally common in shallow and offshore waters. Cape Cod region.

IX

ANNELIDA

PLATE XX

SEGMENTED WORMS

Nereis virens (upper left); *Phylodoce catenula* (middle right);
Harmothoe imbricata (lower middle).

CHAPTER IX

PHYLUM **ANNELIDA**

(THE HIGHER WORMS)

THE annelids, or annulates, as they are sometimes called, are a group of worms in which the body is elongated and composed of numerous ringlike divisions, or segments. There is a distinct head; a coelom, or body cavity, is present, within which is the digestive tube; and very frequently the body segments are provided with paired appendages which are not segmented. In most cases the head is well defined and contains, besides the mouth, a well-organized brain, and also often comparatively highly developed sensory organs such as tentacles and eyes. The remaining segments resemble one another generally, and on these are borne the bristles and other appendages, when present. In most of the marine annelids, these segments partition the body cavity into a series of chambers, thus dividing the body cavity into as many spaces as there are annulations, each separate space also containing a similar set of internal organs.

A mouth, pharynx, œsophagus, intestine, and an anus situated at the posterior end of the body, form the alimentary tract. The pharynx and œsophagus are usually well marked, the former organ in the majority of annelids forming a protrusible proboscis by means of which the animal seizes its food.

In all except the lowest members of the group, the circulatory system is well developed, the blood stream (often consisting of red blood) being transported in tubes, the most important of which are a dorsal and a ventral tube, both longitudinal, and situated respectively above and below the intestine and connected with each other by communicating transverse tubes. A circulatory fluid is sometimes present in the body cavity; in numerous forms this fluid is in direct communication with the tubular system.

The excretory system is composed of numerous nephridia, or kidney tubules, these usually being a pair of coiled tubes in each body segment.

Although the brain is dorsally situated, the rest of the nervous system lies along the ventral region and consists of paired segmental ganglia connected by a pair of longitudinal nerves.

Most marine annelids are unisexual, but except during the breeding season, the reproductive organs are not well marked. Development includes a ciliated trochophore larval stage. The group is a fairly numerous one, consisting of more than 4500 species. The individuals are abundant nearly everywhere, living in water and in moisture-laden areas of the earth. The majority are predaceous; some, however, are vegetarian; others are mud-eaters, ingesting the material of their surroundings for the food particles this may contain; still others are parasitic.

The *Annelida* may be divided into the following five classes:

PHYLUM ANNELIDA

CLASS **Archiannelida**	FAMILIES ..	{ Dinophilidae Polygordiidae
CLASS **Chaetopoda**	ORDERS	{ Polychaeta Oligochaeta (Mostly terrestrial; not listed here.) Echiurida
CLASS **Gephyrea**	ORDER	Inermia

CLASS **Hirudinea** (The leeches. Parastic forms; not described in this work).

CLASS **Myzostomida** (Parasitic forms on echinoderms; no species listed here.)

CLASS **ARCHIANNELIDA**

Members of this group are primitive forms of small size. The external markings of the body segments are not usually well defined, but the internal septa, or partitions, dividing the body cavity are complete. Segmental parapodia and setæ are wanting, but the head sometimes bears bristles or other appendages. The sexes are separate.

Family DINOPHILIDAE

Genus *Dinophilus*

D. conklini. (Fig. 186.) This species has a thick, cylindrical, colorless body composed of 6 segments, each encircled with 1 or 2 bands of cilia. The head bears 2 bands of cilia and is provided with a pair of eyes and tactile bristles. The posterior segment of the body is reduced and modified into a telson, or tail. Length about 1 mm. The development of this worm is quite simple; in many respects the adult form is seemingly not farther advanced in organization than is a polychætous larva, to which it has a certain resemblance. Found among algæ. New Jersey.

Fig. 186 —D. conklini.

Family POLYGORDIIDAE

Genus *Polygordius*

P. appendiculatus. (Plate XXI.) The adult form of this worm has never been found along our coast, but the larva occurs frequently in plankton towings. In this trochophore larval stage the body is much broader anteriorly than it is posteriorly and bears 2 long threadlike anal appendages. Length about 1 mm. The adult attains to a length of 20 mm, and it occurs in the Mediterranean. Larvæ often numerous. All ranges.

CLASS CHAETOPODA

The marine representatives of this class are principally characterized by the possession of setæ, or bristlelike appendages, occurring singly or in groups and segmentally arranged in pairs.

ORDER POLYCHAETA

In this order are comprised those chætopods in which the individuals are provided with *parapodia,* or paired and segmentally arranged fleshy locomotor organs on which the setæ are borne. Certain sedentary forms, however, may have the parapodia much reduced in size or may lack them altogether. The individual parapodium consists of two parts: a dorsal *notopodium* and a ventral *neuropodium,* from which extend both a dorsal and a ventral *cirrus,* or sensory filament. The cirri in some cases may be much enlarged, receiving an extra blood supply; thus functioning also as gills. Nearly all the marine annelids belong to this order.

Family APHRODITIDAE

Genus *Aphrodita*

A. hastata. (Plate XXI.) The sea mouse. This beautiful and colorful worm could hardly be confused with any other, being easily distinguished by its broad elliptical body bearing a thick growth of long, brilliant, iridescent setæ along the lateral margins of the back. These bristles project through a shorter and brownish feltlike coat that covers the entire back of the animal. They are relatively delicate, curving over the back toward the middle line, and among them are numerous coarse setæ barbed near the ends and terminating in sharp points. There are 15 pairs of elytra on the back which are hidden by the furry covering.

On the head there are present 1 short and 2 long sensory tentacles. Length about 12 cm; width about 4 cm. Inhabits mud and sandy bottoms in offshore waters. Vineyard Sound.

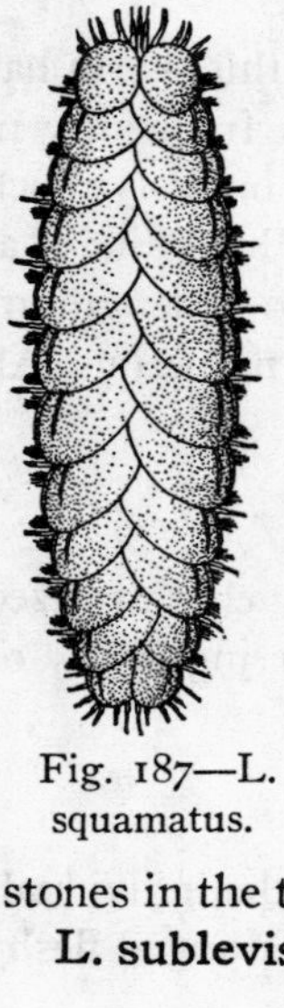

Genus *Lepidonotus*

Fig. 187—L. squamatus.

L. squamatus. (Fig. 187.) A scale worm. The body of this worm is slightly broader at the middle than at the ends, which are blunt. There are 12 pairs of elytra covering the back. These are tuberculated and colored a dark, speckled, sandy brown. The scales are often cast off if the animal is handled or irritated. The head bears well-developed eyes, 3 tentacles, 1 pair of long palps, and 2 pairs of cirri grouped around the mouth. Length about 3 cm; width about 8 mm. Very common. Found under stones in the tidal zone. Labrador to New Jersey.

L. sublevis. (Plate XXII.) This scale worm, though not so common as *L. squamata,* resembles it somewhat and generally occurs in the same situations. It differs from the other in its smaller size and in the texture and markings of the elytra, these being smooth and colored gray or light brown with spots. Length 3 cm; width 7 mm. Southern New England to Virginia.

Genus *Harmothoe*

H. imbricata. (Plate XX.) A scale worm that has the anterior portion of the head produced forward into a pair of pointed tips separate from the tentacle bases. The body is flattened and elongate with 15 pairs of slytra, and is made up of from 42 to 44 segments. The bilobed head carries 3 tentacles, 2 long palps, and 2 pairs of peristomal cirri, and is provided with two pairs of eyes. Color variable, ranging from grayish to brown, and occasionally with a black dorsal stripe. Length about 3 cm. Occurs below the low-tide level to deep water. Greenland to Long Island Sound.

Genus *Pholoe*

P. minuta. (Fig. 188.) In this scale worm, the segments number about 68 in the adult form, young specimens having fewer somites. Forty-four pairs of elytra cover the back, these alternating with the dorsal cirri anteriorly but occurring on all segments posteriorly. The head is provided with 2 pairs of eyes and bears 1 tentacle and 2 short cirri near the mouth. Length about 2 cm. Found in shallow water. Massachusetts Bay.

Genus *Sthenelais*

S. leidyi. (Plate XXI.) This attractive scale worm has a relatively long body consisting of over 100 segments. The back is covered with more than 150 pairs of elytra, those situated anteriorly alternating with the dorsal cirri, while those of the posterior region occur on every segment. It is provided with 2 pairs of eyes and has 1 tentacle. The head is brown with a central red spot on each side of which is a white spot. The remainder of the body is grayish with a dark stripe running along the middle of the back. Length

Fig. 188— P. minuta.

15 cm; width 4 mm. Found in shallow water. Massachusetts Bay to North Carolina.

Family PHYLLODOCIDAE

Genus *Phyllodoce*

P. catenula. (Plate XX.) A paddle worm, so-called from the fact that this and other members of the family have the dorsal and ventral cirri broad and leaflike, giving the individuals a characteristic appearance while in motion. The body is elongate and depressed. The head has a pair of large brown eyes, 4 tentacles on the anterior part, and 4 long cirri. On the proboscis are longitudinal rows of papillæ, or small prominent projections. The color of the worm is bright iridescent green with longitudinal rows of brown spots on the back. Length about 7 cm; width about 1.5 mm. Found in tide pools and in shallow and offshore waters in sandy mud. Rather common. Bay of Fundy to Rhode Island.

Genus *Eulalia*

E. pistacia. (Fig. 189.) The body of this paddle worm is slender and depressed. There are 5 short tentacles on the anterior part of the head, while the posterior part bears 4 pairs of very long cirri. The general color is a bright iridescent yellowish-green. Length 4 cm; width 1.5 mm. Found in offshore waters, usually among hydroids. Maine to Long Island Sound.

Fig. 189—E. pistacia.

Family SYLLIDAE

Genus *Syllis*

S. pallida. (Fig. 190.) The body of this species is slender, tapering at both ends, and with segmented ventral cirri. Another characteristic is the segmentation of the tentacles, 3 of which are borne on the anterior portion of the head. Also in this region are a pair of palps and 2 pairs of eyes. The posterior region of the head bears 2 cirri on each side. Color white. Length of adult variable, being from 15 to 25 mm. This annelid reproduces asexually by terminal budding, the posterior portion of the animal constricting to form individuals which eventually separate from the parent to lead a free life. Often abundant in sand and mud and on dead shells in shallow and offshore waters. Bay of Fundy to Long Island Sound.

Genus *Autolytus*

A. cornutus. (Fig. 191.) Body of full-grown male consisting of 30 segments; of female, 40 to 50 segments. The head has 2 pairs of eyes; but the palps are rudimentary. The tentacles and cirri are filiform, or

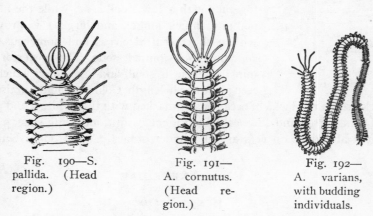

Fig. 190—S. pallida. (Head region.)

Fig. 191— A. cornutus. (Head region.)

Fig. 192— A. varians, with budding individuals.

threadlike. This form is without ventral cirri. Color pinkish, with conspicuous brown eyes. Length 15 mm. Reproduction usually is by asexual budding. Found in shallow and offshore waters, often free-swimming. These worms commonly occur among seaweeds and hydroids, the asexual individuals inhabiting attached tubes which they construct for themselves. Often numerous. Bay of Fundy to New Jersey.

A. varians. (Fig. 192.) Together with the characters of the genus, such as rudimentary palps, absent ventral cirri and filiform tentacles and cirri, a feature by which the species of this syllid may be determined is the intestine with bright-red spots which is visible through the body wall and appears as a median line. Length 15 mm. Occurs usually among hydroids below the low-tide level. Fairly common. Bay of Fundy to North Carolina.

Family HESIONIDAE

Genus *Podarke*

P. obscura. (Fig. 193.) In this species, the body is relatively short, convex above and flat below, with the segments deeply indented at the

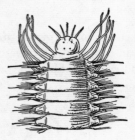

Fig. 193—P. obscura. (Head region.)

sides. There are six pairs of long cirri on the posterior portion of the head and the 2 succeeding somites. The parapodia are elongated and biramous, or forked, the notopodium being short and conical and bearing at the end a long dorsal cirrus together with a few small setæ, while the neuropodium (much longer and larger) is provided with a short ventral cirrus and numerous long compound, or jointed, setæ. The posterior terminal segment is small and has 2 cirri which exceed all others in length. Color variable, but usually dark brown or blackish, occasionally marked with thin whitish and pinkish transverse bands. Length about 4 cm; width, including setæ, 3 mm. Often abundant on eelgrass and under stones. Cape Cod to Gulf of Mexico.

Family NEREIDAE

Genus *Nereis*

N. virens. (Plate XX.) The sand worm. This large, handsome species is the commonest and best known of all our marine worms, being extensively used as fish bait. The body is elongate, convex above and generally flattened below, and consists of numerous segments, which in full-grown individuals may number as many as 200. The head is provided with 2 pairs of eyes, 1 pair of small tentacles, 1 pair of palps and 4 pairs of peristomial cirri. There is a well-developed protrusible proboscis armed with a pair of formidable, serrated, black pincer-jaws. The sexes in this species are separate and may be distinguished by a difference in color, the males being of an intense steel-blue blending into green at the base of the parapodia, these latter having a bright glaucous hue through which runs a brilliant tracery of red blood vessels. The females are of a dull greenish color delicately tinted with orange and red, with the parapodia often having an orange-green cast at the base blending into bright orange toward the ends. In both sexes the entire surface is highly iridescent, exhibiting a variety of fascinating lustrous hues. Length 30 cm or more; width about 1 cm. Found burrowing in sand or mud or under rocks in the lower part of the tidal zone, where by secreting a viscous substance they are enveloped by a loose, flexible tube binding together the sand or other materials of their surroundings. Very abundant. From Labrador to Long Island Sound.

Fig. 194—N. limbata, with proboscis protruded. (Head region.)

N. limbata. (Fig. 194.) Body elongate, rounded above and flattish below, with parapodia and setæ smaller at the anterior than at the posterior region. The head, like that of all nereids, bears 2 small tentacles, 2 palps, 4 prostomial eyes and 4 peristomial cirri. Provided with a pair of sharp, slender jaws, which are light yellow in color. Color of body iridescent brown with light lines on the sides and appendages; posterior end pale red. The male is red in the middle region. Length about 15 cm. Found on sandy shores. Common. Cape Cod to North Carolina.

Fig. 195—N. pelagica. (Head region.)

N. pelagica. (Fig. 195.) The body of this species differs from that of the two preceding forms in that it is widest in the middle region, while the others are widest at the anterior end. The head is provided with a pair of strong curved jaws, and bears a pair of small tentacles, a pair of palps, 2 pairs of eyes on the front part, and 4 pairs of cirri on the hinder part. Color iridescent reddish-brown. The full-grown female is about 20 cm long and 8 mm wide. The male attains to only about half the length of the female. Found under stones, and on shelly and hard bottoms in shallow and offshore waters. From Greenland to Virginia.

FAMILY NEPHTHYDIDAE
GENUS *Nephthys*

N. incisa. (Fig. 196.) The body of this worm is elongated with flattened dorsal and ventral surfaces, thus giving the animal a quadrangular form in cross-section. The head bears on the forward part 2 pairs of small tentacles, the ventral pair being modified palps, while on the hinder region are a pair of short cirri and parapodia with bristles. There is a large proboscis with long fleshy projections in front, and with large dorsal and small ventral papillæ. The lobes of the parapodia are widely separated and are fringed with membrane which functions

Fig. 196—N. incisa.
(Head region.)

Fig. 197—N. bucera. (Head region.)

as gills. Color whitish with red blood vessels showing on the dorsal surface. Length about 13 cm; width 6 mm. Occurs on muddy bottoms in shallow and offshore waters. Bay of Fundy to Long Island Sound.

N. bucera. (Fig. 197.) This species is relatively more slender than the preceding one, the body being composed of a greater number segments. Like that of *N. incisa,* the body is quadrangular in cross-section. There are more than 100 segments, each provided with long parapodial bristles exceeding in length the diameter of the body. Parapodia widely separate. Head provided with appendages similar to those of *N. incisa,* except that forward tentacles are longer than half the width of the head. Color whitish with anterior dorsal mottling of brown, and often a dark median line on the back. Length 20 cm; width 5 mm. Found in sandy mud and under stones in shallow water. Massachusetts Bay to South Carolina.

FAMILY LEODICIDAE
GENUS *Diopatra*

D. cuprea. (Plate XXII.) The plumed worm. A large species in which the body is flattened and bears on the segments posterior to the fifth somite and extending nearly to the end, long red many-branched gills resembling plumes. From the under side of each parapodium projects a small whitish tubercle with a dark spot in the middle. The appendages of the head consist of 1 pair of cirri, 5 tentacles and 1 pair of palps. Color reddish-brown stippled with gray, and with a brilliant opalescent iridescence. Length up to 20 cm; width 10 mm. This annelid constructs and inhabits a parchment-like tube extending nearly 3 feet into the substratum with about 3 inches projecting upward into the water, the whole being thickly covered with adhering shells and other loose material. Occurs on sandy and muddy shores near the low-water mark and in shallow bottoms. Common. Cape Cod to South Carolina.

Genus *Marphysa*

Fig. 198—M. leidyi. (Head region.)

M. leidyi. (Fig. 198.) Body flattened, except at the anterior end where it becomes slender and rounded. Head without cirri, but with a pair of small palps and 5 tentacles in a transverse row. A pair of eyes is present; and it has powerful jaws. Gills begin at or near the 20th somite. Color yellowish or brownish-red. Length 20 cm. Inhabits a tube similar to that of *D. cuprea* but inferior in quality. Found under stones, among rock clefts and in sand in shallow water. Often numerous. Vineyard Sound to North Carolina.

Genus *Lumbrinereis*

Fig. 199 —L. tenuis. (Head region.)

L. tenuis. (Fig. 199.) Slender and relatively very long, the body of this polychæt is very much like a coarse red thread. It has a conical head, without eyes or appendages, the posterior part of which consists of 2 segments. The dorsal cirri are flat, and the parapodia are small. Length about 30 cm. Abundant in sandy mud and under stones near the low-water mark. Massachusetts to Virginia.

Genus *Arabella*

A. opalina. (Fig. 200.) This species has a cylindrical body, thickest in the mid-region and tapering toward the ends, and is with small lateral appendages. The head is somewhat similar to that of *L. tenuis,* except for the presence of 4 eyes in a transverse row on the prostomium, or anterior region. Color iridescent bronze with an opalescent sheen. Length up to 40 cm; width in middle, 3 mm. Common in sandy mud near low-water level. Maine to North Carolina.

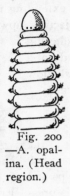

Fig. 200 —A. opalina. (Head region.)

Fig. 201—G. dibranchiata. (Head region.)

FAMILY GLYCERIDAE

GENUS *Glycera*

G. dibranchiata. (Fig. 201.) The cylindrical body is thickest in the middle and tapers toward the ends. Parapodia small, bearing simple, flat gills on the dorsal and ventral sides, and alike on all segments. The head is segmented and there are 4 small tentacles and 2 rudimentary palps on the prostomium. The animal is provided with a proboscis about 25 mm long, which is armed with 4 sharp teeth. With the aid of the proboscis, which it thrusts into the substratum, this creature can burrow with great rapidity. Color reddish. Length 20 cm. Lives in cylindrical passages in sand and mud near the low-water mark. Often abundant. Bay of Fundy to North Carolina.

G. americana. (Fig. 202.) In general, the shape of the body is like that of *G. dibranchiata,* but the parapodia are with 1 lobe on the anterior third of the body and with 2 lobes on the posterior parts. The dorsal gills are retractile and are branched; there are no ventral gills. The head is segmented and with a sharp, conical prostomium, and bears the same type and number of appendages as that of the preceding species. The proboscis is very large and long and is provided with 4 hooklike teeth. It is a rapid burrower, and inhabits the same kinds of localities as does *G. dibranchiata,* although it is not so common as that species. Color dark red or purple. Length 20 cm. Cape Cod to South Carolina.

Family SPIONIDAE

Genus *Spio*

S. setosa. (Fig. 203.) Body flattened dorsally and rounded ventrally, with segments alike throughout, and all bearing gills. Head with a prominent, blunted, median lobe, 4 eyes and 1 pair of very long peristomial cirri, which normally curve over the back. There is a proboscis present, but it is without teeth. Color green with red gills and cirri. Length 8 cm; width 2.5 mm. Sometimes numerous near low-water mark. Vineyard and Long Island Sounds.

Genus *Polydora*

P. concharum. (Fig. 204.) With long slender body composed of about 200 segments. Fifth segment unlike the others, being considerably longer and bearing bristles that differ in their arrangement. Color yellowish or grayish. Length 14 cm; width 1.5 mm. Occasionally found free in younger stages; more often found burrowing in dead shell material. Common. Nova Scotia to Cape Cod.

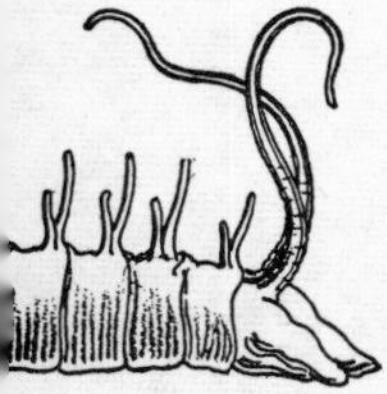

Fig. 202—
G. americana.
(Head region
with extended
proboscis.)

Family CHAETOPTERIDAE

Genus *Chaetopterus*

Fig. 203—S. setosa. (Side view of head region.)

C. pergamentaceus. (Fig. 205.) (See Plate XXII.) Has a short, stout body with the anterior region much flattened. The middle region consists of 1 segment bearing large winglike parapodia, and of 4 thickened segments. Parapodia simple. The head is without tentacles or palps, but there is present a pair of peristomial cirri having a tendency to project backward. Proboscis absent. In this annelid the body walls are transparently thin, revealing the intestinal tract and the sex elements. It is a highly

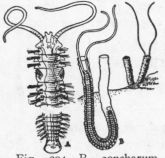

Fig. 204—P. concharum.
A, detail of worm; B, worm
in its tube.

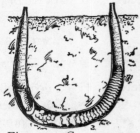

Fig. 205—C. pergamen-
taceus, in its tube.

phosphorescent creature, and lives in a U-shaped, parchmentlike tube
with both ends projecting above the substratum, which tube it forms
in the sand and mud near the low-tide level. The winglike parapodia in
the middle region fit closely to the sides of the tube and are made to
act as a suction-pump whereby the water with its food-laden particles
is drawn within. Length of worm 15 cm. Length of tube up to 50 cm.
Fairly common. Cape Cod to North Carolina.

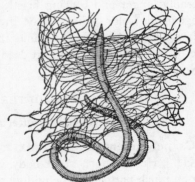

Fig. 206—C. grandis.

Family CIRRATULIDAE

Genus *Cirratulus*

C. grandis. (Fig. 206.) This worm has a cylindrical body tapering
toward the ends. The parapodia are rudimentary, but there are numer-

ous very long filamentous dorsal cirri. These cirri can be extended to nearly the length of the body, and they function as gills. There is a distinct head, though it is without eyes or appendages. Proboscis wanting. Color yellowish-green with orange cirri. Length 15 cm; diameter 6 mm. Very common. Found in sand and gravel near the low-tide level. Cape Cod to Virginia.

FAMILY TEREBELLIDAE

GENUS *Amphitrite*

A. ornata. (Plate XXI.) Body cylindrical with the forward end considerably thicker than the rest. There are 3 pairs of red branched plumelike gills and numerous long tentacular pink filaments at the anterior end. The tentacles are contractile and are constantly performing slow movements. The segments bearing setæ number about forty; these begin with the 4th segment and are confined to those in the anterior region of the body. Color deep pink. Length up to 30 cm. The animal lives in a well-consolidated tube made of sand grains and mud. Common. Found in sandy mud, gravel and under stones. Cape Cod to North Carolina.

GENUS *Polycirrus*

P. eximius. (Fig. 207.) A blood worm. In this species the body is long and slender, and consists of about 100 segments, of which 25 bear small tufts of long bristles. There are no branched gills, but the forward end is provided with numerous, long, crowded tentacles which extend in all directions. Parapodia simple. Color of body and tentacles bright red, the body

Fig. 207—P. eximius.

often being yellowish posteriorly. Length 2 cm or more. Lives in sand

and mud near the low-water mark. Very common. Cape Cod to North Carolina.

Family AMPHARETIDAE

Genus *Ampharete*

A. setosa. (Fig. 208.) Body thickest in anterior region, tapering rapidly backward. Fourteen segments in anterior region bear small groups of long setæ supported at the base by prominent lobes. Anal segments small with a pair of slender cirri. The forward end bears 2 fan-shaped groups of bristles, 40 tentacular filaments, and 4 pairs of unbranched gills, the latter situated on the 3rd and 4th segments. Color of body translucent, light yellowish-green or red. Length 20 mm; width 2 mm. Inhabits rough tubes about 25 mm in length, covered with coarse sand and mud. Found in shallow water. Vineyard and Long Island Sounds.

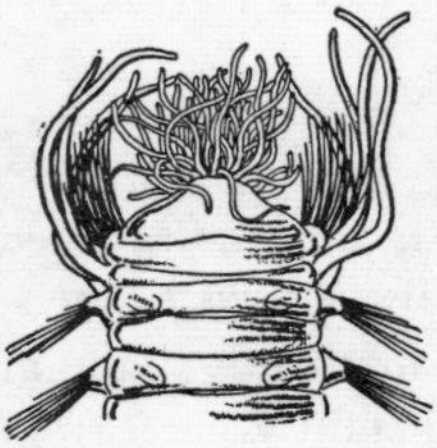

Fig. 208—A. setosa. (Head region.)

Family AMPHICTENIDAE

Genus *Pectinaria*

P. gouldi. (Plate XXI.) Easily identified by the conical tube it constructs of sand grains cemented together in a single layer; and which under a glass reveals a mosaiclike precision in artisanship, thus being not without considerable beauty. The worm itself has a short body with an obliquely flattened head. There are no parapodia on the hinder part of the body, and this terminal portion is folded forward. The prostomium is provided with short filamentous tentacles which are protected by long, golden-yellow, curved bristles arranged in 2 rows, these latter also acting as a sort of operculum to the tube. Color pink or flesh, attractively mottled with deep red and blue. Length 4 cm; width 7 mm. Found on sandy and muddy shores near the low-water level. Rather common. Maine to North Carolina.

Family OPHELIIDAE

Genus *Ammotrypane*

A. fimbriata. (Fig. 209.) Body elongate and smooth, thickest forward of the middle and tapering to both ends. Convex dorsally; ventral surface with a median furrow and rounded margins, and separated from the upper surface by a deep groove. Head acute; without appendages but provided with a pair of small black eyes. Proboscis present. Parapodia rudimentary, bearing dorsal cirri that act as gills. Setæ of forward segments long; those in back much shorter. There is a spoon-shaped caudal appendage, which is fringed with a row of slender papillæ; while at its ventral base arises a pair of slender cirri. Color iridescent purplish pink. Length 7.5 cm; width 3 mm. Found in mud below low-water mark. Not uncommon. Maine to Vineyard Sound.

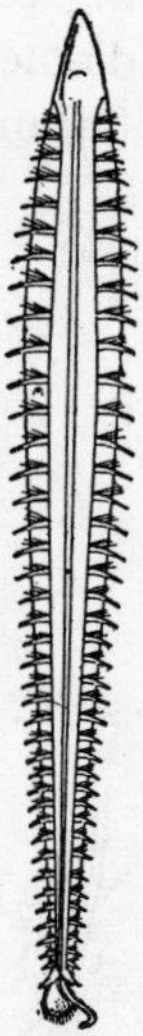

Fig. 209 — A. fimbriata.

Family MALDANIDAE

Genus *Clymenella*

C. torquata. (Fig. 210.) This form has a body consisting of 22 segments, 18 of which bear setæ. Near the middle of the fourth setigerous segment there is a membranous collar. Parapodia rudimentary and without gills. Head obliquely truncated and without appendages. Hinder end funnel-shaped with frilled edges. Color reddish. Length 10 cm. Lives in sand tubes in shallow and offshore waters. Sometimes numerous. Bay of Fundy to North Carolina.

Fig. 210—C. torquata.

Family ARENICOLIDAE

Genus *Arenicola*

A. marina. (Fig. 211.) The lug worm. An elongated cylindrical form with a swollen anterior region, blunt at the end. The skin has numerous annulations obscuring the true segmental divisions. Eight bristle-bearing segments compose the anterior region, 13 the middle region. No appendages are present on the head, but there is an unarmed proboscis. Parapodia rudimentary; those in the middle of the body bearing branched gills. Color dark brownish-green. Length up to 20 cm diameter 8 mm. Lives in deep burrows in sand near the low-tide level Often numerous. Long Island Sound northward.

Fig. 211—A. marina.

FAMILY CHLORHAEMIDAE

GENUS *Trophonia*

T. affinis. (Fig. 212.) This species has a slender elongate body, witl a rough granular skin covered anteriorly with small pa pillæ. The setæ of the anterior segments are very lon, and project directly forward surrounding the head. Ther are 2 small, black eyes present, and the head bears 8 ter tacular filaments of unequal length which act as gills. Th proboscis is without teeth; the whole head is retractile and the blood of this creature is green. Length about cm: width 3.5 mm. Sometimes numerous, burrowing i shallow and offshore bottoms. Vineyard Sound south ward.

FAMILY SABELLIDAE

GENUS *Sabella*

Fig. 212—T. affinis. (Head region.)

S. micropthalma. (Fig. 213.) A form that build parchment-like tubes most often on oyster shells. Th body is somewhat short and composed of about 60 seg ments. Eight setigerous segments form the anterior region, at th end of which are long, slender gill filaments bearing minute eye spot Peristomium raised and reflexed, producing a dorsally notched colla around the gills. Proboscis present. Color greenish-yellow. Length m; diameter 3 mm. The tube occupied by this worm is translucent, and the gills extended from its orifice appear like some dainty flower. The animal is sensitive to light and darkness, and the shadow of a passing hand will cause the gills to retract within the tube. Often abundant near ow-water mark. Cape Cod to North Carolina.

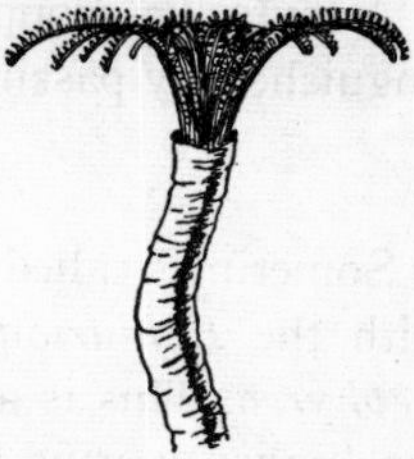

FAMILY SERPULIDAE

GENUS *Hydroides*

H. hexagonus. (Plate XXII.) This species, formerly bearing the name *Serpula dianthus,* lives

Fig. 213—S. micropthalma.

n a contorted, calcareous tube which it forms usually on mollusk shells, both living and dead. The tube is often encircled by ridges indicating the periods of growth. One of the dorsal gill filaments of the worm is flattened and functions as an operculum, or stopper, to close the opening when the animal is retracted! When the worm is extended, a pair of plumelike gills, representing the palps, appear over the orifice; and, like those of *S. micropthalma,* instantly withdraw from a passing shadow. Just below the gill filaments there is a paired membrane composing a collar which is used to smooth the inside surface of the tube. Color of gills variable, commonly purplish-brown. Length 7.5 cm; diameter 3 mm. This species occurs both solitary and in colonies, mollusk shells often being completely covered by the tubes, which are usually fixed for a considerable part of their length, having the open end free. The colonies form clusters in which the tubes cross and recross or are intertwined with one another. Very common near low-water mark. Cape Cod to Florida.

GENUS *Spirorbis*

S. spirorbis. (Plate XXII.) A tube-dweller that forms a small coiled calcareous tube more often on seaweeds, but occasionally on shells, etc. The flat coil is encrusted on one side to its supporting object. There is n operculum, and 9 gill filaments are present, the latter being a flower-like structure commonly extended just without the opening. The gills re colored a delicate greenish-white. Length of worm 3 mm. Often bundant. Nova Scotia to Long Island Sound.

CLASS **GEPHYREA**

A tentative group of animals of obscure relationships; primarily distinguished by passing through trochophore larval stage when young.

ORDER **INERMIA**

Sometimes called the *Sipunculoidea,* under which name, and together with the *Echiurida,* they are usually classed in the *Annelida* as the *Gephyrea.* This is a group of questionable forms whose inclusion with the higher worms is permitted principally by the fact that they pass through a trochophore larval stage. In most instances segmentation never occurs at any period of life. They are elongated forms bearing short, hollow tentacles at the forward end, usually surrounding the mouth. Parapodia are wanting. The body wall consists of a single-layered skin and the muscles. The body cavity is quite large, is lined with cilia, and is filled with blood containing corpuscles. A much-coiled narrow tube forms the intestinal tract, which extends from the mouth at the forward end to the hinder end where it turns upon itself and eventually terminates at the anus situated forward of the middle of the body. A pair of nephridia, called the brown tubes, is usually present on each side near the anus and opening to the outside. There is a well-developed nervous system consisting of a cerebral ganglion situated dorsally to the œsophagus and connects by a pair of nerves with the ventral chord. This chord is a large longitudinal nerve, extending to the hinder end of the body, and from which are given off numerous other nerves. The blood vascular system is made up of a ring canal passing around the œsophagus, a tentacular canal extending into the tentacles, and one or two contractile cæca, or hearts rising from the ring canal and running a short distance along the œsophagus. The function of the tentacles is probably respiratory as well as sensory, as their extension is brought about by movement of the blood. The sexes are separate, have the reproductive organs at the bases of the retractor muscles of the introvert, presently to be described, and discharge their products into the body cavity, whence these escape to the outside through the nephridia.

The introvert (Fig. 214), just mentioned, is a singular, contractile region consisting of the forepart of the body. When the introvert is fully

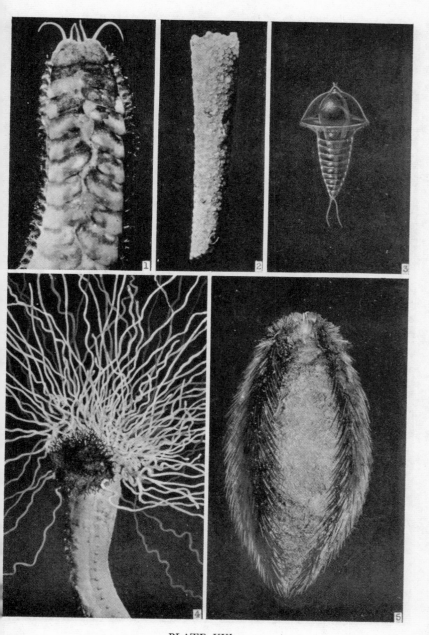

PLATE XXI

1—*Sthenelais leidyi*; head region; 4x.   2—*Pectinaria gouldi*; sand tube; 3x.

3—*Polygordius appendiculatus*; larva; 35x.

4—*Amphitrite ornata*; head region; about natural size.

5—*Aphrodite hastata*; about natural size.

PLATE XXII

1—*Lepidonotus sublevis*; 3x.   2—*Diopatra cuprea*; head region; about natural size.
3—*Spirorbis spirorbis*; coiled tube attached to shell of *Buccinum*; 13x.
4—*Chaetopterus pergamentaceus*; about natural size.
5—*Hydroides hexagonus*; tubes attached to valve of *Mactra*; ¾x.

extended, the circlet of tentacles at the end is plainly visible. In contracting, the introvert invaginates, or turns within itself, until the entire region together with the tentacles disappears within the
body of the animal. In the living animals the action of
invagination and the consequent rolling out of the introvert is constantly repeated. They live in sand and mud
and swallow considerable quantities of this material for
the organic food particles contained therein.

Fig. 214
—introvert
of S. nudus.

Of the more than 100 known species contained in the
order, but 16 have been found on the Atlantic Coast
of the United States.

GENUS *Sipunculus*

S. nudus. (Fig. 215.) The body is long with the introvert, and, except
for a short distance at the end, is covered with numerous small, sharp

Fig 215—S. nudus.

papillæ having their points directed backwards. A fluted tentacular fold
in which the tentacles are not isolated surrounds the mouth. The introvert is smaller in diameter than the rest of the body and has from 30
to 32 longitudinal muscle bundles and 4 retractor muscles. The anal
opening is quite prominent. A horny cuticular covering, which is an
iridescent white or flesh-color, invests the body, and it is divided by
longitudinal and transverse circular markings into a pattern of small
squares, these markings representing the position of the underlying
muscles. Length up to 21 cm. It has been asserted that so much of the
sand and mud in certain areas of the sea bottom has passed through
the sipuncloids that the mineral substances of those areas have been
modified just as the soil of the land has been modified by earthworms.
Sometimes numerous near the low-tide level. Cape Hatteras region and
Key West.

GENUS *Phascoloscoma*

P. gouldi. (Fig. 216.) In this form the longitudinal muscles are separated into about 30 anastomosing bundles, and do not form a con-

tinuous sheath such as is common to most other species
of this genus. The introvert takes up about one fourth
of the body length. At the end are numerous tentacles
arranged in several concentric circles. Opening into the
brain is a pair of pigmented tubes serving as eyes. The
surface of the body is parchmentlike in color and texture, and is marked off into small squares corresponding
to the position of the muscles underneath. Length 18

Fig. 216— cm or more. Lives in sand and gravel near the low-water
P. gouldi. mark. Massachusetts Bay to Long Island Sound.

GENUS *Phascolion*

P. strombi. (Fig. 217.) This species is commonly found living in a
snail shell, though some have been found that form
a thick short tube of mud or sand. There are numerous
varieties of this species; all of them are small, measuring usually not more than 30 mm in length. The
body is long and covered with papillæ. On the long
introvert, just back of the circlet of numerous tentacles, is a band of minute hooks. On the hinder
part of the body occur large, brown crescentric or
triangular hooks pointing forward. The worm plugs
the aperture to its shell with sand which is consoli-
dated with cement, leaving a hole for the extrusion

Fig. 217—P.
strombi.

of the introvert, by means of which it travels from place to place, hermit-
crablike, taking the shell with it. Common in shallow and offshore areas.
All ranges.

ORDER **ARMATA**

Comprised within this order are certain worms in which the segmentation is wanting or obscure in the adult stage. They are usually
thick-bodied, cylindrical animals, and are born as typical trochophore
larvæ. At one stage of their early metamorphosis, however, they possess fifteen rudimentary segments. There are no parapodia or cephalic
appendages, but a pair of setæ are present on the ventral side near the
forward end. Nor is there a true proboscis. What is usually termed
the proboscis in animals of this group is the prostomium, or anterior

part of the head; this region being much elongated, forming a spatu-late or troughlike extension forward from the mouth. The so-called proboscis is very elastic, and its troughlike ventral surface is ciliated, causing the food particles on which the animal subsists to be swept into the mouth. A long alimentary canal (considerably longer than the body), continuing from the mouth, ends with a posteriorly situated anus. The excretory system consists not only of true nephridia, located in the forward part of the body, but also of modified nephridia, which are a pair of cylindrical anal pouches joining the rectum and communicat-ing with body cavity. The vascular system consists of a dorsal and a ventral blood vessel. These run longitudinally and are joined together anteriorly. There are no lateral vessels. A nervous system is present, but a distinct brain is wanting; however, there is an elongated œsopheal ring extending from the prostomium to the ventral chord back of the mouth. The individuals are unisexual, having unpaired reproductive organs with paired ducts. The group is comparatively small, containing fewer than 25 known species, of which only 5 occur on the Atlantic Coast of this country.

Genus *Echiurus*

E. pallasi. (Fig. 218.) Chiefly characterizing this species are the two groups of anal bristles at the hinder end, and which are, in addition to the forward ventrally situated setæ, common to the entire group. The body is marked with 22 faint rings bearing minute spines. The

Fig. 218—E. pallasi.

proboscis is spoon-shaped, cylindrical at the base, and is 6 cm in length. Color of animal variable; may be gray, yellow or orange. Length up to 30 cm. The creature is somewhat protean in character, altering the shape of its body by extending and contracting its length, and by al-ternately constricting and dilating various regions. Found in burrows in the mud near the low-water level. Not uncommon. Maine.

X

ECHINODERMATA

PLATE XXIII

CRINOIDS AND LUMINOUS FISHES IN DEEP WATER

Antedon tenella, stalked form (lower middle); *A. tenella*, free form (lower right).

CHAPTER X

PHYLUM **ECHINODERMATA**

(STARFISHES, SEA URCHINS, ETC.)

THE echinoderms are animals having a radially symmetrical, five-rayed body, and may be distinguished from all other animals by certain well-marked characters. In most forms, the skin is provided with calcareous spines which project out from all sides of the body. In the sea-cucumbers, however, these spines are reduced to minute spicules; in other forms calcareous plates are hidden in the skin. With few exceptions echinoderms have a well-developed water-vascular, or ambulacral, system consisting of a circular canal around the mouth, from which radial branches extend to the periphery of the body. Arising from the radial branches are numerous offshoots leading out into the skin in certain regions of the body, and ending in delicate tubular projections. Such projections are variously modified in different parts of the body for the performance of special functions like locomotion, respiration or feeling. And, so modified, they are variously known as sucker feet, tube feet or pedicels.

Echinoderms are strictly marine animals; they have no relatives, close or distant, living on land or in fresh water. There are, however, a few tropical species that are able to live in brackish water. They are found in all seas, ranging from the tidal zone to the greatest depths.

In the *Echinodermata* are contained approximately 4000 species, which are commonly grouped according to the following classifications:

PHYLUM ECHINODERMATA

CLASS **Crinoidea** (No orders)

CLASS **Asteroidea**

ORDERS........ { Phanerozonia
Spinulosa
Forcipulata

CLASS **Ophiuroidea**   ORDERS........ { **Ophiurae**
                                           Euryalae

CLASS **Echinoidea**   ORDERS........ { **Cidaroida**
                                           Clypeastroida
                                           Spatangoida

CLASS **Holothuroidea**   ORDERS........ { **Actinopoda**
                                           Paractinopoda

CLASS **CRINOIDEA**

The sea lilies and feather stars, as the crinoids are commonly called, get their popular name from the fact that in general their forms suggest the lily or that they have certain featherlike features. All of them are inhabitants of deep water, living in regions of cold and of utter darkness relieved only by the soft luminous flashings of phosphorescent creatures, where throughout a part or the whole of their lives they are attached by a stalk to the bottom.

A typical crinoid has a body consisting of the disk, or calyx, which is somewhat cup-shaped and forms the central part, the arms and the stalk, both of the latter being composed of numerous joints. That side of the calyx which bears the mouth, and which is directed upward, is known as the *oral* surface. On the oral surface near the mouth and situated interradially is the anus. The arms rise from the calyx and are either five or ten in number, and sometimes branch dichotomously, bearing short alternating branches, called *pinnules,* along the side. A ciliated groove, known as the *ambulacral* groove, extends from the central mouth to the tip of each arm and its branches. Through the ciliary action of the grooves, floating food particles are swept into the mouth. Along the edges of the ambulacral grooves are certain appendages which have a tactile, respiratory and excretory function, there being no special sense organs present. It is by the *aboral,* or downward-directed, surface that the animal is attached by a stalk, often bearing whorls of jointed extensions called *cirri,* which has its lower end rooted in the ooze, or other bottom material. This stalk is not always present, in most forms being absent after the animal has reached maturity. The calcareous framework of the calyx is composed of close-fitting or fused polygonal plates; that of the arms and the stalk together with its cirri is composed of series of thick disks stacked one against the other.

In the calyx is contained the main body cavity, but this has extensions reaching throughout the entire skeleton, and in it is contained a mixture of blood fluid and sea water, the latter being drawn in through ciliated pores in the oral surface wall. A tubular digestive tract extends from the mouth to the anus by running once around the main body cavity.

The ambulacral, or water-vascular, system is without terminal suckers or ampullæ. A ring canal, which is part of the ambulacral system, surrounds the mouth, and from it rise a number of so-called "stone canals" which end free in the main body cavity. These stone canals, however, lack the calcareous inclusions in their walls which are typical of the stone canals in other members of the phylum. Also rising from the ring canal and surrounding the mouth are five groups of ambulacral tentacles. There is a fairly well-organized central nervous system which sends large radial trunks into the arms. The sexes are separate in the crinoids, and the reproductive organs extend throughout the arms into the pinnules, in which region they mature, shedding their products through minute pores. Generally present in the oral axis of the calyx is an organ known as the genital stolon with which the five gonads are connected; this organ and the axial organ of other echinoderms are homologous. The early stages of development are similar to those of the starfishes (*see Asteroidea*).

The crinoids are a vanishing host. In former ages they covered vast tracts of the ocean floor, and they left fossil remains which contain about 2000 different known forms. Today this once mighty group is represented by fewer than 500 living species.

As these animals are commonly, with few exceptions, dwellers of the deep sea, they are almost never found cast up on the shore. But to complete the classification of the phylum, and for the benefit of the chance collector who in some unusual and fortunate circumstances may chance to come upon a stray specimen, one of the commoner forms is described in this book.

Family **ANTEDONIDAE**

Genus *Antedon*

A. tenella. (Plate XXIII.) This form is stalked and sessile only during the early period of its life, the adult retaining not more than one or a few of the proximal segments of the stalk, which are fused together and provided with from 25 to 30 cirri, each cirrus having from 10 to 16 segments. This modified appendage is used as a temporary holdfast and is known as the *centro-dorsal*. There are 10 arms, each of which bears smooth pinnules with one or more of the proximal pinnules elongated and flagellate. The arms are very flexible, and by their alternate flexions and extensions the free adult animal is enabled to swim slowly about. This species has a diameter of about 6 cm. It occurs in depths ranging from 35 to 600 fathoms. New England waters.

CLASS **ASTEROIDEA**

The starfishes. As the class name indicates, the animals in this group have a radial structure which is typically star-shaped. Usually this structural plan is that of a conventionally drawn five-pointed star, but some members of the class have normally more than five arms, while certain abnormal individuals may have more or fewer than the normal number.

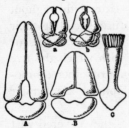

Fig. 219—pedicellariae of starfish. A and a, major and minor pedicellariae of A. vulgaris; B and b, major and minor pedicellariae of A. forbesi; C, a paxilla.

The aboral surface in this group is directed upwards; it is covered with cilia and bears numerous spines which project from the skeletal meshwork. These spines in most cases are movable and are presumed to function as protective appendages. In some forms *paxillae* are present, these being modified spines each consisting of a thick calcareous rod having its summit covered with minute spines. Also usually present in this region are certain peculiar and minute pincerlike appendages known as *pedicellariae*. (Fig. 219.) They occur scattered over the body and in the form of rosettes around the spines. They are modified spines with jaws straight or crossed like scissors, and they are used to keep the body surface clean and to protect the *papulae,* or breathing organs, which are short finger-like projections of the body wall. In an interradius on the aboral surface is a sievelike structure called the *madreporite* which is the external opening to the ambulacral, or water-vascular, system. At the tip of each arm is the only special sense organ these animals possess; this is a red eye spot and is sensitive only to light. The anal opening, when present, is small and is near the center of the back.

The oral surface, which is directed downwards, contains a central mouth and a membranous area surrounding it called the *peristome.* Extending from the peristome along each arm is a median ambulacral groove. In cross-section the ambulacral grooves are roughly V-shaped, and from the walls project either two or four rows of ambulacral, or tube, feet, which are the ograns of locomotion. The tube feet are usually protected by a series of movable spines rising from the ambulacral wall formed by the

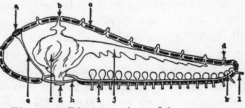

Fig. 220—Diagram of starfish ray. a, madreporite; b, anus; c, calcareous plates; d, papula; e, stone canal; f, stomach; g, mouth; h, ring canal; i, radial canal; j, pyloric caecum; k, ampula; l, tube foot.

ossicles, which are the calcareous structures strengthening the ambulacral grooves. As the tube feet are hollow cylinders of thin skin ending either with or without a sucking disk, they are capable of considerable extension or retraction. Almost invariably the tube foot at the end of each arm has no sucking disk, and it is supposed to function as a feeler and an olfactory organ.

Supporting the body is a framework of calcareous plates of various sizes which are connected by muscles and tissue fibers. And the body structurally consists of two principal regions: the central disk, and the rays, or arms. (Fig. 220.)

The internal structure is comparatively simple and not very difficult even for the inexperienced investigator to determine at first hand by dissection. Occupying almost the entire cavity of the disk is the large folded stomach, which is connected with the mouth by a short œsophagus, and with the anus by a short slender rectum. From the rectum there may rise a small branched cæcum; although both anus and cæcum may be absent, these being of a rudimentary type of structure. The aboral region of the stomach is joined by five pairs of much-branched liver sacs which extend into the arms. Leading from the madreporite, or sieve plate, mentioned in a preceding paragraph, is the *stone canal,* which connects with the *ring canal* lying at the margin of the peristome and surrounding the mouth. The stone canal is so called because its walls are usually encrusted with lime. Both the pores of the madreporic plate and the walls of the stone canal are lined with cilia. Extending from the ring canal are radial canals occupying a position along the ambulacral grooves just external to the skeleton. These radial canals are connected with the tube feet by means of paired side branches, while the tube feet themselves are connected by valvelike structures with bulblike reservoirs, or ampullæ, at their bases which project through pores in the ambulacral plates into the body cavity. The ampullæ are provided with muscles, and by the contraction or distension of these bulbs, water may be forced into or withdrawn from the tube feet, thus extending or shortening them and also operating the sucker disks at their ends. There is evidence that the tube feet are not quite watertight; consequently, in order to make up the loss of fluid through the thin walls, they are obliged to draw upon the contents of the radial canal; and the radial canal in turn draws its supply from the ring canal, which is filled with sea water admitted through the stone canal by the way of the madreporic sieve plate.

There are no special organs of respiration or of excretion, such as a kidney; the functions of such organs are performed by the cilia covering the body together with the papulæ and ambulacral appendages. The nervous system within the skeleton is not well developed; but there is a superficial system of nerves which is epithelial and occurs on the oral surface just without the framework, the radial nerves ending at the eye spots at the tips of the arms.

The sexes are usually separate in the starfishes. The gonads of both males and females look very much alike; they are located at the base of the arms and open to the outside through minute pores in the inter-radii. The sex products in most cases escape into the water where fertilization takes place. However, in some starfishes the female by bending her arms together forms a sort of brood chamber for the eggs.

The larvæ are free-swimming and very unlike the adult in shape.

Most starfishes are carnivorous; and they are voracious feeders, eating almost any nutritious organic material, but their principal food consists of bivalve mollusks. When a starfish attacks some such bivalve as a mussel, it wraps its arms about the shell and, fastening its sucker feet, exerts a steady pull that eventually tires the mollusk, thus causing the shell to open. Whereupon the starfish everts its stomach over the fleshy part of its prey and slowly digests it. Small mollusks are sometimes taken into the stomach whole and the empty shells discarded after the soft parts are digested. Such is the usual manner of feeding in those starfishes having no suckers on their tube feet which would enable them to fasten upon a shell. For the most part, however, those forms without sucker feet are not predatory on shellfish but live on other and minute organic material which they swallow whole.

Starfishes often travel in schools, moving slowly around on the bottom by means of their tube feet, preying on sedantary forms such as barnacles, oysters, and other shellfish that lie in their way. Their bodies are capable of a great degree of regeneration; replacement of rays that are mutilated or lost entire is a common occurrence. They are among the commonest animals of the seashore, and forms are found from the tidal zone to very great depths. About 1000 species are known to occur throughout the entire world.

ORDER **PHANEROZONIA**

In the order *Phanerozonia* the individuals have large conspicuous marginal plates which are in contact and which are usually arranged in two rows, thus making a marginal frame for the aboral area. The pedicellariæ are often wanting, but when these are present they are sessile. Two rows of ambulacral feet occur in each arm.

FAMILY **ASTROPECTINIDAE**

GENUS *Astropecten*

A. articulatus. (Fig. 221.) The body is flat and smooth, and with each upper marginal plate near the tip of the arms bearing a small blunt tubercle. The marginal plates are in two rows and form a conspicuous border on the sharp-pointed rays. Short spines on the lower row of marginal plates fringe the arms. The aboral wall is membranous and bears paxillae. There are no pedicellariae; the anus is wanting; and the ambulacral feet are without sucking disks. Color rich purple. Size over all about 25 cm. Found in tidal zone and offshore waters, usually on sandy bottoms. Not uncommon. New Jersey southward.

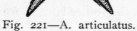

Fig. 221—A. articulatus.   Fig. 222—L. clathrata.

Genus *Luidia*

L. clathrata. (Fig. 222.) Although the family *Astropectinidae,* in which this form is included, is more often than not characterized by the possession of two rows of marginal plates, this species is distinguished by having but one row, the lower. The body is flattened, with a membranous aboral wall, and has long slender arms fringed with marginal spines. Numerous crowded paxillæ occur on the aboral surface. The disk is small and is without an anal opening. Pedicellariae are present. Color very light neutral tint. Size over all about 30 cm. Often common on sandy shores and in shallow water. New Jersey to Florida.

Family **PORCELLANASTERIDAE**

Genus *Ctenodiscus*

C. crispatus. (Fig. 223.) The pentagonal-stellate form of this species is characteristic, although the shape is rather variable. Numerous paxillae cover the aboral wall, which is membranous; and in the center there rises a tubelike prominence. There are two rows of thin porcelainlike marginal plates, giving a flat edge to the rays. Thin folds of integument, called *cribriform* organs, lying in parallel, vertical rows, occur between certain of the marginal plates. The latter are without spines. Madreporite large; anus absent; tube feet without suckers. Color greenish. Size over all about 10 cm. Found from shallow to very deep water, usually on muddy bottoms. Cape Cod northward.

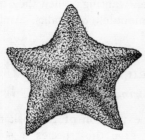

Fig. 223—C. crispatus.   Fig. 224—H. pharygiana.

FAMILY GONIASTERIDAE

GENUS *Hippasteria*

H. pharygiana. (Fig. 224.) Aboral surface of body raised and cushionlike. Principal plates of skeleton arranged mosaiclike and each with a central tubercle; marginal plates in two rows and bearing tubercular spines. Pedicellariae sessile, long and conspicuous. Color of aboral surface bright scarlet. Size over all about 20 cm. Found offshore in deep water. Sometimes common. Cape Cod northward.

FAMILY OREASTERIDAE

GENUS *Oreaster*

O. reticulatus. (See the frontispiece.) (Plate XXIV.) This species, formerly bearing the generic name *Pentaceros,* has a very thick and regularly star-shaped body from the large central disk of which extend short rays. There are numerous prominent blunt spines distributed in a netlike manner over the aboral surface. A row of larger spines projects from the small marginal plates. Both the madreporic plate and the anus are near the center of the disk. Color dull yellow. This is one of the largest American starfishes, measuring about 40 cm over all, and having a thickness of more than 8 cm. Not uncommon. South Carolina southward.

ORDER **SPINULOSA**

Contained in the order *Spinulosa* are those starfishes that have the aboral wall spiny and a skeletal structure commonly imbricated or in the form of a network. Usually the arms are without prominent marginal plates. Pedicellariae are most often wanting; when they do occur they are generally short and not surmounted on stalks. Ambulacral feet provided with sucker disks.

FAMILY **ASTERINIDAE**
GENUS *Asterina*

A. folium. (Fig. 225.) Interradial expansion of the disk connecting the arms give this species a somewhat pentagonal shape, but often distinctly star-shaped forms are found. The body is flattened, often gibbous, or swollen, and bears 3 or 4 minute spines on each aboral plate, the latter being imbricated throughout. Marginal plates small, making the edges rather sharp. Color varying, from blue to greenish. Size over all about 5 cm. Often common. Found in the tidal zone and shallow water, usually on the under side of rocks. Florida and the West Indies.

Fig. 225—A. folium.

FAMILY **ECHINASTERIDAE**
GENUS *Henricia*

H. sanguinolenta. (Fig. 226.) (Plate XXIV.) In the writings of most naturalists, this pretty little starfish goes under the name of *Cribrella sanguinolenta;* but it is known that the generic name *Henricia* must take precedence over the other. The disk is small and the arms are long, slender, and cylindrical. There are no pedicellariae, but the aboral surface is covered with minute, closely set spines, thus giving a comparatively smooth exterior. The madreporite is small and marked by a few wide grooves; and the animal has a relatively conspicuous anal opening. The skeleton is reticulate, forming a fine meshwork of im-

PLATE XXIV

1—*Oreaster reticulatus.* 2—*Echinaster spinulosus.* 3—*Henricia sanguinolenta.*
4—*Solaster endeca.* 5—*Solaster papposus.* 6—*Asterias forbesi.*
7—*Stephanasterias albulus.* 8—*Asterias vulgaris.*

PLATE XXV

bricating plates. There are 2 rows of tube feet arranged in a single row along each side of the ambulacral groove of each arm. Color variable, but uncommonly vivid, rendering the animal rather conspicuous; ranges from scarlet or yellow to lavender on the aboral surface, the oral surface being commonly deep yellowish. Some specimens are mottled red and purple. An aquarium containing numerous individuals of this species is not outdone in colorful splendor by any vase of flowers. Nor are its habits unattractive: the female broods the eggs and protects the young under her disk. Length

Fig. 226—H. sanguinolenta.

about 10 cm over all. Common in offshore waters. Greenland to Cape Hatteras.

Genus *Echinaster*

E. spinulosus. (Plate XXIV.) Disk small; arms long and cylindrical. Aboral plates of the skeleton form a rather large meshwork with coarse scattered spines on the outer surface. No pedicellariae. Tube feet in 2 rows. Color dark violet. Size over all about 10 cm. Common in tidal zone and shallow water on the Florida coast. Cape Hatteras to Florida.

Family SOLASTERIDAE

Genus *Solaster*

S. endeca. (Plate XXIV.) The purple sun star. This form, and most of the members of its family, are distinguished by having more than 5 arms. In the present species, these range in number from 7 to 13; more usually they are from 9 to 11 in number. The arms are long and slender. The aboral plates form a fine network with small meshes bearing paxillae on the outer surface; the latter, notwithstanding, is comparatively smooth in appearance. Pedicellariæ are wanting. Each ray bears two rows of ambulacral feet, a single row extending along each side of the ambulacral groove. Color deep red or purplish-red.

Size over all about 30 cm or more. Rather common in shallow and offshore waters. Cape Cod northward.

S. papposus. (Plate XXIV.) The common sun star. The arms in
this species may number from 8 to 14, but more usually there are either
10 to 11. Each arm has 1 row of marginal plates. The aboral plates are
reticular, but with larger meshes than those in *S. endeca*. There are
no pedicellariae. Aboral surface rough, with numerous paxillae present.
Ambulacral grooves with 2 rows of tube feet, one row along each side of
groove. Color reddish or purple in concentric lines or mottlings, with
lighter tint on the oral surface. Often common. Found in shallow and
offshore waters. Maine to Cape Cod.

Family **PTERASTERIDAE**

Genus *Pteraster*

Fig. 227—P. militaris.

P. militaris. (Fig. 227.) The disk of this
species is large, thick, and arched, from which
extend short, thick arms. A skin covers both
the oral and the aboral surface, passing over the
pines, and encloses a brood chamber which
opens externally by a large central opening on
the upper side, provided with valves. Pedicellariae wanting. Size over all above 9 cm. Found
in offshore waters. Arctic Ocean to Cape Cod.

ORDER **FORCIPULATA**

In the order *Forcipulata* are contained those starfishes having pe-
dunculate, or stalked, pedicellariae, which are with either straight or
crossed jaws. The aboral skeletal structure is netlike; and the spines
are conspicuous.

Family **STICHASTERIDAE**

Genus *Stephanasterias*

S. albulus. (Plate XXIV.) The arms of this starfish are cylindrical
and rather slender; there are 6 and they generally occur in 2 groups,
3 arms being short and 3 being long. The aboral plates are relatively

large and arranged in longitudinal rows. Numerous spines are present on each side of the ambulacral grooves. There are 4 rows of tube feet. Diameter of disk about 3 cm. Found in shallow and deep water. Sometimes common. Arctic Ocean to Cape Hatteras.

FAMILY ASTERIIDAE

GENUS *Asterias*

A. forbesi. (Plate XXIV.) In this form the disk is rather small and the rays are stout, cylindrical, somewhat narrow at the point of attachment, and terminate bluntly. The aboral plates have a reticulate arrangement and bear coarse spines. These spines are encircled with pedicellariae having crossed jaws; however, the pedicellariae occurring among the papulæ have straight jaws. Rows of movable spines are present along the ambulacral grooves. There are 4 rows of tube feet. Color variable, usually a greenish-black, against which the brilliant orange madreporite contrasts conspicuously. Size over all about 16 cm. Very common, being perhaps the most familiar starfish in its principal range, which is from Maine to New Jersey and southwards. Less common in its northern range.

A. vulgaris. (Plate XXIV.) This starfish somewhat resembles *A. forbesi,* but it can be distinguished at once from that species by the more pointed arms and the inconspicuous madreporic plate, which is a pale yellow or more nearly the general color of the body. Also, the rays are flatter, and the spines are more numerous, the latter often being grouped in a distinct row along the middle of the aboral surface of each arm. Color variable, but commonly some shade of purple. Size over all from 15 to 40 cm. Found from the tidal zone to deep water. This is the common starfish of New England. Its occurrence below the eastern end of Long Island is less frequent.

A. littoralis. (Plate XXV.) The body of this starfish is thick. The rays are very broad and with 2 rows of movable spines on each side of the ambulacral groove. On the aboral surface, the bases of the spines are surrounded by pedicellariae with crossed jaws. The pedicellariae among the papulæ are provided with straight jaws. There are 4 rows of tube feet. The aboral plates are arranged in a netlike manner. Color olive-green. Size over all about 5 cm. Found in shallow water, usually on *Fucus,* which it resembles in color. Occasionally numerous. Casco Bay to Cumberland Gulf.

A. tenera. (Fig. 228.) Disk rather small, from which extend the cylindrical, slender, and tapering arms. The skeleton is rather rigid

and with aboral plates arranged in a reticulate manner. Numerous spines occur on the aboral surface, and the bases of these are surrounded by pedicellariae provided with crossed jaws. The pedicellariæ present among the papulæ have straight jaws. Color of aboral surface variable, ranging from whitish to purple; madreporic plate small and white. Size over all about 8 cm. Like several other

Fig. 228—A. tenera.

species of its genus, this little starfish protects its eggs and young beneath the disk, the young being carried about clinging to the partially everted borders of the mother's mouth. Often common in offshore waters. Nova Scotia to New Jersey.

CLASS **OPHIUROIDEA**

The serpent stars. Members of this class, also known as brittle stars, derive their common names from the fact that their greatly elongated rays perform the writhing movements of the serpent, and that they have the habit of breaking off when the animals are seized or roughly handled. It is owing to the serpentine movements of their arms and their covering of plates, which have a superficial resemblance to the scales of a serpent, that they have been given the scientific class name of Greek derivation meaning "snake-tail-form." Locomotion is achieved by the snakelike movement of the rays.

The ophiurans have a general resemblance to the starfishes in form, but the rays, in addition to being very long and slender, are without an ambulacral groove and they are distinctly marked off from the disk. There are no cilia on the exterior; nor are there pedicellariae present; spines, scales or granules occur on the disk, and spines usually project from the sides of the arms. An anus is wanting.

Usually the number of arms is 5, but in some species the normal number is 6, and in other species it is 7 or 8. Projecting from the oral surface of the arms are two rows of ambulacral appendages; these, however, are without terminal sucking disks, and are called tentacles as their function is not locomotor but tactile, respiratory and excretory. Each arm is provided typically with four rows of superficial plates, an oral, an aboral and two lateral rows, which cover an axial row of plates, held together by connective and muscular tissue, and occupying nearly the whole interior space of the ray. It is between the ventral and the abutting lateral plates that the tentacles project. When spines are present these are borne on the lateral plates. (Fig. 229.)

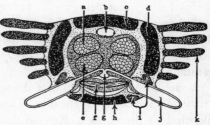

Fig. 229—Diagram of an ophiuran. a, muscle; b, branch of cœlom; c, upper arm plate; d, lateral arm plate; e, vertebra; f, radial nerve; g, radial canal of water vascular system; h, under arm plate; i, tentacle arm scales; j, tentacle; k, arm spine.

As in the starfishes, the oral surface is directed downwards. The mouth is in the center and forms a star-shaped opening surrounded by a rosettelike structure of skeletal parts. The largest and most conspicuous of these parts are the buccal plates in the interradii, one of which is usually larger than the others and marked with a depression and serves as the madreporite. (Fig. 230.) These plates are roughly triangular and have the apex projecting in toward the center of the disk, and form the jaws. The jaws are usually provided with a number of flattened scales or toothlike processes which make up the masticatory apparatus, these processes being known as oral papillæ, tooth papillæ, or teeth, according to their location on the jaw. The toothlike projections along the edge of the jaws are the oral papillæ; similar structures, when present, at the point of each jaw are the tooth papillæ; while the series of plates projecting from the mouth opening into the interior of the body

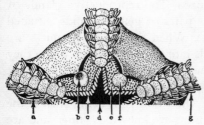

Fig. 230—Diagram of O. brevispina. a, genital bursa; b, madreporic plate; c, mouth papillæ; d, mouth; e, jaw; f, buccal plate; g, arm spines.

are the teeth. In most instances the jaws are not so much used in crushing food as in sorting out and ingesting the softer and smaller particles. And for this purpose they are provided with muscles that allow them to be opened wide or closed tight. The other orifices of the disk occur in the form of slits, there being a pair at the base of each arm. These open into paired pouches, the genital bursæ, which besides being receptacles for the genital products, take in and expel water and probably aid in respiration and excretion. In one genus (*Ophioderma*), the slits opening to the genital bursæ are in two parts.

A saclike stomach is connected with the mouth by a short œsophagus, and it fills the cavity of the disk. This stomach, however, is not eversible like that of the starfish. Nor do the ambulacral canals lie in an ambulacral groove; instead, they are within the skeletal structure. They consist of ring and radial canals, the branch canals terminating in the tentacles, and the stone canal ending at the madreporite.

The nervous system is much the same as that in the starfish, except that the nerves are more highly modified. It consists principally of a central nerve ring with branches having the nerve cells segregated into more definite centers, or ganglia, than is the case in the *Asteroidea*.

As a rule, the sexes are separate; but in the form *Amphipholis squamata,* one of the glands in each of the genital bursæ produces eggs, while the other gland forms spermatozoa. The reproductive glands of serpent stars commonly consist of five pairs of genital glands, situated on the walls of the ten genital bursæ. When the sexual products are ripe they escape from the genital glands into the bursæ, from which they are usually discharged directly into the water through the genital slits; though in the hermaphroditic species just mentioned, the eggs are fertilized and developed within the bursæ, the young thus being born in the form of the adult. The larvæ of ophiurans are known as *plutei,* and are very much unlike the adult. They undergo a complicated metamorphosis before attaining to the typical serpent star form. In some species, asexual reproduction by division of the disk occurs.

Ophiurans are nocturnal in their habits. They are found among seaweeds or corals or on the sea floor, where they feed on small organisms or organic material occurring in their habitat. It is because of their nocturnal habits that many species liberate their genital products between sunset and midnight. Their regenerating powers are considerable, being able to replace lost arms and portions of the disk without apparent difficulty. They form the most numerous group of echinoderms, the class containing more than 1100 species.

ORDER OPHIURAE

Serpent stars included in this order usually have five unbranched arms which have distinct articulating axial plates allowing a great degree of lateral flexion, but which do not permit the arms to be rolled in toward the mouth.

FAMILY OPHIODERMATIDAE

GENUS *Ophioderma*

O. brevispina. (Plate XXV.) The green serpent star. This species is known in the earlier writings of naturalists as *Ophiura olivacea*. It has a pentagonal disk closely covered with granules, and is with 4 genital slits in each interradius. A number of papillæ fringe the mouth. Each arm segment bears 7 or 8 short spines on each side, lying so closely appressed that they are inconspicuous. The color of this species is usually green; however, brownish or blackish, and even mottled, forms occur. Diameter of disk about 1.5 cm; length of arm about 6 cm. Often abundant on sandy shores among eelgrass. Occurs in tide pools and from low-water mark to over 100 fathoms in offshore waters. Cape Cod southward.

FAMILY OPHIOLEPIDIDAE

GENUS *Ophiura*

O. sarsi. (Fig. 231.) Disk pentagonal with several large scales around which are grouped smaller ones. On the indentation of the disk, at the base of each arm, is a fringe of slender spinelets, called the "comb." Each interradius has 2 genital slits. Oral papillæ form a fringe around the mouth. The arm plates are regular and with 2 scales to each tentacle on the middle of the ray. Color dark green or brown. Diameter of disk about 2.5 cm; length of arm 7.5 cm or more. Frequently numerous in offshore waters. Maine and northward.

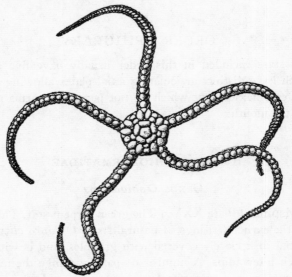

Fig. 231—O. sarsi.

FAMILY **AMPHIURIDAE**

GENUS *Ophiopholis*

O. aculeata. (Plate XXV.) The daisy serpent star. The disk of this handsome species bears a number of radially arranged scales and numerous short blunt spinulose projections, or coarse granular processes. Papillæ fringe the mouth. The upper arm plates are transversely oval and surrounded by a series of smaller plates, while the lateral plates bear small, but stout and conspicuous, spines which project from the surface of the arms. Color exceedingly variable, not only in hue but also in pattern. Red, purple, brown, yellow and green occur in various shades, the disk usually being spotted or blotched in a distinct radially arranged or concentric pattern or both, and the arms being striped or banded. It is only the aboral surface, however, that bears the bright color patterns; the oral surface is commonly some shade of gray or yellowish, except that the dark bands of the arms frequently completely encircle the under side. Diameter of disk about 2 cm; length of arm about 12 cm. Common in shallow and offshore waters, often occurring to great depths. Long Island Sound northward.

Genus *Amphipholis*

A. squamata. (Fig. 232.) The disk of this little ophiuran is covered with scales and is without spines of any sort. The arms are long and slender, are with 3 spines on each side of each segment, and bear 2 scales to each tentacle. An important guide for the determination of the species, however, is the number of oral papillæ in the corners of the mouth, these being 3 tooth-like projections along each edge of each jaw, thus making 6 in each corner. The outer papillæ are very wide and equal to both the inner ones. Color gray or pale brownish, often with numerous lighter markings. The disk in very young specimens is orange.

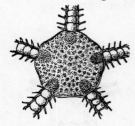

Fig. 232—Disk of A. squamata.

This is the smallest species known to occur on the Atlantic Coast. Diameter of disk about 4 mm; length of arm about 35 mm. In this species the eggs remain in the genital bursæ after they have been fertilized; there, without undergoing the characteristic metamorphosis of other species, they develop into tiny young ophiurans, and these make their escape into the water, similar in form and structure to the parent. Found on eelgrass and hiding in dead mollusk shells and in the small crevices of rocks, etc. Occurs from low-water mark to offshore bottoms. Not uncommon. Long Island Sound northward.

Genus *Amphioplus*

A. abditus. (Fig. 233.) Disk somewhat pentagonal with interradial margins concave and the angles at the base of the arms indented. Arms very long and slender, each segment bearing 3 spines on each side, which stand out almost at right angles to the arm. The middle spine is flatter, less sharp, and considerably stouter than the other two. The principal characteristic for specific determination occurs in the oral papillæ; these are 4 toothlike processes on each side of each jaw, thus making 8 in each corner of the mouth. On the oral surface of the disk at the base of each arm occurs a pair of narrow, curved plates called the radial shields; these are parallel or nearly so, and are not in actual contact, although they are frequently partly obscured by the encroachment of small scales that occur on the disk. Color variable, being commonly gray or brown or both arranged in a patternlike manner. Diameter of disk about 10 mm; length of arm about 175 mm. Found in soft mud in shallow and offshore waters. Sometimes common. Cape Cod to Cape Hatteras and southward.

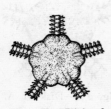

Fig. 233—Disk of
A. abditus.

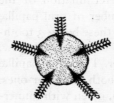

Fig. 234—Disk of
A. macilentus.

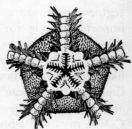

Fig. 235—Disk of O.
bidentata (oral side).

A. macilentus. (Fig. 234.) This form is somewhat similar to *A. abditus,* but differs specifically in that there are 10 oral papillæ in each corner of the mouth; moreover, all the spines on the sides of the arm segments are slender. The disk is nearly circular in outline, and the aboral surface bears numerous small imbricated scales, arranged in a sort of rosette at the center. The radial shields are arranged in pairs; they are long and narrow, and have their outer ends in contact. There are 2 scales to each tentacle. Color light gray. Diameter of disk about 4 mm; length of arm about 60 mm; arms extremely slender. Often abundant. Occurs in muddy localities in offshore waters. Cape Cod southward.

FAMILY OPHIACANTHIDAE

GENUS *Ophiacantha*

O. bidentata. (Fig. 235.) Disk roughly pentagonal and bearing small spines which have the superficial appearance of granules, these more or less obscuring the scales. Long oral papillæ present along the edges of each jaw. Arm plates well defined; lateral plates with hollow spines standing out conspicuously from the sides of the rays. Diameter of disk about 12 mm or more; length of arm about 45 mm or more. Not uncommon. Occurs in offshore waters. Maine and northward.

FAMILY OPHIOTRICHIDAE

GENUS *Ophiothrix*

O. angulata. (Fig. 236.) The pentagonal disk of this species is covered with small spines and granules, and it bears large triangular radial plates. On the slender arms are numerous long brachial spines which stand out almost at right angles to the surface. These spines are serrated along the edges and at the ends. The mouth is without oral papillæ, but at the apex of each jaw occurs a prominent group of tooth papillæ. Color very variable; most often purplish or greenish. There is a longitudinal stripe on the aboral surface of each arm. Diameter of disk about 10 mm or more; length of arm about 50 mm. Found in offshore waters. Not uncommon. Chesapeake Bay and southward.

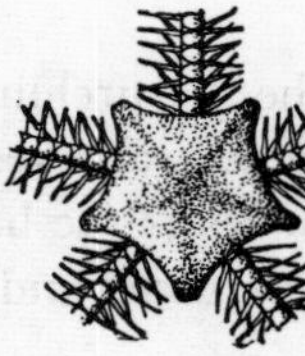

Fig. 236—Disk of O. angulata.

ORDER EURYALAE

In this group of serpent stars are comprised forms that are provided with arms which are usually branched and can be rolled inward toward the mouth. The arms have only axial plates, which are double; the other plates are either rudimentary or wanting entirely. A soft skin, instead of calcareous plates, covers the body.

FAMILY ASTROPHYTIDAE

GENUS *Gorgonocephalus*

G. agassizi. (Plate XXV.) The basket star. This remarkable form has a thick, somewhat pentagonal, granulated disk with radial ridges on the aboral surface marking it into starlike division. Extending from the margin of the disk are five arms, each of which, beginning at the base, branches into pairs, this dichotomous division continuing about a dozen times with the branches becoming increasingly small. The attenuated terminal branches are often in motion, coiling and extending in a graceful manner. When the animal is on the bottom, it supports itself on the tip of the branches with the disk held high; it is, however, able to climb or crawl over seaweeds with the smaller end branches of the rays. It is said to prey upon swimming creatures, which it captures by elevating its basket of branches much in the manner of a purse-net. Color dull yellow or brownish. Diameter of disk about 8 cm; length of arm about 25 cm or more. Found usually in offshore waters; rarely in very shallow areas. Often common. Nantucket northward.

CLASS **ECHINOIDEA**

The sea urchins and the sand dollars; also popularly known as cake urchins, sea cakes, disk urchins, sand cakes, heart urchins, etc. In the sea urchins the body is nearly hemispherical, while in the other forms it is flattened and discoid.

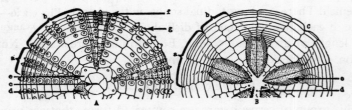

Fig. 237—Diagram of echinoids. A, A. punctata; a sea urchin. a, radius; b, interradius; c, anus; d, madreporic plate; e, genital plate; f, pores for the tube feet; g, tubercles for attachment of spines.

B, a sand dollar. a, radius; b, interradius; c, anus; d, genital pore; e, pelaloid.

In all the echinoderms of this class, the five radii do not extend from the body to form arms, such as are typical of the starfishes; therefore, the body is simpler in general outline. (Fig. 237.) But the anatomical and structural features reflect all the complexities occurring in the other group. The calcareous skeletal plates usually are well developed and fit closely so as to form a compact frame or "test" enclosing the body cavity. In the sea urchins, the test is provided with movable spines articulating with small tubercles by ball-and-socket joints, and with pedicellariae usually having three jaws. Certain pedicellariae, the gemiform type, occurring in some species are provided with poison sacs, while in some other species the tips of certain spines are poisonous. (Fig. 238.) However, none of the sea urchins on our coast are harmful to man. The spines of sea urchins are comparatively long, some of them extremely so, but the spines of sand dollars and other discoid types are minute. A ciliated epithelium, which causes the water to pass over the respiratory tube feet, covers all parts of the sea urchin's body, except the large spines of certain species; in the sand dollar, the spines on the upper surface are ciliated. Minute spherical modified

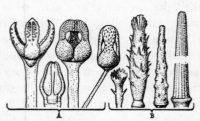

Fig. 238—A, pedicellariae, and B, spines, of sea urchin.

spines (sphaeridia) occur on the oral surface of nearly all echinoids.

In the echinoids the mouth is usually in the center of the under side, and the oral surface is flatter than the aboral. As in the case of the star-fishes, it is on the oral surface that echinoids move about. Surrounding the mouth is a circular area having a certain degree of flexibility, and is known as the peristome. All echinoids, with the exception of the heart urchins, possess a remarkable masticatory apparatus called "Aristotle's lantern," so called from the fact that Aristotle was the first person to liken this structure to an ancient kind of lantern. (Fig. 239.) It is a complicated mechanism, consisting of about forty parts; but its essential features are five long chisellike teeth, the sharp ends of which meet in the center of the mouth opening, where they project slightly and are

Fig. 239—
"Aristotle's
lantern."

visible in the living animal. Each tooth is held in place by a pair of calcareous rods which are so joined as to make a V-shaped structure, thus forming a jaw. The five jaws are connected by other ossicles, and the whole assembly forms a rigid framework, the function of which is to support the teeth and afford attachment for two complex groups of muscles: the group that moves the jaws up and down, and the group that causes the entire lantern to protrude or withdraw. In the sand dollars the lantern of Aristotle is more compressed than in the sea urchins, being in the form of a disk or shield; and the structure is not quite so complicated.

The alimentary canal is a spacious, ample tube that winds once around the body cavity, then folds back on itself and continues in the opposite direction toward the anus. In the sea urchins the anus is in the center of the aboral surface, and is surrounded usually by a system of plates called the *periproct,* one plate of which is modified to form a madreporite; sand dollars have the madreporite in the center, and the anus at or near the edge of the disk; whereas in the keyhole urchins the anus is on the oral surface near the mouth. As an adjunct to the alimentary tract, there is present an accessory tube, or "siphon," which branches off near the mouth, parallels the stomach, and rejoins the tract farther back. It is lined with cilia, and it is presumed to assist peristalsis by keeping a current of water in motion through the alimentary canal.

As in the starfishes, but differing in detail, the water-vascular system consists of sieve plate, ring canal, stone canal and radial water vessels. The ring canal encircles the œsophagus, but is situated above the lantern instead of immediately above the nerve ring, as in the asteroids. The so-called "stone canal" in the echinoids is a misnomer, as this duct is not calcified but has simple, soft walls. The five radial canals extend along the inner surface of the test to the radial plates, through the pores of which the terminal ends of the canals project, forming tube feet. Each tube foot possesses an ampulla, with which it is connected by a pair of minute ciliated canals, the motion of the cilia creating a current through one canal into the tube foot, and the return current reaching the ampulla by the way of the other canal. In the majority of species, particularly among the sea urchins, all the tube feet are provided with sucker disks, thus functioning as organs of locomotion; in every case they are organs of touch and of respiration and excretion. Respiration and excretion are also performed by the entire body surface, also perhaps, by the siphon, and by the peristomal gills, the last-named organs being five pairs of branched outgrowths usually situated at the edge of the peristome.

The nervous system is contained within the skeleton and consists of a ring about the mouth and radial branches lying along the ambulacral vessels. Special sense organs are poorly developed: the use of tube feet as tactile organs has been mentioned; however, although it has not been definitely established as true, it is thought that since the sphæridia contain nerve cells these may be concerned with the sense of balance. And in certain echinoids there are light-sensitive organs present in the form of pigment eyes which are distributed over the test.

There are no hermaphroditic forms, the sexes being separate. In mature individuals, there are five large gonads lying in the upper part of the body cavity, opening to the exterior through the genital pores in the periproct. The males have whitish organs, while those of the females are yellowish. The sex products are discharged into the sea water where fertilization of the eggs takes place. There is a complex development marked by a metamorphosis in which the larva passes through a pluteus stage similar to that of the ophiuran larva.

The class *Echinoidea* contains about 500 species; they are worldwide in distribution, being found in all seas from the tidal zone to very great depths.

ORDER **CIDAROIDA**

Sea urchins without peristomal gills are included in this order. The ambulacral zones are narrow, and the interambulacral zones are broad, both zones continuing over the peristome to the mouth.

FAMILY **CIDARIDAE**

GENUS *Eucidaris*

E. tribuloides. (Fig. 240.) This species has a thick test with the oral and aboral surfaces somewhat flattened. The areas marked by the rows of ambulacral feet are about one third the width of those lying between. Stout, fluted, grayish spines having a length nearly the diameter of the test are distributed over the body. Between the long primary spines occur a number of shorter ones. Diameter of test 4 to 7 cm. Occurs from the tidal zone to deep water. Often common. South Carolina to Florida and Brazil.

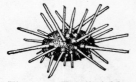

Fig. 240—E. tribuloides.

ORDER **CENTRECHINOIDA**

Sea urchins having peristomal gills and sphaeridia are contained in this order.

Family CENTRECHINIDAE

Genus *Centrechinus*

C. antillarum. (See the frontispiece.) Test circular and with long, slender spines. Color black. Diameter up to 10 cm. Found in shallow water among rocks. Common. Florida.

Family ARBACIIDAE

Genus *Arbacia*

A. punctulata. (Plate XXV.) The purple urchin. This not unhandsome species has a thick test bearing solid, ample-sized fluted spines. The ambulacral zones are narrow, and the ambulacral feet on the aboral surface are without sucker disks; consequently, the animal is obliged to use the spines to aid in locomotion. It is able to move about fairly fast with its spines, which it does in a tilting manner. The aboral pole and its adjacent area, and the interambulacral zones for a distance nearly halfway to the equator of the test are free from spines. Only large primary spines are present, and these are arranged in series of 2 in the ambulacral zones, while in the interambulacral zones they occur in series of from 4 to 8. Color usually purple, but individuals occur in which the test is reddish or purplish-brown; the tube feet are red or brown. Diameter of test 3 to 5 cm; length of spines about 25 cm. Found from low-water mark to offshore bottoms, in shelly or rocky areas. Often numerous. Cape Cod to Yucatan.

Family ECHINIDAE

Genus *Lytechinus*

Fig. 241—L. variegatus.

L. variegatus. (Fig. 241.) Test almost spherical, but slightly flattened near the oral pole. Spines slender and somewhat short. Tubercles supporting the spines about equal in size and arranged in several rows. Ambulacral plates occurring in single pairs at the base of each radius in the peristome, and each plate provided with 3 pairs of pores. Peristome with gills. Periproct consisting of numerous plates. Color green, occasionally with a shade of purple. Diameter from 5 to 8 cm. Found in shallow and offshore waters. Occurs from Cape Hatteras to Brazil. The common species in Southern waters.

GENUS *Tripneustes*

T. esculenta. (Plate XII.) This species, sometimes bearing the generic name *Hipponœ,* is edible, often being used for food by the Negroes of the West Indies. The test is thin and bears numerous small tubercles and spines. Peristomal gills are present and there is a single pair of ambulacral plates at the base of each ray in the peristome. Each one of the plates is perforated with 3 pairs of pores. Color whitish. The test is semiglobular and measures from about 10 to 15 cm in diameter. Often numerous. Florida, West Indies, and the South Atlantic coast.

FAMILY STRONGYLOCENTROTIDAE

GENUS *Strongylocentrotus*

S. drobachiensis. (Plate XXV.) The green sea urchin. The test of this species is circular and somewhat depressed, and is of variable thickness. There are numerous long, slender spines, which are fluted longitudinally. The tubercles bearing the spines are of unequal sizes; they are often crowded, and are arranged in a double series on each ambulacral and interambulacral zone. Thickly scattered among the long primary spines are much smaller secondary spines similar in shape, while a third variety of spines, the miliary spines, which are still smaller and very slender, is also present. This variation in the sizes of the spines is correlated wth that of the tubercles. The anal plates in gull-grown individuals are numerous and very small. Color greenish, but different specimens vary in shade and in markings. There often occur forms in which the color is yellowish, reddish, or purplish, owing to the variations in the hue of the spines; and in some of these the tones are intensified, the spines appearing as if their tips were tinged deeply by some dyestuff of crimson or violet. Diameter of test from 4 to 8 cm; length of spines about 15 mm. Found usually on rocky bottoms in tide pools and shallow water, where it subsists on food consisting mostly of algæ growing on the rocks. It also eats considerable quantities of mud for the diatomaceous and other organic material this contains; and it will even eat the bones of dead fishes, which it chisels with its sharp teeth until they are entirely devoured. This is the common species of the northern New England coast, where it occurs in great abundance. It is also found in south of Cape Cod, but in deeper water and in fewer numbers.

ORDER **CLYPEASTROIDA**

This order includes the sand dollars and keyhole urchins. In these forms there has been secondarily acquired bilateral symmetry of the test, though the internal structure retains its radial features. Thus, there is in the animals of this order a right and a left half. The test is discoidal and much flattened, with the anus situated interradially at the margin or on the oral surface. A characteristic five-rayed figure, composing what is known as the *petaloid areas*, and formed by ambulacral pores, occurs on the aboral surface, the petaloid area being so called because of the arrangement of the pores, which resembles the petals of a conventional flower. The mouth is central, and is provided with a lantern of Aristotle, or dentary apparatus, similar to but more compressed than that which characterizes the members of the preceding order.

FAMILY **CLYPEASTRIDAE**

GENUS *Clypeaster*

C. subdepressus. (Fig. 242.) The test of this species is thick, somewhat elongated, and very roughly pentagonal in outline; short spines cover the surface, and in the center of the aboral surface is the madreporite from which extend the five rays of the petaloid areas, the latter being wide and quite distinct. Five genital pores are present near the madreporite. A groove joins each pair of ambulacral pores. Color greenish-yellow or purplish. Length 12 cm; width 8 cm. Often numerous. Found from the shore to offshore bottoms. North Carolina to Brazil.

Fig. 242—C. subdepressus.

FAMILY SCUTELLIDAE
GENUS *Echinarachnius*

E. parma. (Plate XXV.) The common sand dollar. The body is thin and very flat, and nearly circular in outline with an indentation occurring at one point, indicating the position of the anus. A thick velvety coat of minute spines covers the body, and the petaloid areas are plainly marked by the position of the respiratory tube feet; and at the center of the converging rays is the madreporite. Both the ambulacral and the interambularcral areas bear tube feet, but those of the petals are the larger. Between the petals, near the madreporite, are 4 openings of the genital glands. Color reddish, brownish, or purplish, often with petaloid area darker red than the general color. Diameter about 7 cm. Often very abundant. Occurs on sandy bottoms from the low-water mark to offshore depths. Labrador to New Jersey.

GENUS *Mellita*

M. quinquiesperforata. (Fig. 243.) Keyhole urchin. This form often is described in technical literature under the specific name *testudinata,* which, indeed, was the name given as early as 1734; but as this long preceded the Linnæan code, it has been found more expedient to adopt

the present *Mellita quinquiesperforata.* The shell is very flat and in outline nearly circular, and is perforated with 5 elongated holes, termed lunules, which pass clear through the body. It is from these narrow openings that this and its closely related species derive their popular name. One lunule is located in the posterior interradius, about midway from the central madreporite to the edge, the remaining four being situated in the middle of the lateral radii and distally to the

Fig. 243—M. quinquies-perforata.

petals. The test is covered with small, close-set spines, and has the five petaloid areas of the aboral surface well marked. On the oral surface, the ambulacral grooves have paired branches. In this species the mouth is placed slightly anterior to the center, while the anus occupies a position between the interradial lunule and the mouth. Color pale brown, sometimes with a greenish tinge. Diameter 8 to 12 cm. Found in shallow water. Occurs from Vineyard Sound and southward, but is not common north of Cape Hatteras.

Genus *Encope*

E. michelini. (Fig. 244.) Test flattened orally and arched aborally; 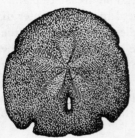 in outline somewhat elliptical and truncated behind. On each lateral margin opposite to the end of the lateral petals are 2 notches which mark the position of lunules that have encroached on the edge; the fifth lunule is in the posterior interradius, occupying a position well toward the madreporite in the center of the petaloid figure. Diameter about 14 cm. Often abundant in shallow and offshore sandy areas.

Fig. 244—E. michelini. Florida and the Gulf Coast.

ORDER SPATANGOIDA

The heart urchins. Like the clypeastroids, the members of this order have the bilateral symmetry of the outer form secondarily acquired. The fundamental structural plan still retains its radial features. No lantern of Aristotle or other dentary apparatus is present, and the mouth is placed near the forward end of the oral surface. The anus is at or near the hinder margin of the body. Five petaloid ambulacral areas occur on the aboral surface, the anterior median area usually having a different proportion from the others, which are similar to one another and occur in patternlike pairs.

Family SPATANGIDAE

Genus *Moira*

M. atropos. (Plate XII.) Test thick and somewhat heart-shaped or ovate with numerous spines. Spines directed backwards. Apical area behind the middle and marking the posterior end. Petaloid areas distinct, and with deep, narrow ambulacral grooves. Color yellowish-white.

Length about 5 cm. Common. Found in shallow and offshore waters. North Carolina to Florida.

GENUS *Metalia*

M. pectoralis. (Fig. 245.) In this large species the test is thin and has a somewhat elliptical outline. The surface is densely covered with long, backward-pointing spines. A well-marked petaloid figure occurs on the test, and the ambulacral grooves are deep and distinct. Color reddish-gray. Length from 12 to 16 cm. Not uncommon. Occurs in shallow and offshore waters. Florida and the West Indies.

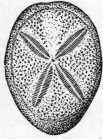

Fig. 245—Test of
M. pectoralis.

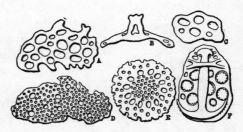

Fig. 246—Calcareous plates of a holothurian. A, plate from tentacle; B, plate from pedicel; C, plate from body wall; D, plate from cloaca; E, terminal plate of pedicel; F, anchor plates.

CLASS HOLOTHURIOIDEA

Sea cucumbers. The holothurians are elongated, cylindrical echinoderms in which the calcareous plates are mostly minute (Fig. 246.) and the body wall in consequence is without rigidity. The skin, however, is tough and leathery or often warty. The oral surface is not directed downwards, as in the asteroids and the echinoids, but, as the great majority of these animals rest on their side and thus have a superficial resemblance to thick-bodied worms, the oral surface is at the forward end, while the aboral surface is at the hinder end in the region of the anus. Only in rare exceptions, such as in the extra-territorial form *Rhopalodina lageniformis,* wherein the mouth and anus are close together on the upper surface, is this typical holothurian character altered. And although in appearance the sea cucumbers are seemingly unlike the star-fishes and sea urchins, examination of the former will often reveal very evident relationships even on the surface in the occurrence of the characteristic radial ambulacral zones. In some forms, two of these zones are on the upper, or dorsal, surface, and three are on the lower (ventral) surface, forming a sort of sole on which the animal creeps; again, the ambulacral feet and papillae may be scattered irregularly over both the radii and interradii or they may be suppressed entirely. The skin is without cilia, and there are no spines or pedicellariae.

The mouth is surrounded by ten or more long, branched tentacles which are continuously in motion when the animal is not contracted, and which serve chiefly as tactile organs. Other sensory organs, such as pigment eyes, otocysts, olfactory cups, and taste papillae are often present. The oral tentacles are in communication with the water-vascular system, and that they are also of aid in respiration is indicated by the fact that they generally are provided with ampullæ through which the water flows during the characteristic movements of the tentacles. (Fig. 247.)

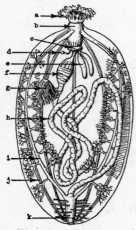

Fig. 247—Diagram of a holothurian. a, tentacles; b, calcareous ring; c, ring canal; d, madreporite; e, Polian vessels; f, stomach; g, reproductive organs; h, intestine; i, respiratory trees; j, longitudinal muscles; k, cloaca.

The water-vascular system in most holothurians receives its supply through an internal madreporite from the interior of the body cavity, the latter normally being filled almost to the point of turgidity with fluid. A stone canal connects the madreporite with the ring canal. The ring canal encircles the œsophagus at the base of the oral tentacles and is in connection with large *Polian* vessels which vary greatly in number according to the species in which they occur. Their function is to hold a reserve supply of fluid. Also connected with the ring canal are the radial canals; and these extend between the two bands of longitudinal muscles in each of the five radii to the hinder end of the body. The tube feet are not so numerous as in the starfish in any instance, but when these are present, the ampullæ appear on the internal surface of the body. In some forms, like *Leptosynapta,* the water-vascular system is very rudimentary; radial canals and tube feet are wanting.

The alimentary canal is nearly three times as long as the body; running from the mouth at the forward end, it generally turns on itself twice before ending at the anus. Just within the anal opening is the cloaca, or rectum, an enlargement of the intestine, which forms a good-sized chamber from which often extend into the body cavity two kinds of long tubular processes: the respiratory trees and Cuvier's organs. The respiratory trees are profusely branched, and extend nearly to the forward end of the body cavity; they expose a vast amount of surface to the fluid of the body cavity, thus functioning as gills. The other organs are unbranched, glandular tubes attached to the base of one of the respiratory trees, which secrete a viscid substance. These organs are extensile, and when thrust out through the anus their sticky surfaces adhere and entangle; thus, in the case of attack by enemies, these organs may be used as weapons of defense. However, neither the "gills" nor the Cuvierian organs are present in all forms. As the cloaca is attached to strong muscles and is capable of considerable distension, it can admit sea water through the anal opening, and then by closing this aperture force the fluid up into the respiratory trees under considerable pressure. A part of the pure water thus supplied to the branches of the respiratory trees is forced through their walls and enters the body cavity; consequently, the body cavity is kept tensely filled.

A superficial nervous system, somewhat like that of other echinoderms, is present, consisting of a nerve ring around the mouth and nerve branches extending along the radial ambulacral canals.

The sexes are, with a few exceptions, separate. The sexual organs consist of a brushlike structure comprising numerous, slender filaments rising from the end of a tube, the genital duct, that opens to the outside on the dorsal surface near the base of the oral tentacles, or within the circlet. The opening of this duct in the males is generally indicated by the presence of a small papilla. In numerous forms, the eggs are expelled into the water, after which fertilization takes place; but in some others the young are carried by the mother, while there are yet others in which the eggs develop in the body cavity, the young finally becoming free by escaping through a rupture of the body wall. The eggs of holothurians, like those of other echinoderms, give rise to free-swimming embryos that at first are similar to the embryos of starfishes, and are called *auriculariae*.

Holothurians are widely distributed, being found in most seas living at all depths. Their movements are usually sluggish, though they respond quickly and energetically to various stimuli. Like all other echinoderms they can regenerate lost or injured parts; indeed, in their regenerative processes, certain forms have capacities far beyond those of any other animals so highly specialized. Some holothurians when disturbed contract their bodies with such violence that the respiratory organs and the greater part of the viscera, including the cloaca and nearly the whole alimentary canal, are ejected, these parts being later restored under favorable conditions. This extraordinary facility, however, is not common to our Atlantic species; nevertheless, our forms are able to withstand considerable mutilation and still regain their normal proportions.

As the holothurians live largely in the mud or sand, moving about from place to place only as the conditions of food or other circumstances require, they remain for the most part with the greater portion of the body buried in the substratum: the paractinopods usually bury deeply the hinder end of the body, leaving only the oral tentacles exposed; but the other holothurians have the cloacal opening connected with the respiratory organs; consequently, in those species the body is usually bent into a U-shape with both ends at or near the surface of the bottom in which they live.

More than 500 species are included in the class *Holothurioidea*.

ORDER **ACTINOPODA**

Holothurians included in this order have the radial canals of the water-vascular system well developed. The radial canals supply the oral tentacles with fluid, the tentacles rising from the radial canal only. Ambulacral tube feet are usually present. Most of the sea cucumbers are in this order.

Family **HOLOTHURIIDAE**

Genus *Holothuria*

H. floridana. (Fig. 248.) This species, formerly known as *H. mexicana,* has an elongated, cylindrical body bearing 20 oral tentacles that are branched only at the tip. The tentacles have ampullæ, and just outside the tentacular ring is the genital pore. Ambulacral appendages are scattered over the entire body surface, those on the under side bearing sucker disks, those on the upper side being either with or without them. The respiratory trees are well developed. In the body wall numerous calcareous plates of diverse shapes occur; but these are so small that a microscope is necessary for their proper study. Color brown or yellowish, sometimes reddish underneath. Body from 20 to 40 cm long. Not uncommon. Florida and the West Indies.

Family CUCUMERIIDAE

Genus *Cucumaria*

C. frondosa. (Plate XXV.) The body in this species is somewhat ovate and decidedly plump, though the animal has the power to alter its shape considerably by greatly lengthening the body or by contracting it almost to a sphere or by constricting it like an hour-glass. There are 10 profusely-branched oral tentacles, inside the ring of which occurs the genital pore. The general surface is smooth with tube feet bearing suckers arranged in double rows along the radii; in the interradii there are a few scattered tube feet without suckers. Respiratory trees are present. Color dark brown or purple above; lighter below. Length 20 to 30 cm; breadth 10 cm. Very common; most abundant along the coast of Maine. Occurs in tide pools and in shallow and offshore bottoms. New England coast and northward.

Fig. 248— H. floridana.

Fig. 249—C. pulcherrima.

C. pulcherrima. (Fig. 249.) Body ovate, often tapering toward the hinder end, and with a distinct curve so that the dorsal side measures more than the ventral side. Ten much-branched oral tentacles are present, of which the 2 ventral are considerably smaller than the others. The tube feet are all provided with sucker disks, and are limited solely to the five radii. Surrounding the anus are 5 groups of tube feet that are made rigid by numerous calcareous deposits. Color white or light yellow. Length about 4 cm; thickness 2 cm. Occurs in shallow water; often found on the beach after storms. Vineyard Sound to North Carolina.

Genus *Thyone*

T. briareus. (Fig. 250.) In shape, the body of this holothurian is

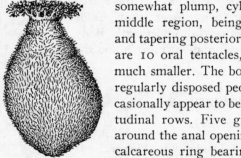

Fig. 250—T. briareus.

somewhat plump, cylindrical, and thickest in the middle region, being rather narrower anteriorly and tapering posteriorly. As in *C. pulcherrima,* there are 10 oral tentacles, of which the 2 ventral are much smaller. The body is thickly covered with irregularly disposed pedicels, or tube feet, which occasionally appear to be arranged in 5 very wide longitudinal rows. Five groups of small papillæ occur around the anal opening, and there is also present a calcareous ring bearing 5 processes known as the anal teeth. Color purplish, brown or black. Length 12 cm; thickness 3 cm. Rather common, being very abundant in the mud of shallow waters in certain regions. Vineyard Sound and southward.

Genus *Psolus*

P. fabricii. (Fig. 251.) This species has a plump body with a flat under side, the surface of which latter forms a sort of sole on which the holothurian moves about. There are 10 profusely branched oral tentacles which, when extended, are about as long as the body. No tube feet occur on the upper surface of the body, but the "sole" is provided with a row of feet on each side and a narrower row running through the middle. The main portion of the body, with the exception of the "sole," has the surface cov-

Fig. 251—P. fabricii.

ered with rounded, overlapping plates bearing granulations. These plates also occur in groups around the oral and anal openings. Color bright red on the upper side; lighter on the lower side; a very attractive form, being an object of considerable beauty when expanded in the water. Length about 10 cm. Found in shallow water, usually among rocks. Not uncommon. Cape Cod and northward.

FAMILY MOLPADIIDAE
GENUS *Caudina*

C. arenata. (Fig. 252.) In this form the body is cylindrical and elongated and terminates posteriorly in a taillike constriction of about one third of the total length. There are 15 short cylindrical oral tentacles; each one is clovelike, bearing 4 small branches at the tip. No tube feet are present. Color pink or purplish. Length about 15 cm including "tail." Lives in sand in shallow and offshore waters. Sometimes common. Gulf of St. Lawrence to Rhode Island.

Fig. 252— C. arenata.

ORDER PARACTINOPODA

These holothurians are without ambulacral feet, respiratory trees, ampullæ, and radial canals. The tentacular canals of the oral tentacles connect with the ring canal.

FAMILY SYNAPTIDAE
GENUS *Leptosynapta*

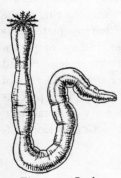

Fig. 253—L. inhaerens.

L. inhaerens. (Fig. 253.) A wormlike, semitransparent form with a long, cylindrical, slender body capable of considerable extension. There are 12 oral tentacles, each of which has several pairs of side branches. Calcareous deposits occur in the tentacles, and the tentacles each have a number of sense organs on the inner surface near the base. However, the calcareous deposits occurring in the body walls are the most characteristic feature of the species and the genus. These occur as miniature anchors, each of which is connected with a perforated plate; they are present throughout the body wall, but those in the forward regions are smaller than the rest, which gradually increase in size toward the hinder end. (Fig. 246.) These may be seen in the living animal even without the aid of a microscope. The anchors help to keep the animal from slipping as it burrows in the sand, being held in an elevated position by contraction of the circular muscles. The contraction starts near the hinder end and moving forward thrusts the anterior end onward. The body is colorless, but often individuals are found in which there is a distinct tinge of yellow or pale red. A thin delicate skin protects the body, through which the 5 whitish bands of longitudinal muscles and certain of the internal organs of the living animal are often visible. The sand-filled intestine and the reproductive organs during the period of the maturity of the sexual products are particularly conspicuous under the translucent integument. Length 10 to 30 cm; thickness 5 to 10 mm. Usually lives in clean sand, but occurs also in muddy situations. Found from the tidal zone to offshore bottoms. Common. Maine to South Carolina.

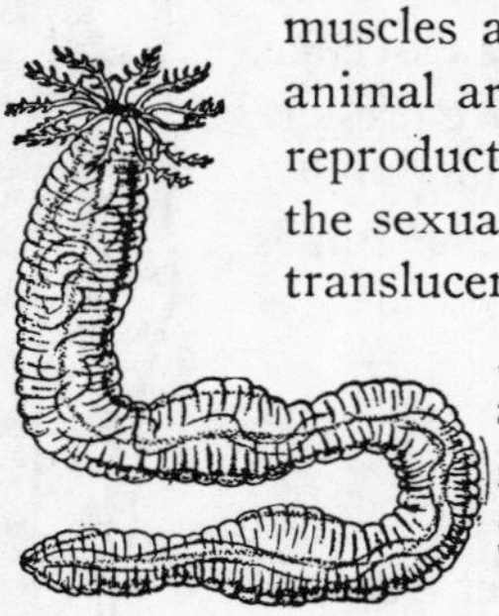

Fig. 254—L. roseala.

L. roseala. (Fig. 254.) Differs from the preceding form in the presence of red pigment granules scattered thickly over the body, thus giving it a delicate rose-pink color. It is smaller than *L. inhærens* and has each of the 12 tentacles provided with but 2 or 3 pairs of side branches. The anchor plates of the body wall are slenderer and the calcareous deposits in the oral tentacle are more irregular. But the chief distinction lies in the calcareous ring at the base of the tentacles, which in this species consists of narrow plates having the radical plates notched on the anterior border, instead of perforated, for the passage of the radial nerves. Length 10 cm or less. Found in gravel and under stones; sometimes in sand; never in mud. Occurs in the tidal zone and shallow regions. Not uncommon. This and the preceding species, when kept in confinement, have a propensity to break into pieces by constricting the body at various points. Cape Cod to Long Island Sound, and Bermuda.

XI

ARTHROPODA

PLATE XXVI

CRUSTACEANS AND ARACHNOIDS

Ovalipes ocellatus (middle top); *Calinectes sapidus* (upper left); *Homarus americanus* (middle right); *Limulus polyphemus* (lower left and lower right).

CHAPTER XI

PHYLUM **ARTHROPODA**

(THE JOINTED-FOOTED ANIMALS)

OF the 500,000 or more species of animals known to occur throughout the entire world, fully four fifths of this number are arthropods. By this token, together with the fact that the number of individuals in the group is vastly greater than that in any other group of the *Metazoa,* they are the most successful in the animal kingdom. Not only have they adapted themselves to all sorts of environments, but also they are in a peculiar sense the product of a diversity of environments. The sea, the fresh water of the land, the air and the soil of the earth have all been factors in molding the destiny of this mighty horde. Nor has the organic world been without a share in the shaping of their biological pattern. Man and other animals and even the plants have provided them with facile methods of existence in the way of parasitic hosts, to say nothing of furnishing other and influential forms of food and shelter.

Marine arthropods include the crustaceans, arachnids and insects. In point of numbers, both of species and of individuals, the crustaceans by far outrank the others; the arachnoids are represented in the sea by such aberrant forms as the horseshoe crabs and the sea spiders; while only a very few insects are identified with salt water, and these few live on the surface. All marine arthropods are easily recognized by the horny, or chitinous, covering of the body and by their division externally into somites, or segments, some of which have jointed appendages. The members of the group are characterized by bilateral symmetry.

Three classes are contained in the phylum *Arthropoda,* which are as follows:

PHYLUM ARTHROPODA

CLASS **Crustacea**   SUBCLASSES .... { Entomostraca
Malacostraca

279

CLASS **Arachnoidea**   SUBCLASSES ....$\begin{cases}\textbf{Xiphosura} \\ \textbf{Arachnida}\end{cases}$

CLASS **Tracheata** (Including the insects and centipedes; no species
described in this work.)

CLASS **CRUSTACEA**

Crustaceans living in the sea are represented by such commonly
known forms as water fleas, barnacles, beach fleas, shrimps, crabs, and
lobsters. They are called crustaceans from the fact that, in the larger
forms at least, there is a hard crust, or armor, encasing the body and its
appendages. In all cases, this crust is composed of chitin, but certain
forms like the larger crabs have a heavy crust, or exoskeleton, which
is partly constituted of calcium carbonate; smaller forms such as some
shrimps and the water fleas have chitinous coverings often so thin and
delicate and transparent that the movements of the internal organs can
be seen. Crustaceans are characterized also by the fact that they are gill-
breathers and bear two pairs of antennæ.

Although crustaceans evidence a distinct segmentation of the body,
usually some of the somites are fused into one piece. The head region,
or cephalic portion, is nearly always fused in this manner, besides which
it is often joined with some or all the somites of the thorax, thus forming a division of the body known as the *cephalothorax*. The hard, shieldlike covering of the cephalothorax is called the *carapace*. Posterior to
this is the abdomen.

The appendages are jointed, and are modified to perform various
functions, chief of which are respiration and locomotion. Typically they
are divided into two branches, which arise from a basal piece attached to
the body. These branches are termed the *endopodite* and the *exopodite,*
that is to say, the inner foot and the outer foot; the basal piece is known
technically as the *protopodite,* the first foot. Actually, however, there is
often one branch poorly developed or wanting entirely. The two pairs of
antennæ, or feelers, in front of the mouth are provided with the sense of
touch and perhaps that of smell. Strictly speaking only the second and
larger pair are the *antennæ:* the first and smaller pair are the *antennules.*

Compound eyes are usually present, and in the higher crustaceans these are borne on stalks.

Considerable variation occurs in the mouth parts, but very generally there is present a pair of short, hard *mandibles* that work from side to side, and are used in biting; then, next to these, on the outside, are the appendages that hold and manipulate the food, the first and second pairs of *maxillae*.

The digestive tract in most crustaceans consists principally of a tube extending from the mouth at the forward end to the anus at the hinder end of the body. In the higher forms there is a well-marked œsophagus, stomach, and intestine, and often a distinct gizzardlike organ which is essentially a part of the stomach and is known as the "gastric mill." Large, tubular livers are frequently present. Excretion takes place through the kidneys, which in most forms are tubular glands opening to the outside near the mouth. Respiration in the lower and smaller crustaceans is carried on through the surface of the body, there being no specialized organs or appendages for this function; but in the higher groups gills are present as modified abdominal appendages or projections of the legs or, in rare instances, as projections of the body wall. The gills are often enclosed in special chambers. In the higher members of the group, a pair of such chambers occurs, one chamber on each side above the walking legs and under a flaplike extension of the carapace.

The nervous system extends backward from the brain, which is a large ganglion in the head above the œsophagus composed of a double knot of nerve tissue; by forming two branches it passes on each side of the œsophagus and runs along a median line on the under side of the body. These parallel branches are connected at intervals by ganglia, one of which occurs in each somite. The ventral nerve chain sends off a number of branches, and other nerves pass from the brain to the eyes, antennules, and antennæ.

There is no closed circulatory system such as arteries, veins or capillaries; the blood passes from the gills to the tissues through the spaces lying between the latter. A heart, however, is usually present on the dorsal side which causes the blood to circulate. The blood is nearly colorless.

In the great majority of crustaceans the sexes are separate, hermaphroditism prevailing only among such sessile forms as barnacles and among parasitic species. The first larval stage is characterized by a typical *nauplius* form (Fig. 255), a minute, unsegmented animal with three pairs of appendages, the two hinder pairs being two-branched. In

the lower crustaceans the nauplius usually leaves the egg as a free-swimmer, but in the higher crustacea the larvæ are born at a later stage of development, usually the stage characterized by the *zoea* form (Fig. 255). In the crabs, the free-swimming zoea passes through a metamorphosis that is marked by the *megalops* stage (Fig.

Fig. 255—Larval forms of crustaceans. A, nauplius; B, zoea; C, megalops.

255) before assuming the typical adult form.

Among the crustaceans, the phenomenon of growth is accompanied by a periodical shedding of the external crust and such chitinous coverings as may occur in the oral passages. Among numerous forms, a crack occurs along a median dorsal line in the crust of the cephalothorax through which the animal emerges; but crabs and lobsters molt by coming out through a transverse dorsal rupture between the carapace and the hind-body. The rapid increase in size which takes place between the time of the casting of the old covering and the subsequent hardening of the new cuticula is due merely to the filling out of tissues that have already attained to a larger growth prior to the process of molting; which is to say, the animal does not grow because it has molted, but, rather, it molts because its rigid and inelastic covering can no longer accommodate the growing tissues within.

The habits of crustaceans are as various as are their species. Some are strictly flesh-eaters, some subsist entirely on plants, and some are omnivorous. Many forms are parasitic. Their food habits are correlated with their diversity of habitat. All are adapted to conform to a certain environment, and therefore they possess organs fitted to cope with their surroundings. Were the various species not so perfectly adapted, or should the environment change too rapidly for the individuals of any given species to adjust themselves, they would become exterminated and their places be taken by others better equipped to maintain themselves in the struggle for existence.

Subclass **ENTOMOSTRACA**

The crustaceans contained in this subclass are mostly minute forms that require for their proper determination the use of either a good strong hand lens or a compound microscope. They are without abdominal appendages, all appendages being restricted to the head and thoracic regions. There are 5 pairs commonly present on the head, while the thorax may bear from 2 to 60 pairs.

ORDER **PHYLLOPODA**

Small entomostracans in which the appendages of the thorax are flattened and leaflike. The majority are fresh-water forms and are variable in shape, but the marine representatives commonly have short, compact bodies. They live chiefly on microscopic algæ.

Suborder **CLADOCERA**

Phyllopods without segmentation of the body. Carapace large and bivalve in type. The second pair of antennæ are very large, with two branches, and serve as swimming appendages.

Family **POLYPHEMIDAE**

Genus *Evadne*

E. nordmanni. (Fig. 256.) Water flea. The carapace of this species is blunted at the forward end and pointed behind. In the female it is somewhat more amply proportioned than in the male and serves as a brood chamber. The 4 legs and the slender abdomen are exposed, the carapace not enclosing these parts. A single spine occurs on the outer branch of the third pair of legs; and a pair of long caudal spines project from the tip of the abdomen. The head is very large and is provided with a single huge compound eye at the front, which in reality is the fusion of a pair of lateral eyes. Colorless. Length 1.15 mm or less. Very common in towings. All ranges.

Fig. 256—E. nordmanni (female).

GENUS *Podon*

P. leuckarti. (Fig. 257.) Except for its rounder carapace, this water-flea has a general resemblance to the preceding species. Its principal difference, however, is in the separation of the head and thorax by a well-marked dorsal depression. Another characteristic feature is the presence of 6 bristles on each branch of the second pair of antennæ. Length 1 mm. Common. Atlantic coast.

Fig. 257—P.
leuckarti.

ORDER **COPEPODA**

Copepods, in contrast to the phyllopods which exhibit a great diversity of form and structure among the various species, are rather uniform in general appearance. They are gracefully and symmetrically proportioned, and are often possessed of a coloration that makes them exceedingly attractive to the amateur collector. Nor is the collecting of them arduous. They swarm the littoral and offshore waters in countless millions. No matter how barren of other forms the tow-net may be, it is almost certain to capture a greater or lesser number of them. Particularly is this so if the towing is done after nightfall; for then the surface is usually literally crowded with these active little crustaceans which migrate upward from the depths. They form the principal food supply of many fishes and other larger marine animals; this, in spite of the fact that individually they are minute; for collectively they constitute a considerable bulk of the total planktonic life. The interest that the group has held for students is indicated by the fact that nearly 2000 marine forms have already been named and described; these compose about nine tenths of all known species. However, it should be added that almost half of these are parasitic forms.

The body is elongate, broad anteriorly, and narrows at the abdomen and toward the tail. Commonly it is distinctly segmented, being made up of fifteen somites, five each occurring in the head, the thorax and the abdomen. But, as the head and thorax always are combined to form a cephalothorax, and as the somites of the head are fused, and often also fused with the first thoracic segment, this latter region may have only four free somites. In the female the first and second abdominal segments are invariably fused; sometimes all of them are united. A single median eye is present in most species. The first pair of antennæ is large, unbranched and usually used for swimming, which act is accomplished in a rapid, jerky manner. On the thorax are present five pairs of branched swimming legs, each pair being connected by a transverse movable ridge, thus causing the opposed legs to work in unison. The last abdominal segment ends in a forked pair of projections, the *furca,* bearing caudal bristles.

The female is often larger than the male; and in certain species during the breeding season she carries a pair of egg sacs or brood pouches attached to the first segment of the abdomen. In both sexes the genital openings are in the first abdominal segment. The young are born as free-swimming nauplii.

Suborder EUCOPEPODA

Copepods contained in this order have elongate bodies. In the free forms the mouth parts are adapted for biting instead of for sucking as in the parasitic forms.

Family CALANIDAE

Genus *Calanus*

C. finmarchicus. (Fig. 258.) The thorax consists of 5 unfused segments. First antennæ as long as the body, and in the female composed of 25 joints, this number being somewhat altered in the male. The second pair of antennæ 2-branched; likewise the 5 pairs of legs. In this species the female bears a single egg pouch. Color yellowish or reddish or colorless. Length about 4 mm. Perhaps the most abundant pelagic species of copepods in the northern range. It oftentimes occurs in such numbers as to color the water a decided yellow or red. Besides serving as food for herring and mackerel, it is a most important article in the dietary of the Greenland whale. Often taken in the tow-net close to shore. New England coast, but widely pelagic.

Fig. 258—
C. finmarchicus (female).

Genus *Calocalanus*

C. pavo. (Plate IV.) This interesting form, which not commonly found close to shore, may not inappropriately be described and figured here as an example of the unique manner in which appendages are adapted to cope with an environment. For, as this is a pelagic free-swimming animal, there hardly can be any question that its large and decorative featherlike tail bristles assist in flotation. The form figured in the plate is a female; the thorax consists of 3 segments, the first somite being fused with the head, while the fourth is fused with the fifth. There are but 2 abdominal segments. The first antennæ are large and bear numerous long bristles, which, like those of the tail, are of use in increasing the copepod's frictional resistance to the water, thus enabling it to float easily. Color pale transparent carmine. Length about 1 mm. Occurs sometimes abundantly in the Gulf Stream, occasionally as far north as New England.

Family CENTROPAGIDAE

Genus *Temora*

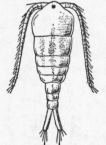

T. longicornis. (Fig. 259.) The thorax in this species has 4 segments. In the male there are 5 abdominal segments, while in the female there are but 3, the last thoracic segment being with backward-pointing lateral projections. The first antennæ are very long and consist of 24 joints, from which arise short bristles. Usually the right antenna in the male is prehensile, functioning as a copulatory aid; the fifth pair of legs, which are fitted for clasping, function likewise. The female bears a single egg sac. The second pair of antennæ and the 5 pairs of legs are 2-branched. Length 1.5 mm. Often abundant. Southern New England.

Fig. 259—T. longicornis (female).

Genus *Eurytemora*

E. hirundoides. (Fig. 260.) In both sexes of this species there are 5 thoracic segments; the male has 5 abdominal segments, however, while

---

the female has but 3, the last thoracic segment ending in lateral points. The first pair of antennæ are about half as long as the body, and each antenna consists of 24 bristle-bearing segments. The second pair of antennæ and the first 4 pairs of legs are 2-branched. The fifth pair of legs is unbranched. A single large egg sac is carried by the female. Color transparent with yellow bands. Length 1.16 mm. This species occurs in fresh as well as in brackish and salt water. Common. Massachusetts and Narragansett Bays.

Fig. 260—
E. hirundoides
(female).

FAMILY PONTELLIDAE

GENUS *Labidocera*

L. æstiva. (Fig. 261.) Thorax in both sexes with 4 segments; abdomen in male with 5, the last occasionally being unsymmetrical. Female with 2 abdominal segments. First pair of antennæ very long, and in female with 23 joints. Right antenna in male prehensile, and fifth pair of legs fitted for clasping. Second pair of antennæ 2-branched; first pair of legs poorly developed. Female with single egg sac. In addition to the single ventral median eye, there is in this species a dorsal pair. Body colorless. Length 2 mm. Southern New England coast.

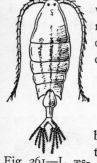

Fig. 261—L. æstiva (female).

ORDER **OSTRACODA**

These crustaceans possess a carapace resembling a bivalve mollusk shell, in which are completely enclosed their usually unsegmented bodies. Traces of segmentation are rare, the vast majority of species having the segments so united that no distinction between the somites is apparent. Seven pairs of appendages are usually present. Of these, the antennæ are protruded from within the carapace —often through a sinus in the lower front margin of the shell— when the animal swims or crawls. Ostracods are bottom dwellers, moving actively about in search of prey; they subsist chiefly on other and smaller living animals. They occur both in fresh and in salt water. There are about 1800 known species, most of which are marine.

Suborder **PODOCOPA**

Ostracods in which the shell is without an antennal sinus. Antennæ simple, with the ends bearing sharp or clawlike bristles. A heart is wanting.

Family **CYPRIDAE**

Genus *Pontocypris*

P. edwardsi. (Fig. 262.) The shell, when seen from the side, is roughly triangular, highest in the middle, and with rounded ends. The

Fig. 262—P. edwardsi (female).

surface is smooth but thickly covered with fine hairs. First antennæ with 7 joints, each joint provided with bristles, those on the terminal joint being the longest. Basal joint with 4 bristles. Second antennæ with 4 slender claws on last joint. Terminal claw of the first pair of feet very long and slender, a little less than the total length of the preceding 4 joints, and provided with pectinations toward the outer end. Length .85 mm; height .47 mm. Occurs in muddy, protected bottoms, usually in August. Southern New England coast.

Family **CYTHERIDAE**

Genus *Loxoconcha*

L. impressa. (Fig. 263.) This form may be recognized by the notch that appears on the anterior dorsal margin of the shell and by the structure of the antennæ. The first pair is 6-jointed and bears on the second joint a comb of fine bristles, while the bristles on the terminal joint are sparse and long. The second and leglike antennæ are 4-jointed, the third joint being much the longest and provided with a number of

Fig. 263—L. impressa (female).

stout bristles. The basal segment of each second antenna gives rise to a long, jointed, hairlike process containing a duct from a poison gland, the distal end of which is in close proximity to a pair of curved claws terminating the antenna. The shell of this species is broadest in the middle when viewed from the side. Both ends and the ventral margin are produced into a flattened rim. Length .82 mm; height .51 mm. This species occurs seemingly more common in the eelgrass and among hydroids of shallow water than it does on the actual bottom. Often abundant. Southern New England and southward.

Genus *Cythereis*

C. arenicola. (Fig. 264.) When seen from the side, the shell of this species is rounded at both ends with the dorsal and ventral margins nearly parallel. It is about twice as long as it is high; that is to say, its length is 1 mm and its height 0.5 mm. There is considerable carbonate of lime in the shell, making it rather thick and opaque. The surface has scattered irregularly shaped areas, each usually with a single hair. These areas are more transparent than the other parts of the shell; and in addition to them, there are a number of scattered papillæ which are pigmented, these being the most numerous near the dorsal margin and the ends. At both ends of the shell is a series of longer hairs. The first

Fig. 264—C. arenicola (male). A, shell; B, antenna; a, poison duct.

antenna is composed of 6 joints bearing 2 groups of hairs on the first joint and 3 groups on the second. On the second joint are also a short bristle near the upper end of the outer margin and some others at the lower end of the inner margin. The second antennæ are 4-jointed, and from the basal segment arises a long, jointed bristle bearing a duct from a poison gland. Length 1 mm; height .5 mm. Fairly common. Occurs on sandy bottoms. Southern New England.

Suborder **MYODOCOPA**

Members of this suborder are characterized by the conspicuous notch in the front margin of the bivalves for the protrusion of the antennæ. The first pair of antennæ are not adapted for swimming but usually function as sensory organs. The second pair of antennæ are 2-branched and fitted for swimming; one of the branches is rudimentary in the female, in the male it is modified for grasping.

FAMILY CYPRIDINIDAE

GENUS *Sarsialla*

S. zostericola. (Fig. 265.) In this species the shell of the female is without the notch, or antennal sinus, present in the carapace of the

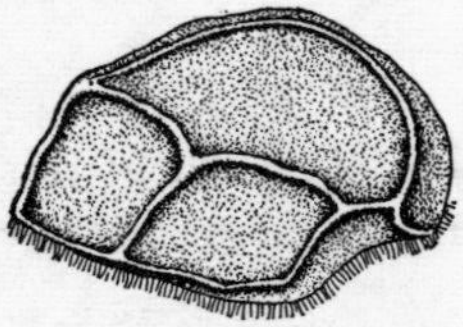

Fig. 265—S. zostericola (male).

male. The shell is marked into well-defined areas by several large, conspicuous and irregularly disposed ridges on the surface. A fringe of hairs extends along the lower margin of the shell. The first pair of antennæ are 5-jointed. In the male, a sense organ is present at the end of the third joint, and consists of a knoblike thickening at the base from which arises a thick cluster of long slender sense hairs. Second antennæ with outer branch 9-jointed; inner branch in male with 3 joints. Length 1.3 mm; breadth 8.6 mm. Found on eelgrass and hydroids. Southern New England.

GENUS *Cylindroleberis*

C. mariae. (Fig. 266.) Shell long and somewhat elliptical with a deep notch at the front end for the protrusion of the antennæ. First antennæ with 8 joints; in the male the antepenultimate joint bears a sense organ consisting of a stout tapering bristle furnished with numerous long slender sense hairs. Second antennæ with 9 joints on outer branch; inner branch is composed of three joints, ending in a recurved claw. Length about

Fig. 266—C. mariæ (male).

2 mm; width about 1 mm. Lives near the surface. Not uncommon. Southern New England.

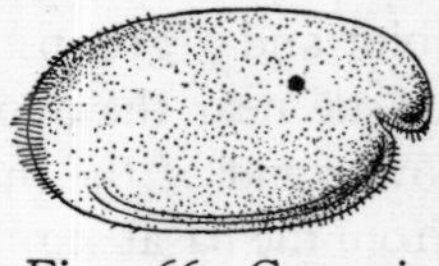

ORDER CIRRIPEDIA

The barnacles. These crustaceans have an imperfectly segmented body which is contained in a calcareous shell; and during their adult life they are either fixed or parasitic, no free forms being known. Barnacles, by reason of their hard shells, were once classed with the mollusks; but when their early life history was traced, the facts regarding their relationship to the crustaceans were established. They are born as nauplii, looking much like the nauplii of other crustaceans. In this stage, they leave the mantle cavity of the parent and swim at the surface for a greater or lesser period, at the end of which they pass through the *cypris* stage, a stage in which the individual acquires a bivalve shell, and attaches itself to some solid support on which it will pass the remainder of its life. In the cypris stage, it possesses three eyes: a single, median eye, and a pair of compound eyes. After making a permanent attachment, it undergoes a metamorphosis in which it loses its eyes and the bivalve shell, and is characterized by the typical hard calcareous shell of a barnacle.

Although cirripeds are often of considerable size, their affinity with the other and smaller entomostracans is evident when one examines the barnacle's body structure. (Fig. 267.) Then it will be seen that, like an ostracod, it is entirely enclosed within a carapace falling in right and left folds over the body, and to which it is attached by the back of the head and the thorax. The appendages consist of one pair of mandibles, two pairs of maxillæ, and six pairs of 2-branched legs. The latter appendages are fringed with fine hairs and are for the most part curled and plumelike; they are used to capture food. This the animal does by thrusting its feet through the aperture between the valves of its shell, and by a continuous extension and retraction causes currents carrying organic food particles and small animals to be swept in toward the mouth. Except for the region that is attached, the body lies usually on its back free in the shell, the cavity of the latter being lined by the mantle. The digestive tract is a simple straight tube extending from the mouth to the anus at the hinder end of the abdomen.

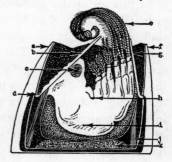

Fig. 267—Diagram of a barnacle. a, rostrum; b, scutum; c, adductor muscle (x-section); d, opercular membrane; e, cirri; f, carina; g, tergum; h, sexual orifice; i, stomach region; j, muscle (x-section); k, ovary.

Most barnacles are hermaphroditic, a condition very frequently prevailing among fixed forms of life. Like other crustaceans, barnacles molt. At each molt they shed the hard covering of the legs and body, and the lining of the shell, but not the shell itself. This increases in size as the enclosed animal grows, and its periodic expansions leave characteristic traces on the surface in the form of markings much like those on the shells of mollusks.

Cirripeds are extremely abundant and are found in all seas. They are among the most obvious and familiar animals of the seashore. Almost any solid object submerged in salt water is likely to be covered with them. They encrust rocks, wharf piles, and other structures below the high-water mark; they attach themselves to floating wood, the bottoms of ships, whales, jellyfishes, and even to one another. They are hardy creatures and easily kept in the aquarium, where, despite their sessile habit, their mode of living affords much of interest and instruction to the beginning naturalist.

Suborder LEPADOMORPHA

In this group are included the goose barnacles. These cirripeds are primarily distinguished by having the shell attached by a peduncle, or stalk. The stalk at the anterior end is provided with muscles and has considerable flexibility. The shell is composed of scalelike plates which are not solidly fused together; one pair of these is usually larger than the others and is known as the *terga,* while another and smaller pair is termed the *scuta,* both pairs corresponding to similarly named structures in the balanomorph barnacles. (Compare Figs. 267 and 268.) Also, in addition to these, there is a median dorsal plate called the *carina* lying along the hinge. The shell opens on the ventral side for the protrusion of the legs. A few species are without any shell; one, *Alepas pacifica,* is pelagic and attaches itself to a large jellyfish.

Family LEPADIDAE

Genus *Lepas*

L. fascicularis. (Plate XXVII.) The stalk in this species is shorter than the shell. The shell consists of 5 plates and these are very thin and brittle with a smooth surface. Scuta with lower part of tergo-carinal edge projecting strongly. Terga flat with apex bent toward carina. Carina bent at right angles, with upper part very narrow, deeply concave within, and appearing as if only central ridge were developed. The shell is translucent and colorless; its length is 4 cm. This form is very widely distributed in most seas, being found attached to various seaweeds and floating material. Common on the bottom of ships' hulls. All ranges.

L. anserifera. (Fig. 268.) In this species the stalk and the shell are nearly equal in length. The shell is composed of 5 plates, a family characteristic. Plates closely approximated and with variably furrowed surfaces, but with terga usually more plainly furrowed than scuta. Carina deeply concave within, and with fine longitudinal furrows without. Color of shell white, edged with bright orange membrane; color of stalk orange. Length of shell 5 cm. Common on floating objects, especially on the hulls of ships from tropical seas. All ranges.

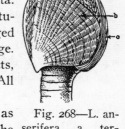

Fig. 268—L. anserifera. a, tergum; b, scutum; c, carina.

L. anatifera. (Fig. 269.) Stalk about as long as or longer than the shell. The 5 plates composing the shell are smooth with traces of fine lines radiating from the umbones, or protuberant processes. Scuta usually more plainly marked than other plates. Terga separated from scuta and carina by a very narrow interspace. Carina occasionally with knobs

or teeth along the ridge. Color of shell bluish-white with scarlet-orange at edges of aperture; interspaces between plates dull orange-brown; stalk purplish-brown. Length of shell about 5 cm. This is the commonest of the goose barnacles, and is almost world-wide in distribution. It is often abundant on ships' bottoms and other floating objects. All ranges.

Fig. 269—L. anatifera.

L. pectinata. (Fig. 270.) The stalk in this species is narrower and shorter than the shell. The latter is made up of 5 plates; these are thin and brittle and are marked with deep radiating furrows on the surface. Very frequently the ridges between the furrows bear a row of prominent curled flat calcareous spines. Owing to the thinness of the plates and the depth of the furrows, the edges of the first-named are wavy. Scuta usually with a very prominent ridge running along outside edge from umbo to apex. Terga with a conspicuous notch to receive apex of scuta. Carina broad, deeply concave within, often bearing stout barbs without. Color of shell bluish or pale purplish; stalk brownish. Length of shell about 1 cm. Sometimes abundant on various floating objects such as wood, cork, bottles, charcoal, seaweeds, bottoms of vessels, etc. All ranges.

Fig. 270—L. pectinata.

Genus *Conchoderma*

C. virgatum. (Fig. 271.) A so-called nude cirriped. This species has the stalk flattened, gradually widened toward the shell, and usually a little longer. The plates of the shell are thin, obscure, imbedded in the membrane, and widely separated. Thus, the animal acquires a flattened shape with a squarish top, the body blending into the stalk often without a distinct demarcation. The scuta are trilobed, while the form of the terga is variable; their position is at right angle to the scuta. Carina nearly parallel to scuta, and concave within. Color gray tinged with blue, and with 6 black longitudinal bands tinged purplish-brown. Length of entire animal 5 cm. Distributed world-wide in the sea; common and often abundant on the bottoms of ships and other floating objects; not infrequently found also attached to large turtles and large fishes. All ranges.

Fig. 271—C. virgatum.

Suborder **BALANOMORPHA**

This group includes the sessile barnacles known as acorn barnacles. These forms are without a stalk. The shell is thickly calcareous and composed of immovably united plates forming a cylindrical structure which is attached at the base to some object and has the free end provided with two pairs of hinged plates, the scuta and the terga. These are furnished with depressor muscles and are homologous with the scuta and terga of lepadomorph barnacles. (Compare Figs. 267 and 268.)

Family BALANIDAE

Genus *Balanus*

B. balanoides. (Plate XXVII.) The shell of this species is extremely variable in shape; usually when isolated it is cylindrical, but sometimes it is much depressed, and, again, it may be conical or it may be greatly elongated, even club-shaped, as the result of close crowding. This barnacle can usually be distinguished from similar forms by the base of the shell which in this species is membranous. There are 6 thick longitudinally grooved plates composing the main structure of the shell, and these are connected by thinner and narrower ones without grooves. The aperture is generally diamond-shaped with a rather deeply toothed margin. The scuta are faintly striated, and the upper ends of the plates are somewhat reflexed, thus allowing the tips to project freely as small flattened points. A conspicuous notch in the margin of the shell often indicates the position of the terga, the latter being a pair of small, blunt plates, each provided with a spur, and usually more or less obscured in the living animal. Color whitish or pale brown. Basal diameter up to 5 cm. This is the commonest and most abundant species occurring in the tidal zone of the North Atlantic coast. It attaches itself to rocks, pilings and other solid objects, often being encrusted in such enormous numbers and so crowded together that the surfaces of these objects are entirely hidden. North Atlantic coast.

B. eburneus. (Fig. 272.) The ivory barnacle. Shell rather low and broad, somewhat obliquely conical; and with a very smooth surface. The 6 heavy plates are covered by a film of epidermis, but the thin connecting plates are naked. The base of this species is calcareous. There is a large aperture, rhomboidal to pentagonal in shape, and moderately toothed around the margin. Longitudinal striations occur on the outer upper surface of the scuta, while the inner upper surface is roughened. The terga each have a spur, the lower end of which is abruptly

Fig. 272—B. eburneus.

truncated. Both the scuta and the terga often project slightly above the margin of the aperture in the form of a pyramid. Color ivory white. Basal diameter about 2 cm. This form is not infrequently found in brackish or even fresh water, though it lives chiefly below the low-tide level. It is a common form, often being abundantly associated with other species of barnacles on the bottoms of ships and attached to other solid objects. Massachusetts Bay to West Indies.

B. crenatus. (Fig. 273.) Shell usually conical, although cylindrical and club-shaped forms are not uncommon. The 6 thick plates composing the shell may be rugged and irregularly folded longitudinally or extremely smooth, and their ends form toothlike projections around the margin of the aperture. The base of the shell is calcareous and very thin. Scuta marked with faint lines and have the surface generally covered with membranous material. Upper ends of scuta often somewhat reflexed,

Fig. 273—B. crenatus.   Fig. 274—B. balanus.   Fig. 275—B. tintinnabulum.

so that tips project freely. Terga small and with short spur having the lower end rounded or somewhat blunt. Color whitish. Basal diameter about 1.5 cm. Very common. Occurs on shells and stones below the low-tide level, and on the bottoms of ships. Widely distributed. All ranges.

B. balanus. (Fig. 274.) Shell conical, somewhat convex, and composed of 6 plates each of which usually has from 2 to 4 distinct longitudinal ribs. Interspaces wide and with smooth surface. Base calcareous. Aperture somewhat ovate. Scuta with prominent ridges crossed by fine striæ that crenulate the ridges. Terga project well above the scuta, and are marked externally with relatively coarse growth lines. Color white or yellowish. Basal diameter up to 3 cm. This species usually attaches itself to pecten, or scollop, shells, probably because these offer the most abundant and solid of supports available in the localities it inhabits. Numerous and common in offshore waters. New England.

B. tintinnabulum. (Fig. 275.) In this species the shell is somewhat conical and is often with fine longitudinal ribs on the 6 plates of which it is composed. The interspaces are indistinctly striated transversely. Base calcareous. The aperture is generally diamond-shaped. Both scuta and terga are marked with distinct transverse ridges, or growth lines. Color reddish or bluish. Basal diameter up to 6 cm. This species is perhaps the most widely known of the sessile barnacles, being recorded for three or four hundred years as a constant visitor to nearly every deep-sea traffic port in the world by virtue of its attachment to the bottoms of vessels. It is common in the collections of individuals and museums, and has been described and figured frequently for the past two centuries. Although its rightful home is in the warmer seas, its occurrence on ships' bottoms makes it available to collectors in all latitudes; thus, it properly may be listed as being found in all ranges.

SUBCLASS **MALACOSTRACA**

These are crustaceans that are usually of considerable size and they are more highly organized than those contained in the subclass *Entomostraca*. The marine members of the group include the familiar sand fleas, crabs and lobsters. With the exception of one division, *Phyllocarida*, the number of segments composing the bodies is constant, 5 forming the head, 8 the thorax, and 7 the abdomen. In all the orders represented on our coast, the head is fused with one or more of the thoracic segments to form a cephalothorax, which in certain orders is covered by a carapace. The last abdominal segment is known as the *telson*. Typically, there are 5 cephalic, 8 thoracic, and 6 abdominal appendages. The thoracic appendages are more highly differentiated than in the *Entomostraca*, in most instances being modified to manipulate food. These accessory mouth parts are known as *maxillipeds*.

DIVISION **PHYLLOCARDIA**

This division comprises certain small primitive crustaceans that have some of the characters of entomostracans and seem to form a connecting link between them and the malacostracans. The malacostracan number of somites are contained in the head and thorax, while the abdomen is composed of 8 segments. The head and thorax are enclosed by a large bivalved carapace and are with leaflike feet. The abdomen is long and partly covered by the carapace, and it bears 6 pairs of 2-branched appendages. A pair of caudal bristles are present on the last segment. Eyes stalked.

Genus *Nebalia*

N. bipes. (Fig. 276.) The body of this form is slender and compressed laterally. Connected by a movable hinge with the front dorsal end of the carapace and partly covering the stalked eyes is a small "rostral plate," or horny beaklike projection. The tail is forked, and in addition to the pair of caudal bristles is provided with lateral spines. On the thorax there are 8 pairs of leaflike feet functioning as gills. These

Fig. 276—N. bipes.

are all alike and resemble those of branchiopod crustaceans. The first four pairs of feet on the abdomen are used for swimming, but the last two pairs are functionless, being reduced to small vestiges. In the male the genital opening is on the last thoracic somite; in the female it is on the last but two. Length 10 mm. This creature is interesting particularly from the paleontological point of view, since the evolution of the entire subclass *Malacostraca* appears to have had its beginning in this type of animal. For the fossils apparently belonging to phyllocaridans are the earliest found of the group, these forerunners of the higher *Malacostraca* being abundant in the Cambrian, Ordovician and Silurian; while it is probable that most of the existing orders date their origin at a period not preceding the Carboniferous era. Often abundant in shallow water and among seaweeds. All ranges.

Division **ARTHROSTRACA**

This division comprises the amphipod and isopod crustaceans. Both types are characterized by the absence of a carapace, by having the head fused by never more than one thoracic segment, by the presence of usually only seven free thoracic segments, a single pair of maxillipeds, and in most cases a pair of sessile, or unstalked, eyes. Other appendages consist of two pairs of antennæ, one pair of mandibles, two pairs of maxillæ, seven pairs of *periopods* (thoracic legs modified for walking, taking of food or defense), and six pairs of *pleopods* (abdominal appendages), also called swimmerets. The antennæ consist of two parts, a basal peduncle composed of a few larger joints, and a flagellum made up of numerous smaller joints. The periopods often are provided with claws of *subchelate* structure, in which the terminal joint closes upon the next one like a jack-knife blade in its handle. (Fig. 283.) The pleopods, besides being locomotor appendages, are also sometimes respiratory in function and may be used for the attachment of the eggs or young. The posterior three pairs of pleopods are sometimes called *uropods*. The young animals resemble the parents, there being no free metamorphosis.

ORDER **AMPHIPODA**

The amphipods may be described as arthrostracans in which the body is elongated and generally strongly compressed from side to side. The abdomen consists typically of six segments and a telson. The first two pairs of periopods are called *gnathopods;* these are generally larger than the other thoracic legs, and are used for seizing and holding food. Gills are present on the inner side of the proximal joints of the thoracic legs. The hinder three pairs of periopods are usually adapted for jumping. The heart lies in the anterior part of the body in a position corresponding to that of the gills.

Suborder **GAMMARIDEA**

Head not fused with first segment of thorax. Maxillipeds with palps having two to four joints.

Family **ORCHESTIIDAE**

Genus *Orchestia*

O. agilis. (Fig. 277.) The common beach flea. First gnathopods subchelate in both sexes. In the male, second pair of gnathopods larger than the first, oval and with notch at hinder end of the palm. First antennæ very short, not reaching to penultimate joint of peduncle of second antennæ. Flagellum of second antennæ shorter than the peduncle and composed of 10 to 15 short segments. Penultimate joint in hinder pair of periopods swollen in male. Color variable but commonly brownish. Length 14 mm. Extremely abundant under débris, seawrack, seaweeds, eelgrass, etc., at the tide marks. When this material is turned over, these

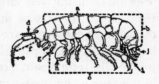

Fig. 277—O. agilis. a, thorax; b, abdomen; c, periopods; d, 1st antenna; e, flagellum of 2d antenna; f, peduncle of antenna; g, 1st gnathopod; h, 2d gnathopod, with subchelate claw; i, periopods; j, uropods.

animals disperse by leaps in all directions to find a hiding place under stones and other objects. As their scientific name indicates, they are very active, and therefore are not easily caught individually. They are easily found, occurring in the situations mentioned almost everywhere along the shores of bays. All ranges.

O. palustris. (Fig. 278.) In this species, the gnathopods are the same as those of *O. agilis,* except that in the male the palm of the second pair is without a notch. The first pair of antennæ extend beyond the tip of the penultimate joint of the peduncle of the second pair. The flagella of the antennæ are slender and longer than the peduncle, and consist of from 18 to 26 joints. Color olive-green to olive-brown or reddish-brown. Length 18 mm. This species is also characterized by its habitat which is different from that of the preceding, being generally found in salt marsh regions often a considerable distance from the shore, sometimes among grass and reeds and in other situations that are comparatively dry. This species is a more active crawler than *O. agilis,* but cannot hop so agilely. Often very abundant. Cape Cod to New Jersey.

Fig. 278—O. palustris (male).

Genus *Talorchestia*

T. longicornis. (Plate XXVII.) First pair of gnathopods subchelate in the male, but not in the female. In the male, the second pair of gnathopods are comparatively large, subchelate and bearing a conspicuous lobe fringed with short setæ on the claw edge of the palm. In the female, the second gnathopods are weak and relatively small, but with a broad basal joint. The first pair of antennæ reaches to the tip of the penultimate joint of the second pair. Second antennæ about one third the body length in the female, but may equal the length of the body in the male. Eyes comparatively large. Color grayish or sand-colored. Length up to 25 mm. Very abundant on sandy beaches near the high-water mark where it burrows beneath the surface to reach moist sand. It is nocturnal in its habits, remaining inactive in its burrow during the day, its presence being indicated only by the small holes it makes in the sand. It feeds at night on the seaweeds and other organic material cast up on the shore;

and as it is strongly attracted by a light, it can be collected in large numbers merely by placing an electric flash lamp in the center of a cloth spread in situations where it is known to live. Cape Cod to New Jersey.

Family **PONTOPOREIIDAE**

Genus *Haustorius*

H. arenarius. (Fig. 279.) The characters of this species are such that it cannot readily be mistaken for any other on our coast. Its shape is peculiar, having a general resemblance to the sand bug, *Hippa,* and like that crustacean is a rapid burrower, perhaps the most rapid along the shore. The first antennæ are slightly shorter than the second pair and have a well-developed accessory flagellum. The second an-

Fig. 279—H. arenarius.

tennæ have the fourth joint of the peduncle broadly expanded and provided with long plumose hairs. Mandibles with rather large palp. Gnathopods comparatively feeble. Posterior periopods very broad and without dactyls, or pointed terminal joints, being modified for digging. Color grayish-yellow, resembling its sandy habitat. Found near the high-water mark. Often numerous. Cape Cod to Cape Hatteras and southward.

Family **AMPELISCIDAE**

Genus *Ampelisca*

A. macrocephala. (Fig. 280.) The head in this species is as long as the first three somites of the thorax, and at its anterior end are 4 small

Fig. 280—A. macrocephala.

red eyes. In the female the first pair of antennæ reach to the peduncle of the second pair ; flagellum of first antennæ twice as long as peduncle. In the male this flagellum is much longer. Second antennæ of female trifle longer than half the body length. Gnathopods simple in structure. Third abdominal segment with an acute upturned projection on the posterior edge at each side. Color whitish. Length 15 mm. This is a mud-burrowing species, inhabiting smoothly lined tubes which it makes, and in the bottom of which it can usually be found. Vineyard Sound and northward.

Family **CALLIOPIIDAE**

Genus *Calliopius*

C. laeviusculus. (Fig. 281.) In the present species, the two pairs of antennæ are almost equal in length. The first pair is without accessory

flagella, but a small lappetlike process is borne on the last pedunclar joint. In both pairs of antennæ, the articulations of the flagella are so strongly marked that the latter has a sort of saw-edged appearance. Eyes large. Gnatho-

Fig. 281—C. laeviusculus.

pods large, equal in size and of subchelate structure. Terminal pleopods slightly larger than those preceding, and are with forked ends fringed with hairs and spines. Color light green. Length about 15 mm. Found in tide pools and among shore algæ. Often numerous. Greenland to Cape Hatteras.

Family **GAMMARIDAE**

Genus *Gammarus*

G. locusta. (Fig. 282.) The scud. This species has the two pairs of antennæ long and slender with the first pair longer than the second pair

and bearing an accessory flagellum which occurs as a small side branch having about 8 joints. On the last 3 abdominal segments there are median and lateral tufts of small hairs. Telson small and somewhat deeply cleft. Gnathopods comparatively small, subchelate and equal in size, those of the male being generally stronger. Mandibles

Fig. 282—G. locusta.

are with palps. Three posterior pairs of periopods large and strong and gradually increasing in size. Terminal pleopods generally extending beyond the others, and with forked, leaflike tips provided with a fringe of hairs and spines. Color brownish or greenish. Length about 20 mm. Found under stones that are covered at high tide and under sea wrack. This is one of the commonest and most abundant of the amphipods. Arctic Ocean to New Jersey.

G. annulatus. (Fig. 283.) Similar to *G. locusta* but with first antennæ shorter than the second. Gnathopods with rather narrow and uneven palms armed with short heavy spines along the lower edge and bearing a fringe of hairs. Terminal pleopods elongated and furnished with a fringe of hairs and spines. Fifth and sixth abdominal segments with both median and lateral tufts of hairs; third segment with only a median tuft. Telson cleft and provided with short spines along the lateral edge and at end. Color lighter than that of the preceding species, but usually has

Fig. 283—G. annulatus. A, 1st gnathopod; B, 2d gnathopod; C, uropod; D, telson.

the abdomen marked with dark bands having red spots. Length 15 mm. Found under stones and seaweeds near the high-water mark. Often very abundant. Bay of Fundy to Long Island Sound.

Genus *Carinogammarus*

C. mucronatus. (Fig. 284.) This form has many of the characters of *G. locusta* and others of that genus. It differs, though, in having the first three abdominal segments produced dorsally into acute backward-projecting spines. Also the two pairs of antennæ are of the same length. However, it is in the character of the dorsally produced spines that this form can be distinguished not only from *Gammarus,* but also from all other amphipods commonly occurring on our coast. Color greenish with minute

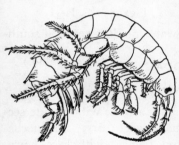

Fig. 284—C. mucronatus.

black or brown spots. Length 15 mm. Found in brackish tide pools, salt marshes and on mud flats at low water. Cape Cod to Florida.

Genus *Melita*

M. nitida. (Fig. 285.) The principal distinguishing characters are the terminal pleopods which have the outer branch elongate and the inner branch considerably reduced in size. First antennæ longer than second and two thirds the length of body. Eyes not kidney-shaped, as in *Gammarus,* but round and black. First gnathopod in male small, ending with minute dactyl, or simple claw; second gnathopod large and subchelate. Color dark greenish-slate. Length about 9 mm. Found occasionally in the tidal zone under rocks and sea wrack. Cape Cod southward.

Fig. 285—M. nitida.

Genus *Elasmopus*

E. laevis. (Fig. 286.) First antennæ two thirds as long as body, peduncle and flagellum the same length, and with accessory flagellum. Second antennæ one half as long as first pair. Last three segments of abdomen bent underneath. First gnathopods small; second gnathopods considerably larger; both pairs subchelate. Last 3 pairs of pleopods broad and flat and provided with tufts of hairs. Color brownish. Length 10 mm. Found frequently under rocks and among seaweeds near low-water mark. Cape Cod southward.

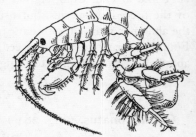

Fig. 286—E. laevis.

Family **PHOTODAE**

Genus *Leptocheirus*

L. pinguis. (Fig. 287.) Body thick. First antennæ with minute secondary flagellum. Both pairs of gnathopods large; first pair subchelate and fringed with long hairs; second pair narrower and longer than first, not chelate, and with basal joint having the anterior margin densely fringed with very long slender plumelike hairs. Terminal pleopods 2-branched and bearing bristles. Color variegated with dark bands and spots. Length 13 mm. Very abundant on muddy bottoms in shallow and offshore waters. Labrador to New Jersey.

Fig. 287—L. pinguis.

Family **AMPHITHOIDAE**

Genus *Amphithoe*

A. valida. (Fig. 288.) In this species the antennæ are all about the same length, and are less than half as long as the body. The first pair is without a secondary flagellum. In the second pair, the flagellum is longer than the last peduncular joint. Gnathopods subchelate, the second pair being larger than the first. Terminal pleopods end in hooks,

Fig. 288—A. valida.

the latter feature being a family characteristic. Color green with numerous starlike spots. Length about 13 mm. Lives under stones and among seaweeds in shallow water. Long Island Sound and New Jersey.

Family **COROPHIIDAE**

Genus *Corphium*

C. cylindricum. (Fig. 289.) A form in which the body is depressed. It may be easily recognized by the enormous development of the second

Fig. 289—C. cylindricum.

pair of antennæ in the males and by its occurrence in the soft tubes it makes in sponges, seaweeds, etc. It is usually common in the red sponge, *Microciona prolifera,* and can readily be collected by tearing apart such fragments as have been cast up on the beach. The gnathopods in both sexes are relatively weak. Mandibles with 2-jointed palp. Color light with gray mottling. Length 5 mm. Common all along the shore. Maine to New Jersey.

Suborder **CAPRELLIDEA**

The skeleton shrimps. An aberrant group in which the body is elongate and usually cylindrical, and with a very rudimentary abdomen. The thorax has but six free segments.

Family CAPRELLIDAE

Genus *Caprella*

C. geometrica. (Fig. 290.) This odd form has the head fused with the first thoracic segment, but the line of fusion is marked with a suture;

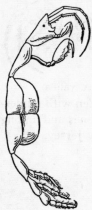

from the head projects a short horizontal spine. The first pair of antennæ are longer than the second pair and are generally bare except for minute hairs on the flagellum; the second pair are conspicuously fringed with hairs of moderate length. The gnathopods are subchelate, the first pair being much the smaller. The next two pairs of periopods are wanting, but the segments on which they normally would be present bear small leaflike gills in the male, while in the female there is a brood pouch. On the last three thoracic segments are borne the walking legs. In the male a pair of rudimentary legs occur on the abdomen, which is reduced to merely a tiny knob. Color vari-

Fig. 290—C. geometrica (female).

able, often resembling the objects to which they are attached; colorless forms not uncommon. Length up to 15 mm. This and other members of the family are generally found clinging to hydroids, starfishes, seaweeds, etc., which they do by the aid of their hinder legs, extending their bodies out in a rigid and motionless posture awaiting their chance to seize smaller prey or grasp particles of floating organic food material with their chelate gnathopods. Very common. Cape Cod to Virginia.

Genus *Aeginella*

A. longicornis. (Plate IV.) A species very much like the preceding, but attaining to a larger size and bearing antennæ of relatively greater length. The first antcnnæ are twice as long as the second pair; and the mandibles are with a palp. The form may be further identified by the characters of thc second pairs of gnathopods, which characters consist essentially of an elongate palm and the presence of two conspicuous teeth on the lower edge. The surface of the body is genrally smooth; but individuals covered with short, slender spines are not uncommon.

Color variable, generally resembling environmental objects. Length up to 25 mm. Lives among hydroids and algæ. Labrador to New Jersey.

ORDER **ISOPODA**

The isopods may be described as arthrostracans that have the body flattened dorsoventrally instead of laterally. The abdominal segments are smaller than those composing the thorax, and they are often more or less fused. Usually the under sides of first five segments of the abdomen bear delicate leaflike appendages, the pleopods, which function as respiratory organs; that is to say, they function as gills or as lungs, according to whether they are present on species that live in the water or on the land. The last pair of appendages are modified into uropods having a distinct expodite (outer branch) and endopodite (inner branch). The isopods are probably the most widely distributed of any group of crustaceans. They are found in fresh water as well as in the sea; and the most familiar of them are strictly terrestrial. These latter are the common pill bugs and sow bugs found in damp earthen cellars and under logs and bark. The marine forms are found under rocks and seawrack or burrowing in the sand and timber. Some forms are free, living in the open sea or in the shallow waters along the shore. Others cling to floating seaweeds, eelgrass or the hydroids attached to piles. Many are parasitic.

Family **TANAIDAE**

Genus *Leptochela*

L. savignyi. (Fig. 291.) Body elongate and somewhat semi-cylindrical. In the species the male is provided with large chelæ on the first pair of gnathopods, while those of the female are small; also, the first antennæ of the male are much longer than those of the female. There are 6 free thoracic segments. The uropods are 2-branched, the outer branch being composed of 1 joint, the inner branch of 6 joints. There are 5 pairs of well-developed pleopods. Color white. Length 2 mm. Occurs among seaweed, eelgrass and on piles. Cape Cod to New Jersey.

Fig. 291— L. savignyi. A, male; B, female.

Family ANTHURIDAE

Genus *Cyathura*

C. carinata. (Fig. 292.) The elongate and somewhat cylindrical body has 7 free thoracic segments. The first 5 segments of the relatively short abdomen are fused so as to resemble an eighth thoracic segment. The 2 pairs of antennæ are nearly alike, being short and stout and with a rather small flagellum. The uropods are laterally situated, are large and leaflike, and have the outer branch arching over the telson. Color brownish, mottled with yellow. Length about 18 mm. Found among seaweeds and in sandy and muddy bottoms of shallow water. Greenland to New Jersey.

Fig. 292—
C. carinata.

Family LIMNORIIDAE

Genus *Limnoria*

L. lignorum. (Fig. 293.) The gribble. Body somewhat broad and flattened with parallel sides. Thorax with 7 free segments, the first segment being longer than the others. Abdomen composed of 6 distinct segments, the first 5 being short, while the sixth, or terminal, segment has the posterior end widely rounded. Antennæ short. Uropods laterally placed with small, rudimentary outer branch and with inner branch reaching to tip of abdomen. All the legs are of simple structure and used for crawling. Color light gray. Length about 4 mm. This species is destructive to wood and submerged timber in which it makes burrows about 12 mm in length. It is very common and is widely distributed. All ranges.

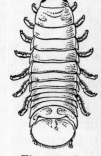

Fig. 293—L.
lignorum.

Family SPHAEROMIDAE

Genus *Sphaeroma*

S. quadridentatum. (Fig. 294.) Body broad and ovate. Abdomen with 2 segments; telson rounded and with leaflike uropods extending beyond end of body. Antennæ inserted on under side of head; second pair longer than first pair. Legs ambulatory and provided with a stout dactyl at the terminal

joint. This isopod, like the land pill bugs, has the
ability to roll itself up into a round ball when disturbed.
Color extremely variable; some forms are slaty-gray
with dorsal patches of creamy white or pinkish mar-
gined with black. Length 8 mm; width 4 mm. Lives
abundantly under stones or on seaweeds in the tidal
zone. Cape Cod to Florida.

Fig. 294—
S. quadriden-
tatum.

FAMILY **IDOTHEIDAE**

GENUS *Idothea*

I. baltica. (Fig. 295-A.) This species, often bearing the names *I. marina* or *I. irrorata,* has an elongate, somewhat ovate body with the sides
nearly parallel. The first antennæ are short and are composed of 4 joints;
the second pair are much longer with a flagellum longer than the peduncle.

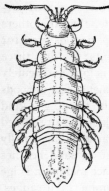

There are 3 complete and 1 partial abdominal segments, the last, or telson, being long and terminating with 3 toothlike projections. Uropods laterally placed, forming shields for the other pleopods
under abdomen. All the legs are alike in structure.
Color very variable; often green with dark brown
stipples. Usually the females are darker than the
males, the latter often having a median dorsal
stripe. Length of male up to about 35 mm. Length
of female seldom exceeding 20 mm. Occurs on
seaweeds and among rocks at low water. It
also is found on the surface, on floating seaweeds
and occasionally living free far from shore.
Very common. Nova Scotia to North Carolina.

Fig. 295-A—I.
baltica.

I. phosphorea. (Fig. 295-B.) A species similar in many respects to
I. baltica; however, it may be distinguished at once from that form by
the telson, which tapers to a point. Another distinguishing character can
be found in the configuration of the thorax, which has the segments
marked by deeper lateral indentations. Color generally darker than in

Fig. 295-B—I. phosphorea.   Fig. 296—I. metallica.

I. baltica. Length about 25 mm. Found usually in the same situations as the preceding species but more rarely, as its range is more restricted, being a more northern form. New England.

I. metallica. (Fig. 296.) Similar to *I. baltica,* but differing principally in the shape of the terminal abdominal segment which is truncated at the tip. In this form there is also a decided difference in the outlines of the thorax, the segments projecting laterally to form a conspicuous serrate margin. Color bluish or greenish, often with metallic iridescence. Length 18 mm. Pelagic. Found free or creeping over floating seaweeds. Not uncommon. All ranges.

Genus *Chiridotea*

C. caeca. (Plate XXVII.) Thorax ovate, broad and flattened; abdomen tapering to a sharp, pointed tip. Eyes dorsally placed, small and inconspicuous. Antennæ nearly equal in length. First three pairs of periopods subchelate, the first pair being somewhat shorter than the others but having the palm more robust. Fourth to seventh periopods used for walking. In this species there are 4 abdominal segments; the first three are rather short, but the fifth, which forms the telson, is long and acutely pointed and has lateral sutures indicating a fused segment. The uropods are laterally situated and arch over the other pleopods, forming a kind of cover on the under side. These shieldlike appendages are provided with hairs along the hinder outside margin. Color variable but usually dark grayish. Length about 15 mm. Often common on sand beaches in the tidal zone where it burrows just below the surface making a ridge like a miniature mole-hill. It is, however, a good swimmer. All ranges.

Genus *Edotea*

E. triloba. (Fig. 297.) Thorax elliptical with lateral margins of segments smooth and rounded and with 2 depressed laterally-running lines extending from the hinder angles of the head across all the segments to the telson. Abdomen consisting of a single segment, the telson, which is long and pointed posteriorly. A lateral indentation occurs on each forward margin of the abdomen, marking a fused segment. First pair of antennæ longer than second pair, but both pairs relatively short; first pair with 4 joints; second pair with peduncle of 5 joints and minute flagellum consisting of 1 article. All the legs are fitted for grasping. Color dull neutral uniform tone, often obscured by adhering dirt particles. Length 7 mm. Lives in mud, under stones, on piles, among eelgrass and decaying algæ in sheltered shallow waters very close to the shore. Often numerous. The lateral depressions of the thoracic segments, which give this crustacean the trilobed appearance whence its name is derived, should make it easy of identification; a further aid to this end is the

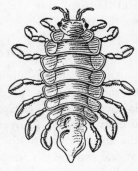

Fig. 297—E. triloba.

presence of the rudimentary flagellum of the second antennæ. Maine to New Jersey.

E. filiformis. (Fig. 298.) Abdomen shield-shaped and consisting of but a single segment. Second antennæ twice as long as first pair, and composed of 6 joints. All legs ambulatory. Color neutral, reddish or brown. Length 8 mm. Lives in sand and seaweeds near the shore. Common. Cape Cod northward.

Fig. 298—E. filiformis.

FAMILY **JANIRIDAE**

GENUS *Jaera*

J. marina. (Fig. 299.) Body flattened and oval, being a trifle more than twice as long as it is broad. First pair of antennæ very short; second pair longer than half the body length. All the legs are similar and are used for creeping. In the female the first pair of pleopods form a conspicuous opercular plate; the corresponding plate in the male bears the laterally joined copulatory organs. The uropods in this species are small and project from a notch at the hinder end of the semicircular telson. Color very variable; slaty-gray forms mottled with yellow or reddish-brown are common; also forms in which the color is of various

Fig. 299— J. marina.

shades of green; even whitish individuals having conspicuous transverse bands are not infrequent. Length 5 mm. Very common in tide pools and the tidal zone clinging to rocks and seaweeds. New England and southward.

DIVISION **THORACOSTRACA**

Animals in this division of the *Malacostraca* have three or more thoracic segments fused with the head forming a cephalothorax which is covered with a carapace. In most species a rostrum is present. This usually is in the form of a spikelike projection pointing forward from the anterior end of the carapace. With the exception of the members of one order, the *Cumacea,* the crustaceans comprised in the *Thoracostraca* have stalked eyes. The exopodite of the second antennæ in this group is known as the "scale." In nearly all species, the young pass through a metamorphosis before acquiring an adultlike form.

ORDER **SCHIZOPODA**

The opossum shrimps. Malacostracans in which the body is elongate and in which the thin carapace does not entirely cover the thorax. The thoracic feet 2-branched and usually bear gills. The foremost pair of legs are modified to a slight extent to form maxillipeds. About 300 species are known, most of which are marine.

FAMILY **MYSIDAE**

GENUS *Mysis*

M. stenolepis. (Fig. 300.) The somewhat cylindrical, translucent body of this species has the carapace produced into a short, blunt rostrum, while the lower anterior margin forms a sharp toothlike projection. Antennal scale longer than carapace. A bend in the body occurs between the first and second segments of the abdomen. Maxillipeds shorter than the following 6 tho-

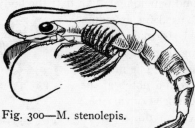

Fig. 300—M. stenolepis.

PLATE XXVII

1—*Lepas fascicularis.*   2—*Balanus balanoides.*   3—*Talorchestia longicornis*; 4x.
4—*Chiridotea coeca*; 4½x.   5—*Palaemonetes vulgaris*; 2x.
6—*Hippa talpoida*; 1¼x.   7—*Pagurus pollicaris.*   8—*Libnia emarginata.*

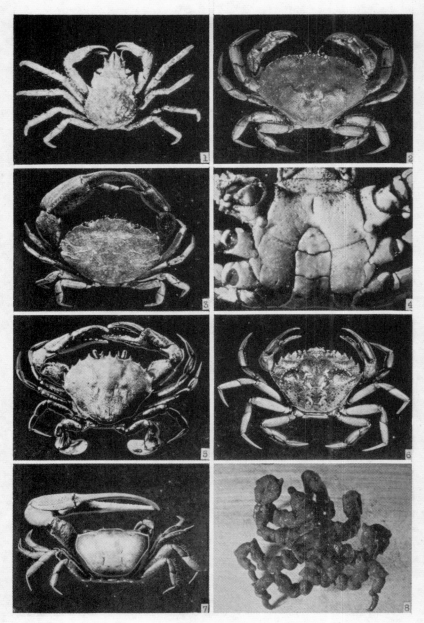

PLATE XXVIII

racic appendages. Gills are wanting in this species. In the male the
fourth pair of abdominal appendages are modified into long copulatory
stylets; in the female, marsupial plates within which the young develop
are present under the thorax. Color white, marked with somewhat stellate
black pigment spots. Length: male 23 mm, female 30 mm. Common. Occurs along quiet shores, especially in eelgrass, often in swarms. New
England and southward.

ORDER **STOMATOPODA**

The mantis shrimps. In the members of this order the carapace is
small and does not cover the posterior segments of the cephalothorax.

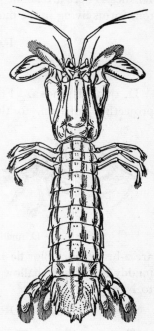

The abdomen is long and broad and proportionately very large. Gills are borne
on the abdominal appendages. The thoracic appendages consist of five pairs of
maxillipeds, the second pair being much
larger than the others and provided with
powerful subchelate claws, and three
pairs of periopods. About 90 species have
been recorded, all of which are marine.

Genus *Squilla*

S. empusa. (Fig. 301.) The carapace
in this species does not cover the last five
thoracic segments. The first thoracic segment is very small, the second segment
bears a spur on each side, while the three
last segments bear the periopods. First
antennæ with 3 flagella; second antennæ
shorter and with a comparatively large
flat scale. The second and large subchelate
maxilliped has a serrated claw, or dac-

Fig. 301—S. empusa.

tylus, which folds back on the palm, or propodus, like a knife-blade. The
tail fin, or telson, is rugged and provided with spines; these latter can be
used with telling effect if the creature is incautiously handled. Color
variable; usually greenish-gray. Length up to 25 cm. Often very common in the soft mud of shallow water and in the tidal zone, where it

inhabits burrows of its own construction, the burrows generally having 2 or 3 openings a few feet apart. Cape Cod to Florida.

ORDER **CUMACEA**

Small malacostracans in which the cephalothorax is large and the abdomen long and slender. The carapace is small, never covering the last four or five thoracic segments. Some forms are without eyes; when these are present they are unstalked and set close together or occur singly. Only a single pair of gills are present, these being borne on the first pair of maxillipeds. Except for uropods, pleopods are wanting in the females. There are about 300 known species, all marine. They are bottom forms living in the mud or sand.

FAMILY **DIASTYLIDAE**

GENUS *Diastylis*

D. quadrispinosa. (Fig. 302.) Carapace with triangular forward-projecting rostrum, to the rear of which, on each side, extends a

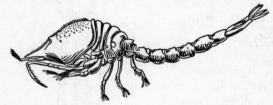

short spine. There are 7 abdominal segments with a long, pointed, well-developed telson. In the female the 3 anterior pairs of periopods are 2-branched; in the male the 5 anterior pairs

Fig. 302—D. quadrispinosa.

are 2-branched. Color flesh or brownish. Length 10 mm. Found on soft muddy bottoms in shallow and deep water. Not uncommon. Nova Scotia to New Jersey.

ORDER **DECAPODA**

These are the ten-footed crustaceans and they are the most highly organized of the group; they include such forms as the shrimps, crayfishes, lobsters, and crabs. The head and thorax are fused into a cephalothorax that is entirely covered by a carapace. There are five pairs of periopods, or thoracic legs, of which at least the first is very commonly provided with strong chelæ, or pinching claws, the remaining pairs being fitted for locomotion. The gills occupy special chambers at the sides of thorax under the carapace; their attachment, however, may be to the legs, the leg-joints, or the body wall; and according to their base of attachment they are termed respectively *podobranchs, arthrobranchs* and *pleurobranchs*. About 6000 species are recorded as occurring throughout the world, most of which are marine.

SUBORDER **MACRURA**

The animals of this suborder are primarily distinguished by their elongate and somewhat cylindrical bodies. The abdomen is well developed and commonly terminates in a horizontal fin made up of the telson and uropods, the latter appendages being the sixth pair of pleopods. Prominent antennæ are present, the first pair having two or more flagella, while the second pair are usually with an expodite, or outer, branch, known as the antennal scale. With few exceptions the young are born in an advanced larval stage.

TRIBE **CARIDEA**

The shrimps and prawns. Body generally compressed and more or less transparent or translucent. Carapace smooth and unmarked by sutures. Antennæ with large scale. It may here be observed that the terms "shrimp" and "prawn" are somewhat loose as generally used. The real difference between these two types of crustaceans lies in the fact that the true shrimp is not provided with a well-developed rostrum, whereas in the prawn the rostrum is long and conspicuous and often with a pronounced saw-toothed dorsal margin.

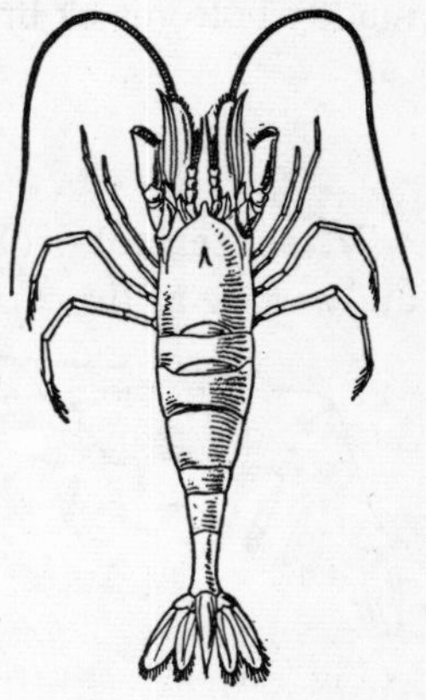

FAMILY **CRANGONIDAE**

GENUS *Crangon*

Fig. 303—C. vulgaris.

C. vulgaris. (Fig. 303.) The sand shrimp. The dorsal surface of the carapace is flattened and has a very small flattened rostrum. First antennæ with 2 flagellas. Second antennæ about as long as the body and with large prominent scale. The first pair of legs are the largest and are subchelate in structure, whereas the second pair are the smallest and of simple type; the remaining 3 pairs are increasingly longer and rather delicate. Color light, with gray spots causing the animal to resemble the sandy or muddy bottoms in which it often buries itself with only the eyes and antennæ exposed. Length 5 cm. Common in sheltered shallow bays. All ranges.

FAMILY **PALAEMONIDAE**

GENUS *Palaemonetes*

P. vulgaris. (Plate XXVII.) The glass prawn. This form may be identified by the long straight, slender rostrum bearing 8 or 9 teeth on the dorsal side and 4 on the ventral side. The first antennæ are with 2 flagella and are two thirds the length of the second pair, which latter are a trifle longer than the body. The first two pairs of feet are provided with pincer claws, those of the second pair being the larger. Colorless and nearly transparent, except for a scattering of very minute brownish pigment spots. Length about 40 mm. Occurs in great abundance, often in swarms, in shallow water. Lives among eelgrass, rockweeds, and on muddy bottoms of brackish salt marshes. All ranges.

GENUS *Virbius*

V. zostericola. (Fig. 304.) The rostrum in this species is straight and as long as the carapace, and bears 2 or 3 teeth on the upper edge near

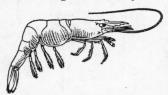

the base, while on the lower side near the tip there may be from 2 to 4 teeth. On the anterior part of the smooth carapace there are present 3 spines. A sharp downward bend occurs in the abdomen at the third segment. No mandibular palp present.

Fig. 304—V. zostericola.   Color translucent greenish with specks of brown or red and a broad median band of dark brown. Length up to about 25 mm. Lives among eelgrass (*Zostera*) just below the low-water level. Common. Vineyard Sound southward.

Family PENEIDAE

Genus *Peneus*

P. setiferus. (Fig. 305.) The common edible prawn. Rostrum long and slender and bearing about 9 or 10 teeth on the upper edge, of which the forward 6 are on the rostrum proper, whereas the others are on the carapace. On the lower side 2 teeth are present. Each side of the forward half of the carapace is marked with a lateral groove, while the anterior margin below the base of the eye stalk bears a strong spine from which a sort of keel extends toward, but does not reach, an-

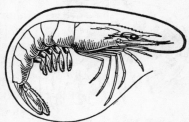

Fig. 305—P. setiferus.

other well-marked spine. Abdomen curved, not sharply bent. First antennæ with short flagella. Second antennæ long and with large scale. Color translucent, almost transparent, having a bluish-white cast with dusky areas consisting of minute pigment spots. Length up to 16 cm. This is the commercial form of the southern markets; the larger-sized individuals being called "prawns," and the smaller ones "shrimps." It occurs in both shallow and deep water, traveling in large shoals. Virginia southward.

P. brasilensis. Another edible form, very similar to the preceding species. It may be distinguished from *P. setiferus* by the lateral grooves which extend the whole length of the carapace instead of part way. It also travels in schools, often together with the preceding described form. Its range, however, extends farther northward. Cape Cod to Cape Hatteras and southward.

Tribe THALASSINIDEA

Ghost shrimps. Mud lobsters. A tribe of burrowing crustaceans characterized chiefly by a shrimplike form having the cephalothorax compressed and the abdomen large with the appendages of the sixth segment usually fitted for swimming. The covering of the body, instead of being hard, is usually soft and more or less membranous in character. An antennal scale is not generally present; and the first pair of thoracic legs are provided with pincer claws of unequal size.

FAMILY **CALLIANASSIDAE**
GENUS *Gebia*

G. affinis. (Fig. 306.) The cephalothorax of this species is compressed, deeper than broad, in the forward region, and the rostrum is

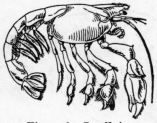

triangular in shape with a hairy covering above and a smooth under surface. Eyes globular and surmounted on thick stalks set close together. First antennæ small with 2-branched flagella; second antennæ comparatively long and with the first two joints of the peduncle shorter and stouter than the others. The mouth parts are hairy with the third pair of maxillipeds modified into a footlike appendage. Chelipeds quite large, not differing greatly in size. In this species, only the first pair of periopods are with pincer claws, the other pairs not being chelate. All the legs, however, are quite hairy toward the end, and the pleopods are well fringed with hairs. Color reddish-brown. Length 10 cm. Inhabits burrows of its own construction in muddy situations between the tide marks and in shallow water. The burrows are roundish with smooth walls and of about the diameter of one's little finger. These descend obliquely for the distance of one or two feet, then divert laterally or downward another two or perhaps three feet, and are generally tortuous in character. The animal is a very active swimmer. Common. Long Island Sound to South Carolina.

Fig. 306—G. affinis.

GENUS *Callianassa*

C. stimpsoni. Resembles *G. affinis* very closely in general appearance, and is frequently found in the same situations. It can, however, be readily distinguished from that form by the first two pairs of periopods which in *C. stimpsoni;* are both provided with large pincer claws. The claws of the first pair, however, are the larger. Then, also, the eye stalks and the third pair of maxillipeds are flattened.

TRIBE **ASTACIDAE**

Lobsters and crayfishes. The animals of this group often attain to a considerable size. All of them are usually provided with a strongly calcified shell giving them an efficient protective covering. The first antennæ have two flagella; the second pair are very much longer than the first and may be either with or without a scale. Usually there is present a transverse suture in the forward region of the carapace which marks the separation of the gastric regions from the posterior regions; longitudinal sutures are wanting. The external maxillipeds are nearly always leglike in form, and shorter than the first pair of periopods. All the periopods are generally strongly developed. The males can be distinguished from the females by the presence of a genital pore in the basal joints of the fifth pair of thoracic legs, and by the appendages of the first abdominal segment, when present, which are modified for copulation.

FAMILY NEPHROPSIDAE

GENUS *Homarus*

H. americanus. (Plate XXVI.) The American lobster. Rostrum conspicuous, narrowly pointed and with 3 teeth on each side. First antennæ comparatively short and with 2 flagella; second antennæ when turned back reach to a distance slightly beyond the telson. First 3 pairs of periopods chelate. It is, however, characteristic of this animal that the pincer claws on the first pair of periopods are enormously developed and bear very strong tubercles along the inner edge of the palm and the movable pincer. The tail, which is formed of the telson and the uropods of the last abdominal segment, is quite broad and makes a powerful swimming fin. It swims forward and backward with equal ease. Color variable, but usually greenish with darker areas and yellowish underneath. Length up to 60 cm or more, or about 2 feet, with an average length of about 35 cm; greatest weight 13 kg, or about 28 lbs., with an average weight of less than 1 kg, or about 2 lbs. It occurs in depths of from less than 1 to more than 100 fathoms, living in shallow water in summer and in deeper water in winter. This is our most commercialized crustacean, the fisheries of which are chiefly located off the coasts of Canada and the New England States. In former years the annual catch has amounted to more than 100,000,000; but at the present time, owing to the constant fishing and the pollution of the waters by large industrial centers and by oil-burning steamers, its numbers are greatly diminished. Labrador to North Carolina.

Family PALINURIDAE

Genus *Panulirus*

P. argus. (Plate XII.) Spiny lobster. Florida crayfish. This species is characterized by the presence of short sharp spines scattered over the carapace, and by the absence of pincer claws on any of the appendages. A rostrum is wanting, nor are there any antennal scales. First antennæ with a long, prominent basal joint. Color purple, reddish and brownish with pairs of yellow, eyelike spots on sides of second and sixth abdominal segments. Length of body up to 40 cm. The larval form of this and its closely related species is peculiar and is known as a *phyllosoma* (leafbody) larva from the flattened condition of the body. At the time of hatching, the larva is very fragile and is devoid of yolk material; only the appendages of the head, and some belonging to the thorax, are developed, while the abdomen is considerably reduced in size. The adult form is edible and is of great commercial importance. Common on coral reefs along the Florida coast.

Tribe ANOMURA

The crustaceans in this group are aberrant forms which at one time were included in a class by themselves but are at present united with the *Macrura*. They are characterized principally by the reduction of the last pair of thoracic legs and their position, these extending backward and upward.

Family HIPPIDAE

Genus *Hippa*

H. talpoida. (Plate XXVII.) The sand bug. The body of this curious species is greatly convex, with the carapace as viewed from above strongly ovoid in shape. The abdomen is bent under and terminates in an elongate, triangular telson. The eyes are peculiar in that they are mounted on long and very slender stalks. A conspicuous feature of this creature, however, is the second pair of antennæ, each antenna being like a long, gracefully curved plume. All the legs are stout and very broad basally and rather hairy; and none is chelate. Color whitish, tinged with lavender above; yellowish below. Length 25 mm. This is a very common form on exposed sandy beaches near the low-water mark, where it burrows under the surface or moves about over the sands shifted by the action of the waves. Besides being a rapid burrower, it is an excellent swimmer. The species is gregarious and often occurs in congregations composed of hundreds of individuals. Cape Cod to Florida.

FAMILY **PAGURIDAE**

The hermit crabs. The members of this family occurring on our shores are readily recognized by seashore visitors, owing to their odd habit of living in empty gastropod shells which they carry around as a protective covering for their unarmored hind-bodies. It is from this habit that they have a fancied resemblance to a hermit in his cell. The cephalothorax is flattened, broadened posteriorly, and, though protected by a crust, it is feebly calcified on the sides. The soft abdomen is asymmetrical, elongate and spirally bent under the thorax; and is with appendages, excepting those of the sixth segment, which are more or less atrophied or wanting entirely. The first three pairs of thoracic legs are well developed, the first pair especially so and provided with strong pincer claws.

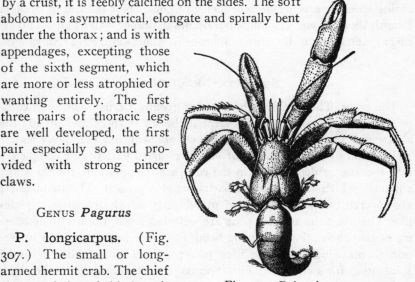

GENUS *Pagurus*

P. longicarpus. (Fig. 307.) The small or long-armed hermit crab. The chief characteristics of this hermit

Fig. 307—P. longicarpus.

crab are its small size and elongated, smooth chelæ with the right claw usually the larger. This smoothness, however, is merely comparative, for under a glass the chelæ will be seen to have a finely granulated to minutely spinescent surface. First antennæ short; second antennæ long and serving as organs of feeling. Color variable, usually shades of light yellowish-brown with iridescent chelipeds. Length about 25 mm. This is the commonest hermit crab within its principal range, which is from Maine to North Carolina, occurring in countless numbers in tide pools and shallow water. The shells it inhabits consist largely of *Nassa* and *Columbella,* and these are frequently covered with colonies of tubularian hydroids (*Hydractinia*), this being a case of commensalism. Maine to South Carolina.

P. pollicaris. (Plate XXVII.) The large or warty hermit crab. In this form the chelæ are broad and rugged, and are covered with tubercles. One claw is always the larger; usually it is the right claw. The animal uses both chelæ together to close the opening of its shell when retreating in time of danger, the shells it occupies being commonly those of Fulgar and Natica, although other large shells may also be used. First antennæ short; second antennæ long and, like those in the preceding species, serve as efficient tactile organs. Color reddish or orange. Length about 10 cm. Common, but occurs in deeper water than *P. longicarpus,* appearing to be more numerous in sounds and bays. All ranges.

Suborder BRACHYURA

The crabs. This is the highest major group of crustaceans and comprises the true crabs only. The cephalothorax is short and broad; the abdomen is much reduced in size and closely bent under it. The eyes are stalked and often are capable of being retracted partially or completely into sockets, or depressions in the carapace. Depressions for the short antennæ, which can be folded back, are also present. The mouthparts are covered by the third pair of maxillipeds, which are flat and platelike. The five pairs of periopods are well developed, the first pair bearing strong chelæ, the other pairs being fitted for walking, except in certain forms which have the fifth pair provided with oarlike terminal joints used for swimming. The appendages of the abdomen are considerably reduced, those in the male consisting of two pairs, which are used in reproduction, while those in the female consist of four pairs modified for the attachment of the eggs. In the adult animals, the sexes can be readily distinguished by the shape of the reflexed abdomen; in the male it is narrow, fitting in a groove on the under side of the cephalothorax; in the female it is very broad, often covering the entire surface of the under side between the paired legs. The young are usually hatched as zoea larvæ, passing through the megalops stage before attaining to the adultlike form. (Fig. 255.)

Division **OXYRHYNCHA**

The spider crabs. The carapace, when viewed from above, is usually narrow in front, broad and rounded behind, and generally with a prominent rostrum. First antennæ longitudinally folded. Nine pairs of gills are present with incurrent channels opening in front of bases of chelipeds and excurrent channels opening at sides of mouth region.

Family **MAIIDAE**

Genus *Libnia*

L. emarginata. (Plate XXVII.) When seen from above, the carapace is broadly ovoid, tapers to a pointed, bifid, or forked rostrum at the front, and has the surface covered with wide-set spines or tubercles and a dense growth of chitinous hairs. A distinguishing feature of this species is the presence of 9 spines or tubercles on the median dorsal surface of the carapace. In the adult male the chelipeds are more rugged than the ambulatory legs and longer than the first pair. Color dark ochrous or brownish. Length of carapace 7 cm; breadth 6 cm. Male larger than female. Female with shorter and smaller legs and relatively weak claws. This species is very common, especially in bays and sounds on muddy or shelly bottoms. Not infrequently, however, it occurs near the low-water mark and along sandy beaches. It is often found covered with algæ, hydroids, barnacles, etc., which, in addition to the dirt and mud it has the habit of collecting on the velvety covering of its body, serve to conceal it. Its movements are slow and it can be easily captured. All ranges.

Fig. 308—L. dubia.

L. dubia. (Fig. 308.) A species very similar to but not so common as the preceding. It may be distinguished from *L. emarginata* by the fewer spines on the carapace and the number of those in the median row, which consist of 6 instead of 9 as in the other species. Moreover, it grows to a larger size, its carapace measuring 13 cm in length and about 10 cm in breadth. Color, habits and habitat much the same as the preceding. Cape Cod to Florida.

Genus *Hyas*

H. coarctatus. (Plate XXVIII.) The toad crab. Viewed from above, the carapace of this species is somewhat lyre-shaped, with the upper surface bearing some conspicuous tubercles and a scattering of short, stout, hooked bristles. In the typical form, the rostrum is expanded into two long, narrowly triangular projections with the inner edges separated. First antennæ small and situated in sockets below the bases of rostral projections; second antennæ larger and with basal joint not having the large tubercle present on the outer front angle, a feature characteristic of certain other spider crabs. Eyes in depressions but not entirely capable of concealment. Chelipeds and walking legs slender. Color reddish or muddy brown or olive. Length of carapace 8 cm; greatest width 6.4 cm. Occurs on sandy and gravelly bottoms in offshore waters. This form is common locally and is usually taken with considerable foreign material attached to the hooked hairs on its upper surface. Greenland to Virginia.

H. coarctatus, variety *alutaceus,* is a form almost identical with the foregoing described species, except that the rostrum is much shorter and broader. It occurs in the same situations and is found in the same ranges.

Division **CYCLOMETOPA**

The crabs in this division do not have a well-developed rostrum. The carapace is somewhat circular or elliptical in outline, generally broader than long, and has the front margin more or less rounded, or arched in outline.

Family **CANCRIDAE**

Genus *Cancer*

C. irroratus. (Plate XVIII.) The rock crab. Carapace about two thirds as long as wide and with granulated dorsal surface. Front margin divided into 9 toothlike projections, the edges of which are roughened and not minutely toothed, as in the following described species. The notches between the teeth are continued onto the carapace, thus giving to the teeth a pentagonal appearance. Chelipeds not so long as second pair of legs. Color yellowish or reddish-brown, with stippling of dark purplish-brown. Length of carapace 7 cm; breadth 10 cm. This is a very common form and is found in tide pools and from well up to the high-water mark to deep water. It is frequent on rocky or stony bottoms, although it also occurs in muddy and sandy situations where it has the habit of burying itself in the substratum often with only the eyes exposed. All ranges.

C. borealis. (Fig. 309.) Northern crab, Jonah crab. This species has a striking resemblance to the one just described, but it may be distin-

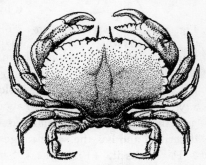

guished from that one by its more massive appearance and the rougher surface of the carapace. The front margin has the same number of teeth as in *C. irroratus,* but the edges of these are again produced into teeth, which are relatively minute. Chelipeds about as long as second pair of legs. Color brownish above, yellowish below. Length 7.5 cm or more; breadth 10 cm or more.

Fig. 309—C. borealis.

Found among rocks on exposed shores, rarely in muddy or sandy regions. Common locally, but less abundant than the preceding form. All ranges.

FAMILY PILUMNIDAE

GENUS Panopeus

P. sayi. (Fig. 310.) A mud crab. This and the following described form are very much alike, but there should be no confusion in their identity after studying the abdominal flap of the male. In the present species the terminal segment of the abdomen is broader than the one preceding it. The terminal segment is about a third broader than it is high, with margins somewhat concave and tip bluntly triangular. There are 5 teeth on each forward lateral margin of the carapace, of which the first 2 are fused and separated only by an indistinct sinus. Rostral region slightly arched and with small median notch. Color dark slatish-green or dull brownish; fingers, or ends, of claws black, extending well onto the palm. Length of carapace about 16 mm; breadth about 25 mm. Common and often abundant in oyster beds and in muddy localities under stones below the high-water mark. Massachusetts to Florida.

Fig. 310—A, P. sayi; B, abdomen.

P. depressus. (Plate XXVIII.) Although closely resembling the preceding form it can readily be distinguished from it by the shape of the terminal abdominal segment of the male which is narrower than the preceding segment, about one fourth as broad as it is long, the margins convex, and has the tip broadly rounded. (Fig. 311.) Rostral region nearly straight and with median notch indistinct or wanting. Color similar to that of *P. sayi*. Length of carapace up to about 20 mm; breadth about 30 mm. This species is usually associated with the one just described and has the same range.

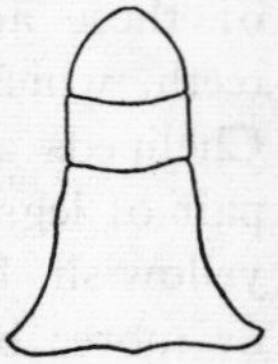

P. herbstii. (Fig. 312.) Similar to the preceding species, but with a tubercle immediately beneath the first tooth on each forward margin of the carapace, and with a prominent tubercle at the base of the movable finger of the large claw. Color similar to that of *P. sayi*. Length of carapace 40 mm; breadth 60 mm. Common in shallow waters and in muddy situations, and sometimes found in burrows associated with the fiddler crab *Uca minax*. Long Island Sound to Florida.

Fig. 311—Abdomen of P. depressus.

Fig. 312—P. herbstii.

P. harrisii. (Fig. 313.) This little mud crab, although rarer than the previously described forms, is not uncommon and might easily be mistaken for one of them. It differs, however, in its generally smaller size and the shape of the carapace, which is more of a quadrilateral form. Also, the large claws are more slender and without the black color that characterizes the tips of the preceding species; besides which, the walking legs are compressed and not so stout as in the other forms, and the carapace has more prominent transverse lines. Color dull brownish. Length of carapace about 11 mm; breadth about 14.5 mm. Found near the high-water mark and in salt marshes or muddy situations generally. Cape Cod to Cape Hatteras and southward.

Fig. 313—P. harrisii.

FAMILY PORTUNIDAE

GENUS *Callinectes*

C. sapidus. (Plate XXVI.) The blue or edible crab. The conspicuous character of this crab is the shape of the carapace, which is twice as broad as it is long and is produced at each side into a stout spinelike projection. The front margin between the eyes and the lateral spines bears a row of 8 strong teeth. Chelipeds not masive but elongated and very strong, these being formidable weapons of offense and defense. The last pair of legs have the terminal joints flattened, each leg forming an oarlike appendage used in swimming. Color of carapace deep olive above, whitish or yellowish below; chelipeds with claws and spines tipped with red; walking and swimming legs blue. Length of carapace 7 cm; breadth 13 cm. This is the common form sold in the markets. It occurs on the bottoms of bays, in shallow, brackish and even rivers emptying into the sea. It is a strong swimmer and is often found near the surface or clinging to the seaweeds on wharves. Cape Cod to Louisiana.

GENUS *Ovalipes*

O. ocellatus. (Plate XXVIII.) The lady crab. The carapace in this species is nearly as long as it is broad. On each side of the front margin there are 5 sharp-pointed teeth. The chelipeds are stout and strong, while the last pair of legs are with flattened terminal joints fitted for swimming. The name "lady crab" was given to this attractive species because of its striking coloring, which is yellowish with purplish-red spots. Length of carapace 5 cm; breadth 6 cm. This crab is an excellent swimmer. It is usually abundant along sandy beaches in shallow water, also on sandy bottoms offshore, where it often lies buried with all but its stalked eyes unexposed watchfully awaiting the approach of its prey or enemies. Cape Cod to the Gulf of Mexico.

Genus *Carcinides*

C. maenas. (Plate XXVIII.) The green crab. In this species the carapace is slightly broader than it is long and has 5 acute teeth on each side of the forward margin. The chelipeds are robust and the pincer claws are provided with stout molarlike tubercles. On both edges of the terminal joints of the first and the last pairs of walking legs are ridges of close-set, brushlike hairs. Color greenish above marked with yellowish blotches; lighter below. Length 4 cm; breadth 5 cm. This species lives in the tidal zone and in shallow water. At low water it usually conceals itself under seaweeds or rocks. It is a good runner and is active and pugnacious by nature. Cape Cod to New Jersey.

Division **CATOMETOPA**

The crabs included in this division are characterized by a carapace which is somewhat rectangular and which has the front margin straight and more or less parallel with a narrower hind margin. A rostrum is wanting.

Family **PINNOTHERIDAE**

Genus *Pinnotheres*

P. ostreum. (Fig. 314.) The oyster crab. A commensal species in

which the female inhabits the mantle cavity of the oyster; the male is free. The carapace of the female is slightly broader than it is long, membranous, and with a smooth surface. The front margin is rounded, slightly projecting and concealing the eyes. The eyes and antennæ are much

Fig. 314—P. ostreum.   reduced, while the chelipeds and walking legs are

slender and weak. Abdomen very large. Color whitish or salmon pink. Length about 8 mm; breadth about 10 mm. The male of this species is seldom collected, being rare; it is distinguished by its smaller size, firmer carapace, and dark brown color with dorsal stripe and 2 prominent spots. The female is abundant. Cape Cod to South Carolina.

P. maculatus. (Fig. 315.) The mussel crab. This is a commensal species that lives in the mantle cavity of the common mussel *Mytilus edulis* and other lamellibranchs. The carapace is membranous but firmer than that of the preceding form. In shape it is somewhat roundish, being a trifle broader than long. The surface is covered with a dense, but very short, growth of small hairs. Abdomen very large and broad. The chelipeds are comparatively stout, but the walking legs are slender as in *P. ostreum*. Color dull brownish. Length of female about 10 mm; breadth about 11 mm. The male is less numerous than the female, which is abundant. Occasionally he is found in company with her in the shell she is occupying, but generally he is taken free, swimming in the water. He is about half the size of the female. Cape Cod to South Carolina.

Fig. 315—P. maculatus.

Fig. 316—O. albicans.

Family OCYPODIDAE

Genus *Ocypode*

O. albicans. (Fig. 316.) The ghost crab. Carapace nearly square with distinct raised margins at sides. Surface finely granulate on middle and hind regions, coarsely granulate toward the sides. Eye stalks large, club-shaped, and each occupying an elongate, groovelike socket. Chelipeds well developed, the larger chela with a vertical stridulating ridge of tubercles on the inner side close to the base of the movable finger. Walking legs with numerous tufts of bristles along their edges. On each side of the ventral surface, between the basal joints of the third and fourth legs, is a breathing slit. Color pale sandy gray or brown. Length of carapace about 35 mm; breadth about 40 mm. This is also known as the "sand crab," and is a very alert and active creature. It is a swift runner, intelligent, and perhaps the most vigilant of all our crustaceans; to capture it without the aid of a hand net or other equipment is very difficult. Like the fiddler crab, it inhabits a burrow, its hole extending to a depth of 2 or more feet. Large numbers of the ghost crab's burrows often occur near the extreme high-tide mark; but they also occur, though in smaller numbers, far beyond the highest reach of the tides. The ghost crab is nocturnal in its habits, and at night will prowl over the beach in droves. New Jersey to Florida and southward.

Genus *Uca*

U. minax. (Plate XXVIII.) The large or red-jointed fiddler crab. In this species the carapace is quadrilateral in form and about one and one third times as wide as it is long. In the female the chelipeds are small and of equal size; in the male these are very unequal, one chela, usually the right, attaining to a relatively enormous dimension. It is tuberculate above and in front, while across the inner surface of the palm is a central, oblique, tuberculate ridge. The inner opposable edges of the claw are tuberculate with a few larger teeth at irregular intervals. The coloration of this species is a characteristic feature; it is brownish becoming grayish in front with red spots at the 3 articulations of the large joints of the cheliped; chelæ yellowish or ivory-white, walking legs olive or grayish-brown. This is the largest of the fiddler crabs commonly occurring along our shores, the carapace being about 25 mm long and 38 mm wide. It is usually found farther from the salt water than other species, often where the water is quite brackish or even fresh. It is abundant in sandy localities and in salt marshes, sometimes well above the tide level, where it excavates burrows measuring a foot or more in depth. These burrows are often distinguished by the presence of excavated material in their immediate neighborhood, and many of the holes are surmounted by an archway, or ovenlike structure, in which the fiddler crab stations itself. Cape Cod to Florida.

U. pugnax. (Fig. 317.) The marsh or mud fiddler. This species differs slightly in its proportions from the preceding, the carapace being one and one half times or more as wide as it is long. The large cheliped of the male is somewhat similar to that of *U. minax* in respect to the central, oblique ridge on the palm of the great claw, but the cutting edge of the movable finger is often evenly toothed, while the edge of the immovable finger is with 1 larger tooth near the middle. Color of carapace deep olive or nearly black, becoming bluish in front, chelipeds brownish-yellow; walking legs brownish. Length of carapace about 15 mm; width about 23 mm. As its popular name indicates, it is a mud-loving species, and its burrows often completely honeycomb the muddy banks of salt marsh runnels. It occurs in prodigious numbers from Cape Cod to Florida.

Fig. 317—Large claw of U. pugnax.

U. pugilator. (Fig. 318.) The sand fiddler. The carapace in this species is less than one and one half times as wide as it is long. It may be distinguished from the 2 preceding forms by the absence of the central, oblique ridge on the palm of the large chela of the male. Color of carapace purplish-gray or olive; great claw purplish or olive with fingers whitish, pinkish or yellowish-brown. Length of carapace about 15 mm; breadth about 21 mm. This is the smallest of our commoner species. It occurs in the mud or sand near the high-water mark, often in countless numbers. It usually selects as sites for its burrows, however, somewhat sheltered locations where the shore is compacted or somewhat hardened. This and its related species subsist on the minute algæ growing in the mud or moist sand. They scrape the surface for their favorite algæ, often swallowing some of the sand with their food. Cape Cod to Florida.

Fig. 318—Large claw of U. pugilator.

CLASS **ARACHNOIDEA**

In this class are contained those arthropods in which the body generally is composed of two principal parts, the cephalothorax and the abdomen. Antennæ are never present.

SUBCLASS **XIPHOSURA**

The horseshoe crabs. There are but five known species of these curious animals, four of them occurring on the eastern coast of Asia and its islands, the other on the Atlantic Coast of North America. They are here grouped in the class *Arachnoidea,* a division of the *Arthropoda* that also contains the scorpions and spiders, partly as a matter of convenience, but mainly because of their internal anatomy, which is arachnid in character. They cannot properly be grouped with the *Crustacea,* for, despite their crablike and other crustaceous characters, the evidence of embryology and of fossil-bearing rocks strongly indicates that their nearest relatives died out ages ago.

Genus *Limulus*

L. polyphemus. (Plate XXVI.) The body of the horseshoe crab is composed of a horseshoe-shaped cephalothorax, a roughly triangular abdomen, and a spinelike telson, or tail, which is about as long as the rest of the body. The cephalothorax is unsegmented but it represents the coalescence of a head and a thorax.

The dorsal surface of the cephalothorax is domelike and bears a pair of compound eyes and a pair of simple eyes. The compound eyes are large and are situated laterally and about midway between the front and the hind margins of the cephalothorax; the simple eyes are small and inconspicuous and occupy a position close to a small median spine near the forward margin.

The dorsal surface of the abdomen is somewhat flattened and has the lateral margins armed with a row of movable spines. On the ventral surface of the cephalothorax are a central mouth around which are grouped six pairs of legs, and a small tuberculate sense organ (supposedly olfactory in function) located in front of the first pair of legs. This first pair of legs, called the mandibles, are very small and are situated in front of the mouth. The five anterior pairs of legs have pincer claws, whereas the sixth pair have the terminal joints provided with *flabella,* or leaflike expansions wherewith the animal can effect a purchase in the sand or mud, thus aiding it to push itself, molelike, through the substratum in search of food. In the males the second pair of legs bear comparatively large and rugged chelæ; these are used for clinging to the abdominal tips of the females during the mating period. On the basal joints of the last five pairs of legs are sharp spines which are employed in chewing the food.

On the ventral surface of the abdomen are six pairs of broad plate-like appendages, the first pair together being protective in function and known as the *operculum,* the other five pairs having on their hinder surfaces numerous thin gills. The opercular and gill plates together constitute what is termed the "book." At the base of the operculum are the paired openings to the genital ducts. There is a well-organized digestive, nervous, and blood-vascular system, the blood being bluish in color. The color of the carapace is usually dark brown, but sometimes it occurs with a grayish or greenish cast. The total length may be up to 50 cm.

Limulus is also known as the "king crab"; it lives along the shore in shallow sandy or muddy bottoms where it moves about just beneath the surface exploring for worms and other food material. During the summer months it comes ashore to deposit its eggs in the sand. On this occasion the female, which is the larger, is attended by her mate who clings to her with his great claws, and crawls up the beach near to the high-water mark, where she makes a hole, or depression, in which the eggs are laid. The male then fertilizes the eggs by covering them with spermatozoa; after which they are left to develop by themselves without further care by the parents. When the larva hatches from the egg it is without the spinelike tail and the abdominal appendages of the adult form, resembling more the ancient trilobite. Although the horseshoe crab is edible, it is not commercially valuable in this respect; nevertheless, it is destroyed in large numbers for its use as a soil fertilizer. *Limulus* occurs in all ranges and is one of our most interesting and best known seashore animals; but owing to its wholesale slaughter within the past half-century, it is far less abundant than formerly; indeed, unless the present rate of decrease in its numbers is altered, the time may not be far distant when it will become extinct on these shores.

Subclass **ARACHNIDA**

In this subclass are contained the spiders, scorpions, ticks, mites, etc., most of which are terrestrial, air-breathing forms or inhabitants of fresh water. In all of them the eyes are simple, that is to say, they are not faceted like the compound eyes of crustaceans; the mouth parts are without true jaws, and, with a few possible exceptions, are modified for sucking instead of for masticating, although numerous forms are provided with sharp or piercing mandibles. The sea spiders, or pycnogonids, are the only members of the subclass that are within the scope of this work.

ORDER **PYCNOGONIDA**

As to the exact position that the sea spiders should occupy in the zoölogical system, not all naturalists are in agreement. By some they are referred to the *Crustacea,* by others they are classed with the *Arachnida,* whereas still others place them in a group apart from both. Most pycnogonids have a superficial resemblance to the terrestrial spiders, but in anatomical structure they are quite different. The body consists of a large cephalothorax which is segmented, and a minute, undeveloped abdomen. The cephalothorax usually bears seven pairs of appendages of which the forward three pairs consist of the *mandibles,* the *pedipalps* and the *ovigerous legs.* The mandibles are long and bear chelæ, or pincer claws; the pedipalps are slender, antennalike sensory organs; the ovigerous legs are small leglike appendages used for carrying the incubating eggs, and are absent in certain females. These three pairs of appendages, together with four eyes surmounting a dorsal tubercle, and the first pair of walking legs, are borne on the second body segment. The four pairs of walking legs are nine-jointed, a pair being borne on the second, third, fourth and fifth, or last, segments of the cephalothorax. The first body segment is produced usually into a conical or cylindrical proboscis terminating with the mouth. At the end of the rudimentary abdomen is the anal opening. There is a rather complex digestive tract. A masticatory apparatus is present just beyond the sucking mouth, and the intestine sends out long branches into the pedipalps and walking legs. These branches are, in many species, quite visible through the body wall. No special respiratory or excretory organs are present.

The sexes are separate, and the reproductive organs, like the intestinal diverticula, extend far out into the walking legs. The genital openings in the male are usually on the second joint of the last two legs; whereas, in the female they are on all the legs. The male incubates the eggs, usually cementing them together with a glandular secretion from the legs, and carrying them about with his ovigerous legs until they hatch. In most cases the larvæ are with only three pairs of walking legs, undergoing a metamorphosis before achieving an adultlike form, but the larvæ of some species resemble the adult at the time of hatching, while those of others have long tendrillike appendages by which they attach themselves to the hydroids on which they live.

Most sea spiders live among marine algæ, hydroids and bryozoans. Their small size, immobile habits, and the protective coloration that renders them inconspicuous have all contributed to their relative remoteness from the average person's acquaintance with seashore life. They are, however, fairly common and easily obtainable, and their interesting habits and grotesque forms will repay careful study on the part of the beginning naturalist who would essay a venture into the realm of morphology or comparative anatomy.

Family **PYCNOGONIDAE**

Genus *Pycnogonum*

P. littorale. (Plate XXVIII.) Body broad and flat with surface covered with very small rounded tubercles. The lateral extensions of the cephalothorax are so closely spaced that there is scarcely any interval between them. Eyes small, black, widely separated, and on a prominent tubercle. Proboscis large with basal half somewhat swollen. Abdomen club-shaped, truncated at the end. Mandibles and pedipalps wanting. Walking legs stout and bearing tubercles and hairs on many of the joints. Ovigerous legs 10-jointed and present only on the male. Color various shades of brown. Length of body 16 mm; legspread 38 mm. Found in shallow and offshore waters under stones and sometimes clinging to sea anemones. Not uncommon. Gulf of St. Lawrence to Long Island Sound.

Family **AMMOTHEIDAE**

Genus *Tanystylum*

Fig. 319—T. orbiculare.

T. orbiculare. (Fig. 319.) Body rounded in outline as viewed from above, and deeply incised between the lateral extensions of the cephalothorax, which are closely set together. The tubercles bearing the eyes are large and rounded. Proboscis large and conical. Mandibles rudimentary and consisting of 1 joint; pedipalps 6-jointed. Ovigerous legs 10-jointed, present in both sexes, but about half as large in the female as in the male. Walking legs stout and sparsely hairy. Color brownish. Length of body 1.5 mm; legspread 6.4 mm. Found among hydroids and on ascidians occurring on wharf piles, etc. Not uncommon. Southern New England.

GENUS *Achelia*

A. spinosa. (Fig. 320.) Body circular with lateral extensions of the cephalothorax separated by a conspicuous interval. The tubercle bear-

ing the eyes prominent and pointed. Eyes ovate and black. Proboscis large, thickest in the middle. Abdomen very long and slender, cleft at tip. Mandibles not well developed, consisting of 2 joints. Pedipalps 8-jointed. Ovigerous legs in both sexes, but larger in the male. Walking legs stout and rather long. The entire surface of the legs and the body is covered with hairs which are mounted on

Fig. 320—A. spinosa.

tubercles. Color brown. Length of body 2.6 mm; legspread 8.4 mm. Occurs on hydroids, ascidians and under stones near the low-water mark. Not uncommon. New England southward.

FAMILY **PALLENIDAE**

GENUS *Pallene*

P. empusa. (Fig. 321.) Body rather robust, distinctly segmented and with smooth surface. The lateral extensions of the cephalothorax are rather widely separated. The tubercle bearing the eyes is small but conspicuous. Proboscis nearly hemispherical; abdomen very small and short. Mandibles 3-jointed and with chelæ; pedipalps wanting. Ovigerous legs 9-jointed and present in both

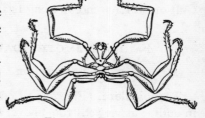

Fig. 321—P. empusa.

sexes. Walking legs very long, slender at the base, stouter toward the ends, and with a few scattered hairs on terminal joints. Colorless with red eyes. Body length 1.5 mm; leg spread 13 mm. Found on hydroids in shallow water. Not uncommon. Southern New England and southward.

Genus *Phoxichilidium*

P. maxillare. (Fig. 322.) Body stout with smooth surface. The cephalothorax is distinctly segmented, and the lateral extensions are well separated. The tubercle bearing the eyes prominent and pointed. Eyes ovate. Proboscis stout, nearly cylindrical, and rounded at end. Abdomen small and rounded. Mandibles 3-jointed and bearing chelæ; pedipalps wanting. Ovigerous legs 5-jointed; absent in female. Walking legs relatively slender, with smooth surface bearing a few scattered hairs. Color variable, ranging from blackish or brown to almost pure white. Body length from 2 to 4.75 mm; leg spread from 15 to 30 mm. Occurs in shallow and offshore waters under stones and on hydroids. Common. Cape Cod northward.

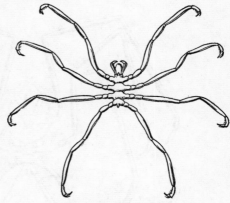

Fig. 322—P. maxillare.

Genus *Anoplodactylus*

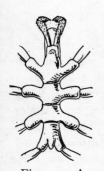

Fig. 323—A. lentus.

A. lentus. (Fig. 323.) Body slender, segmented, and with rough surface. Lateral extensions of cephalothorax widely separated. Tubercle bearing the eyes prominent and pointed. Proboscis large and somewhat constricted at base. Abdomen small and slightly cleft at tip. Mandibles 3-jointed and with chelæ; pedipalps wanting. Ovigerous legs 6-jointed; wanting in the female. Walking legs very long and slender, and with a few scattered hairs which are more numerous on the outer joints. Color purple, grayish, or brownish. Length of body 7 mm; leg spread 30 mm. Found on hydroids and ascidians in tide pools and shallow water. Common. Southern New England.

Family NYMPHONIDAE
Genus *Nymphon*

N. strömii. (Fig. 324.) Body comparatively slender, distinctly segmented, and with nearly smooth surface. Lateral extensions of cephalo-

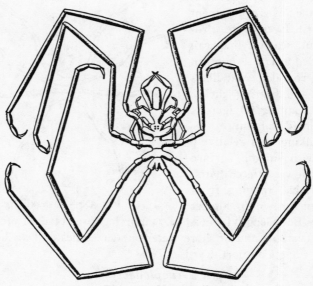

Fig. 324—N. strömii.

thorax widely separated. Tubercle bearing the eyes prominent and smoothly rounded. Proboscis large, nearly cylindrical, and slightly enlarged in middle. Abdomen small, tapering toward end. Mandibles 3-jointed and bearing chelæ; pedipalps 5-jointed. Ovigerous legs 11-jointed and present in both sexes. Walking legs very long and slender. Color light pinkish-yellow with the legs often ringed with reddish bands. Body length 15 mm; leg spread 140 mm. Found on all bottoms, though oftenmost on muddy bottoms in offshore waters. Not uncommon. Cape Cod northward.

N. longitarse. (Fig. 325.) Body very slender, and with smooth surface. Lateral extensions of cephalothorax very widely separated. The tubercles bearing the eyes are rounded. Eyes ovate. Proboscis slender, cylindrical, and rounded at end. Abdomen small and tapering toward tip. Mandibles and pedipalps as in *N. strömii*, except that the chelæ in the present species are very much longer and have their tips crossed when closed. Ovigerous legs very slender, and present in both sexes. Walking legs like those of preceding species, but more slender. Color light salmon or whitish, sometimes banded with purple. Length of body 7 mm; leg spread 65 mm. Habitat as in preceding form, but usually in deeper water. Common. Nova Scotia to Cape Cod.

Fig. 325—N. longitarse.

N. grossipes. (Fig. 326.) Body slender, and with smooth surface. Lateral extensions of cephalothorax very widely separated. Tubercle bearing the eyes prominent, conical and pointed.

Proboscis large, slightly expanded at end. Abdomen small, tapering, and often with the tip directed upward. Mandibles slender and with chelæ somewhat like those of the preceding species, but shorter and stouter. Ovigerous legs slender and 11-jointed. Walking legs long and slender and sparsely hairy. Color light pinkish-yellow, often with purple-banded legs. Length of body 10.5 mm; leg spread 90 mm. Found on all bottoms, but usually among rocks and gravel in offshore waters. Common. Nova Scotia to Long Island Sound.

Fig. 326—N. grossipes.

XII

MOLLUSCA

PLATE XXIX

1—*Loligo pallida*, marooned in a pool by the falling tide.
2—Egg-capsules of *Loligo*.   3—Wood with burrows of *Teredo chloritica*; ¾x.
4—Egg case of *Natica heros*; 1/3x.   5—Egg cases of *Nassa obsoleta*; 7x.
6—Egg cases of *Fulgur canaliculatum*; 1x.
7—Egg cases of *Fulgar carica*; 1x.   8—Radula of *Natica heros*; 50x.

CHAPTER XII

PHYLUM MOLLUSCA

(CHITONS, SNAILS, CLAMS, AND DEVIL-FISHES)

A MOLLUSK is a bilateral animal chiefly characterized by a soft, unsegmented body. The body, in most cases, is protected by a calcareous outer covering, or shell; in some forms, however, the shell may be internal, while in other forms it may be wanting entirely. The shell is usually secreted by the mantle, an outer skin or flaplike fold that more or less completely invests the body, and generally it consists of three layers, of which the inner one is often composed of mother-of-pearl. When the shell is in one piece it is known as a univalve; when it is composed of two halves it is known as a bivalve; in certain forms, as in the chitons, the shell is made up of eight similar plates.

The body of some mollusks, such as snails and slugs, has a very distinct head which is featured by a mouth, tentacles and eyes; but that of others, such as clams and oysters, is without a head, the mouth being placed at the hinder end of the visceral mass, and the tentacles usually at the margin of the shell. Very often there is a muscular foot; this in the amphineurans and gasteropods forming a flat, solelike, creeping organ, in many of the pelecypods a hatchet-shaped or a wedge-shaped digging organ, and in the cephalopods a modified structure forming long tentaclelike, sucker arms.

The visceral mass, just mentioned, forms a considerable part of the body, and its components are closely compacted. It is separated from the mantle in certain regions by a space called the *mantle cavity*. Except in the case of pelecypods, the mouth opens into a muscular pharynx containing a characteristic rasping organ known as the *radula*. This is a chitinous ribbonlike structure moving over a muscular tongue, and it is provided with minute calcareous teeth arranged in regular transverse and longitudinal rows. (See Plate XXIX.) Each transverse row is made up typically of a single middle tooth, on each side of which are one or more lateral teeth, and flanking these, one or more marginal

343

teeth, the teeth being raised or depressed by means of muscles. Each species has its own characteristic radula, the teeth varying in number and shape; consequently, this organ is often of diagnostic value in determining the various forms. There is an œsophagus; a stomach with a large liver, and an intestine making up the remainder of the alimentary canal. The openings of the anus and the ducts from the kidneys, the respiratory organs, the olfactory organ, and a mucous gland are generally situated within the mantle cavity in those forms having a mantle. The circulatory system consists usually of a heart and an arterial and a venous system. There is a well-developed nervous system, and the special sense organs, such as those of sight, taste and smell, when present, are often rather highly organized. The sexes in most cases are separate, but hermaphaphroditism is not uncommon in the group. Some mollusks are live-bearing, but most usually the young animal is born as a "veliger" larva, a characteristic form having wormlike affinities, and undergoes a metamorphosis before assuming the adultlike form.

The phylum *Mollusca* is a very large group of animals, being second in size only to the phylum *Arthropoda*. More than 60,000 species are known, the great majority of which are marine. They may be arranged in five major divisions, or classes, containing the following orders or families:

PHYLUM MOLLUSCA

CLASS **Amphineura**
ORDERS ... { Aplacophora (No forms listed in this work.)
Polyplacophora

CLASS **Scaphopoda**
(No orders.)
FAMILIES .. { Dentaliidae
Siphonodentaliidae (No forms listed in this work.)

CLASS **Gastropoda**
ORDERS ... { Pulmonata
Opisthobranchiata
Prosobranchiata

CLASS **Pelecypoda**

ORDERS ... $\left\{\begin{array}{l}\text{Protobranchiata}\\\text{Filibranchiata}\\\text{Pseudolamellibranchiata}\\\text{Eulamellibranchiata}\\\text{Septibranchiata} \quad (\text{No forms}\\\text{listed in this work.})\end{array}\right.$

CLASS **Cephalopoda**

ORDERS ... $\left\{\begin{array}{l}\text{Tetrabranchiata} \quad (\text{No forms}\\\text{listed in this work.})\\\text{Dibranchiata}\end{array}\right.$

CLASS **AMPHINEURA**

The chitons are the best known representative of this class, and these may be described as bilaterally symmetrical mollusks in which the mouth is at the forward end of the body and the anal opening at the hinder end. The body is covered with eight scalelike plates, which compose the shell. The mantle consists merely of a small fold surrounding a narrow mantle cavity and projecting from the sides and hinder end of the body. No well-developed sense organs are present, the nervous system being of a very primitive character, consisting generally of only four nerves connecting the œsophagal ring with the foot and the mantle tissues. Ganglia are very often wanting, but when they do occur they are usually small and scattered throughout the two longitudinal nerve cords which pass along the mantle furrow.

The class according to the modern arrangement consists of two orders, the *Aplacophora* (*Solenogastres*), and the *Polyplacophora*. The species included in the first-named group are relatively few, are inhabitants of remote areas or deep water, and, although possessing a distinct brain and having a more highly developed nervous system than obtains in the second-named group, are generally characterized by a more primitive body than prevails in the Polyplacophora. *Solenogastres* are not commonly found by the average seashore collector; for that reason no species are described here.

ORDER **POLYPLACOPHORA**

The chitons. These can not easily be mistaken for any other group of mollusks; their flattened, elliptical bodies, protected dorsally by the longitudinal row of eight calcareous plates, surrounded by the marginal girdle, usually plainly indicate their identity. Each of the eight plates usually partially overlaps its posterior neighbor, and consists of two layers, the upper, exposed *tegmentum* and the under, porcelanous layer, the *articulamentum*. The anterior margin of the lower layer of all plates except the first extends under the hinder margin of the plate in front. Oftentimes this lower layer projects into the marginal girdle, these concealed portions being called insertion plates. The girdle is a muscular integument and may be protected by spines or scales or a horny coat; it is of variable width, in some cases entirely covering the plates.

Chitons creep about or attach themselves by a sucker foot. This is usually broad and flat, though there are some forms in which the foot is small and narrow. Extending around the sides and hinder end of the body between the girdle and the foot is the deep, furrowlike mantle cavity, in which lie the gills and the genital and kidney pores, all of which are paired. (Fig. 327.)

Fig. 327—Diagram of a chiton (x-section). a, body cavity; b, plate; c, pericardium; d, ventricle; e, auricle; f, gonad; g, nephridium; h, pleuroviscera connective; i, branchial artery; j, intestine; k, gill; l, mantle cavity; m, pedal nerve chords; n, gill tip; o, girdle; p, foot.

The mouth opens into the pharynx, which is provided with a radula and a pair of salivary glands. There is a short œsophagus, a stomach into which open the ducts from the liver-mass, and a long coiled intestine. The heart is made up of an elongated, hollow ventricle and paired auricles, and is contained in the pericardium, the cavity of the latter being situated dorsally in the posterior third of the body. Such special sense organs as may occur are present in the shell, these being in the form of buds composed of definite nerve groups and called "æsthetes," or they may be modified buds which are sensitive to light.

With few exceptions, the sexes are separate. Certain species protect their eggs with a chitinous case; others lay them in gelatinous strings or masses attached to various objects, or the eggs may even be attached separately; still others retain the eggs within the mantle cavity until they hatch; and one species develops the eggs within the oviduct. The young chiton hatches as a trochophore larva.

SUBORDER **MESOPLACOPHORA**

Chitons characterized by the presence of insertion plates with all the transverse plates. The insertion plates of either the first plate alone or the first seven transverse plates are slit.

FAMILY **ISCHNOCHITONIDAE**

GENUS *Chaetopleura*

C. apiculata. (Fig. 328.) The shell of this species is oval and is produced into a ridge along the middle. The girdle is leathery and provided with numerous hairs. Tubercles arranged in longitudinal rows occur in the central area, while in the lateral areas and on the end plates they are irregularly scattered. There are 24 gills present on each side of the body, extending nearly to the forward end. Color grayish or light brownish-yellow. Not infrequently southern specimens are reddish. Length 17 mm; breadth 10 mm.

Fig. 328—C. apiculata.

This is a common shallow-water species, and it occurs from Cape Cod to Florida.

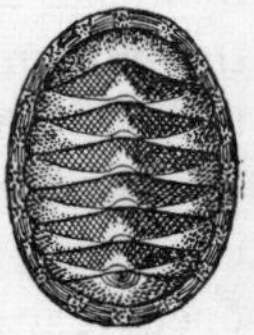

GENUS *Trachydermon*

T. ruber. (Fig. 329.) Shell oval with median ridge and minutely roughened girdle. The plates are smooth in appearance, but under a glass show a reticulated surface. Color buff with reddish marbling. Length 12 mm; breadth 7.5 mm. Circumpolar, Sitka to Long Island Sound, but not common south of Cape Cod.

Fig. 329—T. ruber.

CLASS **SCAPHOPODA**

The tooth shells. These molluusks have a shell resembling somewhat a miniature, elephant's tusk. The valve is an elongated, slender cone, slightly curved, and open at each end. The dorsal side of the animal's

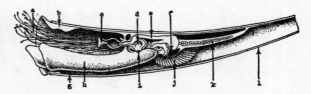

Fig. 330—Diagram of a tooth shell. a, captacula, or cephalic filaments; b, mantle; c, mouth; d, radula; e, intestine; f, kidney; g, mantle cavity; h, foot; i, anus; j, liver; k, gonad; l, shell.

body is represented by the concave side of the shell. (Fig. 330.) These mollusks are probably very primitive forms, being unlike the rest of the phylum in many particulars. There are no heart, gills, eyes or tentacles, but there are a radula, a foot, and a mantle that secretes the shell. The foot and certain cephalic filamentous structures, called the *captacula,* extending out beyond the mouth project from the larger and anterior opening of the shell. The captacula are threadlike organs with club-shaped ends, and are presumed to aid in the capture of the minute plant and animal forms on which the mollusk lives. The animal works its way into the bottom with the foot and buries itself in an oblique position with the smaller and posterior end projecting upward into the water, the opening at this end serving to convey both the incoming and the outgoing currents. As there is no heart, a circulatory blood system is wanting; however, an organ somewhat suggestive of a heart is present in the form of a slightly contractile blood space near the anus. No special sense organs, except lithocysts, are present, but the nervous system is well developed, there being paired cerebral, visceral, pedal, pleural, and buccal ganglia. The sexes are separate. Not all zoölogists are agreed as to exact systematic position of the *Scaphopoda* among mollusks.

Family DENTALIIDAE

Genus *Dentalium*

D. entale. (Plate XXX.) The common tooth shell. The shell of this species is only slightly curved. The anterior margin is jagged; the posterior opening is oblique and notched on the side. The surface is glossy and ivory-like, and shows growth lines, or annulations. At the posterior end are traces of longitudinal striæ. In this and the following species the foot is divided into 3 distinct lobes, instead of being elongate and wormlike. Color of shell ivory-white. Length 50 mm; largest diameter 5 mm. Lives usually in clean sandy bottoms in offshore waters. The dead shells are often cast up on the shore. Arctic Ocean to Cape Cod.

D. occidentale. (Fig. 331.) Shell distinctly and gracefully curved, resembling an elephant tusk. Surface with about 20 longitudinal striations. Glossy. Color ivory-white, Foot trilobate. Length 25 mm; largest diameter 8 mm. Occurs along the New England coast in offshore situations and very deep water.

Fig. 331—D. occidentale.

CLASS GASTROPODA

The sea slugs, sea hares, limpets, snails, etc. In those forms possessing a shell, this structure is usually spirally coiled, with the first-formed end of the coil representing the dorsal side of the animal. The animal is unsymmetrical, has a distinct head and a broad flat foot that usually is used for locomotion.

Although in some forms, such as the sea slugs, the shell is rudimentary or wanting, it is perhaps the most conspicuous feature of the group as a whole. Gastropods that are without a shell during their adult life are usually provided with one in the younger stages, which later disappears. In a few forms, such as in the limpets, the shells are not spirally coiled, being simple broad cones. The normal twist of the shell in spirally coiled forms may be either to the right, as in certain species, or to the left, as in others; left-handed individuals, however, sometimes occur in species in which the normal twist is to the right. Many of the snails are able to close the aperture of the shell, when the body is withdrawn, by a horny plate, the *operculum,* attached to the posterior dorsal side of the foot. The shell is not only formed by certain cells in the mantle, especially those at the margin, but it also can be thickened through the deposition of successive layers by all parts of the mantle, this last-named structure in the gasteropods being a single fold covering the visceral mass much in the manner in which a mitten fits the hand. The beginning of the spiral coil. represented by the extreme top of the shell, is called the apex. Thereafter, each coil is known as a *whorl;* but as the last and largest whorl is commonly called the *body whorl,* the others are usually referred to collectively as the *spire.* The division lines between the successive whorls are referred to as *sutures.* The axis around which spire and body whorl are coiled is known as the *columella.* In certain instances the columella is a hollow tube, and this open space is called the *umbilicus;* the latter, however, may in some cases be partially obscured by a thickened process, the *callus.* The margin of the aperture of the body whorl is termed the *lip.* Not infrequently gasteropod shells exhibit various sculpturings in the form of ribs, spines, striæ, growth-lines, and varices, the last being ridges or thickenings on the whorls indicating the former positions of the lip during the intermissions between the growth periods. (Fig. 332.)

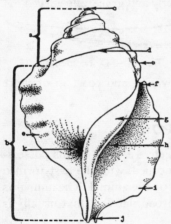

Fig. 332—Diagram of an idealized gasteropod. a, spire; b, body whorl; c, apex; d, suture; e, varix; f, posterior canal; g, columella; h, aperture; i, outer lip; j, anterior canal; k, umbilicus.

Marine gastropods are provided with one or two pairs of tentacles, which are borne on the head; and usually a pair of eyes, the latter being situated near the base of the tentacles or halfway out on them, but never at the ends as in the land snails. Nor are the tentacles of salt-water snails retractile like those of the species living on land. The mouth is at the forward end of the head, often, however, being contained in a long proboscis. Just within the lips is the pharynx, or muscular throat, which is provided usually with one or two jaws and invariably with a radula, or lingual ribbon, the jaws serving to bite off the food and the radula to reduce it into smaller particles. The radula in certain carnivorous species may also be used as a drill for boring through the shells of mollusks and other armored animals on which they feed.

The greater part of the body consists of the visceral mass. It is covered with the mantle which extends beyond it and leaves only the head and foot projecting from underneath. A space on one side (usually on the right side or the front side) between the mantle and the visceral mass forms the mantle cavity. Because of the spiral twisting of the body and the concomitant displacement of certain organs, one member of the usual pair is generally wanting, there ordinarily being but one gill, kidney, auricle, and olfactory organ. The gills, also known as *branchiae,* or as *ctenidia* (meaning "little combs"), are characteristic featherlike or comblike organs.

Not all marine gastropods are of the foregoing type. Certain forms, such as the sea slugs, or nudibranchs, are without any mantle whatsoever; nor are they provided with gills as these structures are commonly known. The respiratory organs are adaptive structures formed by simple projections of the dorsal integument; and in many cases even these are wanting, the animal breathing through the surface of the body wall, which may be smooth or covered with papillæ. Moreover, nudibranchs are entirely without a shell during their adult life.

There is a well developed nervous system with cerebral, pedal, pleural, visceral, and perietal ganglia. The special sense organs consist of a pair of eyes, one or two pairs of tentacles which in certain forms have the posterior pair functioning as *rhinophores,* or olfactory organs, one or two *osphradia* near the gills, also, presumed to be olfactory in function, and a pair of *lithocysts* in the foot, these being the organs of equilibrium.

With the exception of the opisthobranchs, which are usually hermaphroditic, most marine gastropods are unisexual. The eggs as a rule are deposited in protective capsules of a chitinous or gelatinous nature; in some cases, however, snails are live-bearing, there being no free incubation period. The young are usually hatched as "veliger" larvæ, a free-swimming stage in which they are provided with locomotor cilia.

ORDER **OPISTHOBRANCHIATA**

This order contains those gastropods that are characterized mainly by the position of the gills (when these are present) in relation to the heart; which is to say that the animals have these breathing organs in a position posterior to the heart. Many of them are without a mantle and shell, and where the shell is present, it is usually relatively small. All the forms are hermaphroditic.

Suborder **TECTIBRANCHIATA**

Opisthobranchs in which there are a shell, mantle, and ctenidium usually present, the latter being situated on the right side.

Division **BULLOIDA**

The bubble shells. Shell well developed, but may be either internal or external. Head with a broad cephalic disk on the dorsal side. Tenacles usually present, which are obscured by cephalic disk.

Family **ACTEONIDAE**

Genus *Acteon*

Fig. 333—A. punctostriatus.

A. punctostriatus. (Fig. 333.) Shell bulbous with conical spire. Aperture narrow and extending over about one half the length of shell. Strong tortuous fold produced by inner lip of aperture margin. Horny operculum present. Cephalic disk with posterior cleft. Color of animal white; shell with 2 or 3 color bands. Length 6 mm. Occurs in offshore waters. Cape Cod to Florida.

Family **AKERIDAE**

Genus *Haminea*

Fig. 334— H. solitaria.

H. solitaria. (Fig. 334.) The shell in this species is a fragile ovate structure, its delicacy making it deserving of its popular name "bubble shell." The spire is very small and sunken; the aperture is very large, being as long as the shell, and is with a sharp outer lip. The surface is shining, and is with minutely impressed revolving grooves crossed with irregular growth lines. Color bluish-white or light brownish or greenish. Length 9 mm; width 6 mm. Abundant locally in shallow water. Common from Massachusetts Bay to South Carolina.

FAMILY BULLARIIDAE

GENUS *Bullaria*

B. occidentalis. (Plate XXX.) A southern species in which the shell can be easily recognized by the small hole, or pit, that occurs in place of the absent spire, and by the brown mottlings on the white, glossy surface of the whorls. The aperture is large and extends the full length of the shell. The surface bears minute growth lines which are crossed near the lower end of the aperture with revolving grooves; these markings, however, are not readily revealed without a glass. Length about 20 mm; width 12 mm. Lives in offshore waters, but the shells are often strewn along the beaches of southern Florida.

FAMILY LIMACINIDAE

GENUS *Limacina*

L. arctica. (Fig. 335.) A pteropod. Pteropods in general are featured by two large fins, which in reality are the *epipodia,* or lateral projections of the foot extending around the dorsal side of the head. A shell and mantle are present, but ctenidia and eyes are wanting. All of them are small pelagic forms living usually in the open sea; not infrequently, however, they are carried inshore by wave action and currents. In the present described species, the shell is transparent and decidedly snaillike, being a low, left-handed spire consisting of

Fig. 335—L. arctica.

6 whorls, and is with an operculum. The umbilicus is rather wide. Width 4 mm. Arctic Ocean to New Jersey.

FAMILY CAVOLINIIDAE

GENUS *Creseis*

Fig. 336—
C. conica.

C. conica. (Fig. 336.) A pteropod having all the described characters of the preceding species, except those pertaining to the shell. In this species the shell is in the form of a straight slender cone, not coiled. Found along the Atlantic Coast.

DIVISION APLYSIOIDEA

In this division of tectibranchs, the shell is often wanting; where it is present, it is much reduced. The head bears two pairs of tentacles, but there is no cephalic disk. The epipodia, or fins, are not extensions of the foot, but are projections from the sides of the body.

FAMILY APLYSIIDAE

GENUS *Aplysia*

A. protea. (Fig. 337.) A sea hare. This mollusk has a shell, but it is internal. The shell is rudimentary and is hardly more than a thin horny plate buried in the mantle. The animal is sluglike, larger posteriorly, and has a long neck and head, the latter being produced at the anterior angles into 2 elongated tentacular folds. A pair of eyes are present behind the tentacles, and behind the eyes are the rhinophores. The epipodia are rather large lobes and extend upward from the sides of the body, turning over the back; these are used for swimming. Color variable, but commonly yellow or green with ring-shaped spots of green, red and black. Length 16 cm. This animal, when disturbed, ejects a reddish-purple fluid for the purpose of concealment. This fluid is popularly thought to be poisonous, but it is in fact harmless. Occasionally abundant on the Florida coast; common in the West Indies.

Fig. 337—A. protea.

FAMILY CLIONIDAE

GENUS *Clione*

C. limacina. (Fig. 338.) A pteropod. This form belongs to a family in which the shell and mantle are absent. The body is somewhat elongated, tapering to a point behind, and is with a distinct head which is provided with 2 pairs of tentacles, the hinder pair being with eyes. From the base of the neck project a pair of broad fins. Gills are wanting, the animal breathing through the general surface of the body. The body is hyaline with a pale bluish cast. Length 35 mm. Found in offshore waters and the open sea swimming near the surface in enormous shoals. It is a common food for whales. Arctic Ocean to Long Island.

Fig. 338—
C.limacina.

SUBORDER NUDIBRANCHIATA

The naked sea slugs. In this suborder the adult animals are without a shell, ctenidium, and osphradium, although a coiled shell is always present in the embryo. There are two pairs of tentacles. As no true gills are present, the animals are often provided with adaptive gills in the form of outgrowths of the body wall, or respiration may take place directly through the unspecialized general surface. When adaptive gills are present, these are usually grouped around the anal opening when it is in the middle of the back, or they may occur dorsally and arranged in longitudinal or transverse rows, forming a waving fringe along the back. These substitute dorsal gills are known as *cerata,* and they commonly contain hollow extensions of the liver. The sea slugs have perhaps the most gorgeous color patterns among marine invertebrates.

DIVISION TRITONIOIDEA

Cerata branched and usually occurring in two longitudinal rows. The anal opening is situated on the right side.

FAMILY DENDRONOTIDAE

GENUS *Dendronotus*

D. arborescens. (Fig. 339.) n this species the 2 pairs of tentacles are forward-projecting, branched, antlerlike appendages situated at the

forward end of the elongate body. Along the entire back occur 2 rows
of cerata, each row having from 5 to 7 cerata, these structures being

branched and treelike. The liver is
large and extends into the cerata,
the latter being retractile. Color variable, but commonly reddish or rose
with pale or brownish spots; cerata
transparent and sometimes brightly

Fig. 339—D. arborescens.

colored. Length up to about 8 cm. Found among rocks and algæ. Common locally. Circumpolar; south to Rhode Island.

Division **DORIDIOIDEA**

Gills encircling the anal opening which is median and dorsal in position and situated posteriorly. The liver does not extend into the gills.

Family **POLYCERIDAE**

Genus *Ancula*

A. sulphurea. (Fig. 340.) The body of this species is smooth with
the marginal ridge bearing long appendages, those on the head being
in the form of tentacular projections at the base of the rhinophores, or
second pair of tentacles. There is one of these tentacular projections at
the side of the head in front, as well as 8 to 12 surrounding the median
dorsal gills. The rhinophores are club-shaped, while the gills are 3-
branched, featherlike structures. Color light brownish and transparent
with yellow-tipped appendages. Often common. Found under stones.
Occurs on New England shores north of Cape Cod.

Fig. 340—A. sulphurea.   Fig. 341—D. bifida.

Family **DORIDIDAE**

Genus *Doris*

D. bifida. (Fig. 341.) The body in this species is broad and elongated,
somewhat flattened, and provided with protective integument bearing

calcareous spicules. The back is convex and covered with papillæ of nearly uniform size. The forward tentacles are small; the hinder tentacles, or rhinophores, are conspicuous and retractile. The gills consist of 7 retractile plumelike structures forming a rosette in the middle of the back posteriorly. Color purplish-brown with speckles of white. Length 25 mm; width 12 mm. Often common. Found under stones at low-water mark. Occurs along the entire New England coast from Maine to New York.

Genus *Onchidoris*

O. bilamellata. (Fig. 342.) This form, which may be confused with the preceding, may be distinguished from that by the papillæ covering the back, which are short and unequal in size, and by the gills, which are arranged in an oval and number from 20 to 25. The gills are not retractile. Color pale rose or brownish. Length 25 mm; width 12 mm. It is frequently common north of Cape Cod, usually in the same situations as *D. bifida.*

Fig. 342—O. bilamellata.

Division **AEOLIDIOIDEA**

Gills simple and unbranched, and arranged dorsally on each side of the body. The liver projects into the cerata.

Family **AEOLIDIDAE**

Genus *Aeolis*

Fig. 343—A. papillosa.

A. papillosa. (Fig. 343.) The body in this species is broad anteriorly, tapering posteriorly, and depressed dorso-ventrally. The cerata are placed close together in 12 to 20 overlapping oblique rows on each side, leaving a bare space along the middle of the back. In each row there are 10 to 12 cylindrical cerata, these often containing stinging cells derived from ingested hydroids. There are no spicules in the body covering as in *Doris;* nor is there a mantle. The anal opening is on the right side. Tentacles not retractile. Color variable, usually gray or orange with brown or white spots. Length 7 cm; width 2.3. Common. Lives among stones, sea-weeds, hydroids, and eelgrass in tide pools and shallow water. Arctic Ocean to Long Island Sound.

Fig. 344—A. pilata.

Genus *Aeolidia*

A. pilata. (Fig. 344.) In this species the body is slenderer than in *Aeolis,* a form with which it is likely to become confused. The cerata are crowded and arranged in 5 clusterlike rows on each side, and with about 6 somewhat spindle-shaped cerata in each cluster. Color pale drab with the middle of the back marked with a broken carmine stripe edged with silver stippling. Length 38 mm; width 6 mm. Occurs from Massachusetts Bay to Long Island Sound.

Family **DOTONIDAE**

Genus *Doto*

D. coronata. (Fig. 345.) The cerata in this species are club-shaped, tuberculated structures arranged in a single row with 5 to 8 on each side of the animal's back. The body is slender and pointed behind, but the anterior end is forked, each fork flaring trumpetlike at the end whence arises a slender retractile tentacle. Color variable, usually flesh or brownish, with red spots on cerata. Length 12 mm. Not uncommon. Found on bryozoans and hydroids in shallow water. Labrador to New Jersey.

Fig. 345—D. coronata.

Division **ELYSIOIDEA**

In this division the cerata may either be present or absent. But one row of teeth occurs in the radula.

Family **ELYSIIDAE**

Genus *Elysia*

E. chloritica. (Fig. 346.) A species without cerata. The body is elongated, rather slender, and is with a ciliated surface. The sides are produced into broad leaflike ridges which overlap on the back when the animal creeps. These leaflike expansions are used in swimming. Color bright green with white and red dots. Length 4 cm; width .8 cm. Found in shallow and brackish waters. Occasionally abundant. Occurs from Massachusetts Bay to New Jersey.

ORDER **PULMONATA**

This order is composed mostly of fresh-water and land snails, but the group contains a few forms that are marine. They are air-breathing mollusks; intsead of ctenidia, or gills, they are provided with a respiratory sac, or "lung." Except in the case of the young of certain species, which inhabit the bottom of deep lakes and can obtain oxygen from the water that is taken into the mantle cavity, the aquatic members of this order are obliged to come to the surface of the water to get air. The pulmonates are without an operculum; and all of them are hermaphroditic animals.

Fig. 346—E. chloritica.

Suborder **BASOMMATOPHORA**

Pulmonates with but one pair of tentacles. The shell is somewhat fragile, and generally is a simple spire.

Family **AURICULIDAE**

Genus *Phytia*

P. myosotis. (Fig. 347.) This species has a conical shell with a pointed spire, the exterior being covered with a horny periostracum. The aperture is elongated, and the columella is toothed. The tentacles are cylindrical and placed well back on the snoutlike head. Color brownish-yellow with a reddish sutural line. Length 8 mm; diameter 4 mm. Found near the high-water mark and living in the crevices of wharfs. Nova Scotia to Chesapeake Bay.

Fig. 347—P. myosotis.

Genus *Melampus*

M. lineatus. (Fig. 348.) Shell with a large body whorl and a very short, blunt spire; in outline, broadest about the upper third portion. Structure of shell thin and translucent. Aperture elongate, widest below, and with a white callus and a conspicuous tooth on the columella. Color dark reddish-brown, reddish-gray or brownish-yellow. Length 13 mm; diameter 7 mm. Common near the high-water level among the grasses (*Spartina*) in salt marshes. New England to Texas.

Fig. 348—
M. lineatus.

ORDER **PROSOBRANCHIATA**

The prosobranchs with few exceptions are gill-breathers, being provided with a ctenidium that lies in the mantle cavity forward of the heart. Both a shell and an operculum are generally present, the latter being calcareous or horny. The shell is usually spiral and in the higher forms it is produced at the aperture into characteristic prolongation known as the anterior, or siphonal, canal; which, as its name indicates, is the passage through which the siphon is extended, in those forms possessing this organ. The head bears a single pair of non-retractile tentacles. There is a well-organized nervous system. Eyes are present, these consisting of a pair at or toward the base of the tentacles, never at the tips, although they may surmount stalks of their own. The sexes are separate, and often can be distinguished by the shell, which is generally slenderer in the male than in the female. The great majority of gastropods are contained in the order *Prosobranchiata,* and most of these are marine.

Suborder **SCUTIBRANCHIATA**

The prosobranchs in this suborder are characterized by *bipectinate* ctenidia, when gills are present. Bipectinate gill structure may be likened to that of a typical feather; that is to say, smaller lateral extensions are arranged on *each* side of a longer axis. These animals are without a proboscis, nor is there a siphonal canal present in the shell.

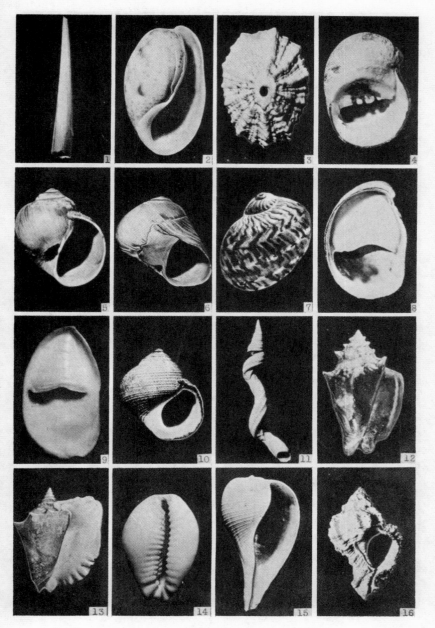

PLATE XXX

1—*Dentalium entale*; 1x.   2—*Bullaria occidentalis*; 4x.

3—*Fissurella barbadensis*; 1½x.   4—*Nerita peleronta*; 1¼x.

5—*Natica heros*; ¼x.   6—*Natica duplicata*; 1/3x.   7—*Natica canrena*; 1x.

8—*Crepidula fornicata*; 1x.   9—*Crepidula plana*; 1x.

10—*Littorina litorea*; 1¼x.   11—*Vermetus radicula*; 2/3x.

12—*Strombus pugilis*; ½x.   13—*Strombus gigas*; 1/6x.

14—*Cypræa spurca*; 2½x.   15—*Pyrula papyratia*; ½x.   16—*Murex pomum*; ¼x.

PLATE XXXI

1—*Urosalpinx cinereus*; 2x 2—*Eupleura caudata*; 2x. 3—*Nassa obsoleta*; 3x.
4—*Nassa trivittata*; 3x. 5—*Buccinum undatum*; 2/3x.
6—*Fulgur carica*; ¼x. 7—*Fulgur canaliculata*; 1/4x. 8—*Fulgur perversa*; 1/3x.
9—*Oliva litterata*; 2/3x. 10—*Conus floridanus*; ¾x.
11—*Leda tenuisulcata*; 1½x. 12—*Anomia ephippium*, lower valve; ¾x.
13—*Arca pexata*; ½x. 14—*Arca ponderosa*; ½x.
15—*Arca noae*; ½x. 16—*Mytilus edulis*; variety *pellucidus*; ½x.

DIVISION DOCOGLOSSA

This division includes certain limpets that have but one auricle in the heart. The shell is a simple flat cone, not spiral, and is without an operculum. In the gill-breathing forms, but one ctenidium is present. The jaw is unpaired.

FAMILY ACMAEIDAE

GENUS *Acmaea*

A. testudinalis. (Fig. 349.) Shell oval in outline when viewed dorsally; seen from the side it is a depressed cone with the apex inclined toward the anterior end. The shell is solid and with a smooth surface. Ctenidium present. Color gray or greenish with dark brown stripes radiating from the apex; inner surface white with a large brownish blotch. Length 35 mm; width 25 mm; height 9 mm. A common species occurring on rocks in the tidal zone and shallow water. Is. found from the Arctic Ocean to Long Island Sound, but is less common south of Cape Cod.

Fig. 349—A. testudinalis.

A. testudinalis, variety *alveus,* is a smaller form. (Fig. 350.) The shell is narrow, being adapted to occupying the slender leaves of eelgrass which it commonly inhabits. Color whitish with reddish-brown spots. Length 15 mm; width 8 mm; height 5 mm. Common along the New England Coast.

Fig. 350—A. testudinalis, var. alveus.

DIVISION RHIPIDOGLOSSA

The gastropods in this division usually have depressed shells which may be spiral in form or not. When spiral the shell is only slightly coiled. The jaw is paired. The radula is provided with one central tooth, 3 to 9 laterals and numerous closely set marginals. Ctenidia are present, either singly or in pairs; and the heart is with 2 auricles.

SUBDIVISION ZYGOBRANCHIATA

Grouped in this subdivision are those forms having the shell and mantle with a slit at the apex or the margin, or with marginal holes. Two ctenidia are present.

Family FISSURELLIDAE

Genus *Fissurella*

F. barbadensis. (Plate XXX.) Keyhole limpet. Shell low and conical. This form can be recognized by the nearly circular perforation in the apex of shell. The shell is further marked with heavy radiating ribs, about 11 of which are considerably larger than the others. Margin crenulated. Color variable, but commonly gray or whitish with brown markings; interior greenish and with a pink area around the apical perforation. Length 37 mm; width 24 mm; height 9.5 mm. This is a southern form, being common along the shores of Florida, Bermuda, and the West Indies.

Subdivision AZYGOBRANCHIATA

The gastropods in this group have a shell that is usually distinguished by a large body whorl and an elevated spire. The animals have only one ctenidium.

Family TROCHIDAE

Genus *Margarita*

M. obscura. (Fig. 351.) In this species, the shell is globose with a rather small conical spire and an open umbilicus. In structure it is somewhat thin. There are 5 rounded whorls, and the surface is marked with 2 or 3 revolving ridges crossed with coarse growth lines. There is a horny operculum. Color brownish. Length 10 mm; diameter 12 mm. A common species occurring in shallow water in its northern range, but lives in deep water toward the southern limits of its range. Labrador to Long Island Sound.

Fig. 351—M. obscura.

M. undulata. (Fig. 352.) A species with a shell having some of the characters distinguishing the preceding, but which can be determined by the 6 whorls with short wrinkles near the suture, and the numerous uniformly spaced revolving ridges sculpturing the dull surface. Color brownish or reddish. Length 7.5 mm; diameter 10 mm. Common in shallow and offshore waters. Arctic Ocean to Cape Cod.

Fig. 352—M. undulata.

Family NERITIDAE

Genus *Nerita*

N. peleronta. (Plate XXX.) The bleeding-tooth shell. This species has a shell that is not likely to be confused with that of any other gasteropod. Two prominent white teeth occur on the columellar lip, which are surrounded by a bright orange-red blotch strongly suggesting the popular name of the shell. The body whorl is very large and with a greatly flattened spire. The operculum is calcareous. There is no umbilicus. The animal has a broad, short snout, long, slender tentacles and stalked eyes. Color yellowish with wavy purple bands. Length about 3 cm; diameter about 2 cm. Occurs along the shore in southern Florida.

Suborder PECTINIBRANCHIATA

This, the second suborder of prosobranch gasteropod mollusks, contains the largest number of forms. They are characterized by having one auricle in the heart, one osphradium and one ctenidium. The ctenidium is of the *monopectinate* type; it may be compared to a feather that has the lateral extensions, or vanes, on but one side of the axis, or shaft. A proboscis, siphon, and anterior, or siphonal, canal are generally present.

Division PTENOGLOSSA

This is a small group of prosobranchs in which the shell is with a small, simple aperture and without a siphonal canal. Although the radula is well developed, being provided with many small lateral teeth, the central tooth is sometimes wanting.

Family JANTHINIDAE

Genus *Janthina*

J. fragilis. (Fig. 353.) The shell of this mollusk is transparently thin and very fragile. The body whorl is large, while the spire is low and considerably reduced. Color violet at base, blending into blue toward spire. Length 20 mm; diameter 25 mm. This is a pelagic species which occurs in schools in the warm waters of tropical seas. It floats at the

Fig. 353—J. fragilis.

surface by means of a long, many-chambered, raftlike structure secreted by the small and almost rudimentary foot. The animal also attaches its eggs to this vesicular float. Despite its pelagic existence, it is often available to collectors, as it is driven ashore in large numbers to perish on the beaches. It is frequently especially abundant along the Florida coast; and it is occasionally thrown up on the shores of New England.

FAMILY SCALARIIDAE

GENUS *Scalaria*

Fig. 354
—S. groen-
landica.

S. groenlandica. (Fig. 354.) The ladder shell or wentletrap. The shell is an elongated spiral cone with 10 whorls, each whorl being with 6 or 8 revolving ridges and crossed at regular intervals with strong, prominent elevated ribs. The aperture is nearly round and with a continuous lip. There is a horny operculum. A short siphon is present. Color whitish or brownish with white ribs. Length 25 mm; diameter 9 mm. Not uncommon. Occurs in offshore waters from the Arctic Ocean to Cape Cod.

Fig. 355—
S. lineata.

S. lineata. (Fig. 355.) This wentletrap has a shell similar to that of the preceding species, but, although smaller in size, it is proportionately longer and more slender. There are 11 whorls in the shell, with strong but not greatly elevated ribs. Color light brownish with white ribs, and sometimes marked with a few dark revolving lines. Length 12 mm; diameter 5 mm. Lives in offshore areas. Not uncommon. Cape Cod to Florida.

Fig. 356—
S. angulata.

S. angulata. (Fig. 356.) The shell of this wentletrap can be distinguished from others by the fact that the whorls touch one another only by the elevated ribs. There are from 6 to 11 whorls, each of which has about nine ribs, the latter being slightly angled next to the suture above. Color whitish. Length 20 mm; diameter 8 mm. Found in offshore waters. Common in southern areas. Long Island Sound to Texas.

S. multistriata. (Fig. 357.) A wentletrap in which the transverse ribs are very small and numerous. There are about 9 whorls and these are marked with many fine revolving lines. Color white. Length 13 mm; diameter 10 mm. Occasionally common. Cape Cod southward.

Fig. 357—S. multistriata.

Division TAENIOGLOSSA

In this the second division of the pectinibranch mollusks, the radula is a long narrow ribbon with seven teeth in each row arranged in the following order: one central, having on each side one lateral and two marginals.

Subdivision PLATYPODA

The mollusks in this subdivision have the foot in the form of a broad creeping sole. There is a well-developed shell.

Family NATICIDAE

Genus *Natica*

N. heros. (Plate XXX.) The moon snail. This large gastropod often bears the scientific name *Polynices* or *Lunatia heros*. The shell is globose and with 5 convex whorls. It may be distinguished from the following described species by the large open umbilicus. A yellowish periostracum covers the shell, which is ashy-white or brownish in color and fairly thick and solid in structure. There is a horny operculum. Length 9 cm; diameter 8 cm. The curious large "sand collars" (Plate XXIX) found in the shallow water and tide pools and strewn along the beach are the egg cases of this animal. Also found in these situations is a small shell of similar shape to the one described, but marked with 3 revolving rows of bluish or brownish spots, a shell formerly classified as a separate species (*N. triseriata*) but now known to be the young of *N. heros*. *N. heros,* as well as the other members of the genus, is a predatory mollusk, living chiefly on other mollusks which it kills by enveloping with the large, fleshy foot and by boring a hole through the victim's shell with the radula and tearing out the soft parts with the sharp jaws in the proboscis. On occasion, however, *Natica* will devour dead fishes or other non-living animal remains. The present species is very common, especially in Long Island Sound. It lives in sandy regions, occurring in shallow and offshore bottoms where it burrows usually after bivalve mollusks. Labrador to Virginia.

N. duplicata. (Plate XXX.) Similar to the preceding but with a more flattened spire and a large, thick callus, the latter almost completely closing up the umbilicus. Color ashy-white below and blending into chestnut-brown above; inner surface chestnut-brown. Length and diameter 5.5 cm. Very common. Found in shallow water. Massachusetts Bay to Mexico.

N. clausa. (Fig. 358.) A species similar to *N. heros,* but much smaller and with a calcareous instead of a horny operculum. The operculum is white, and there is a white callus which completely fills the umbilicus; it is by the color and the characters of these structures that the species may be recognized at once. Color of shell whitish or brownish. Length 15 mm; diameter 13 mm. Found in offshore waters. Circumpolar. Arctic Ocean to Cape Cod.

Fig. 358—
N. clausa.

N. canrena. (Plate XXX.) This species is easily determined by its coloration, the shell being white with revolving bands of chestnut which are crossed with zigzag lines of purple. The shell is white beneath, while the interior of the aperture is purplish. A callus partially closes the umbilicus. The operculum is calcareous and is sculptured with deep eccentric grooves. Length and diameter about 3.5 cm. Lives in sand near the low-water mark. Occurs from Cape Hatteras to Florida, but is commoner toward the south.

Genus *Sigaretus*

S. perspectivus. (Fig. 359.) The ear shell. The shell of this gasteropod is much flattened. This feature, together with the fact that the spire is very small and that the aperture is comparatively enormous, gives it the ear-shaped appearance whence is derived its popular name. The operculum, however, is peculiarly minute, and is horny. There is no umbilicus. The surface of the shell is smooth and polished. Color white. Length 37 mm; width 30 mm. This animal lives in the sand and mud in shallow water, where it burrows, molelike,

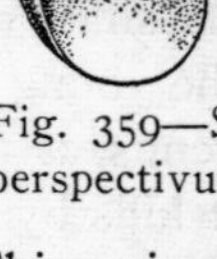

Fig. 359—S.
perspectivus.

under the surface seeking other mollusks for food. The forepart of its foot is enormously developed, making an efficient organ both for burrowing and for enveloping its prey, the latter consisting largely of bivalves. Common, especially toward the south. Occurs from New Jersey to Florida.

FAMILY CAPULIDAE

GENUS *Crucibulum*

C. striatum. (Fig. 360.) The cup-and-saucer limpet. The shell of this species is flattened and without a spire. Viewed from above it is ovate; when seen from the side it is conical with the apex directed toward the hinder end. Radiating from the apex to the margin are numerous elevated lines crossed by circular lines of growth. However, a distinctive feature of this shell is the cup-shaped shelf

Fig. 360—C. striatum.

extending from one side of the inner wall near the apex. This shelf is for the attachment of the adductor muscles. There is no operculum. Color white or yellowish. Length about 20 mm; height about 12 mm. Lives in shallow water, clinging to stones and shells with its broad foot. It is inactive in its habits. Common. Occurs from the Bay of Fundy to Florida, but is more abundant toward its northern range.

GENUS *Crepidula*

C. fornicata. (Plate XXX.) The boat shell. In this species the shell is somewhat obliquely oval in outline, when seen from above, and has one side less curved than the other. The spire is inconspicuous and is at the hinder end where it is deflected to one side and is depressed against the large body whorl. Within the shell, the hinder half is covered by a concave diaphragm or shelflike partition, extending from the body of the shell. Thus, the interior of the shell has some likeness to a boat, the space below the diaphragm at the pointed end representing the forecastle, and the rounded end representing the stern. Color whitish or brownish. Length 45 mm; width 33 mm; height 15 mm. This is a very common species and is found attached to stones, shells and other objects in shallow water. Often several individuals will be found superimposed on the shells of one another. The animal is inactive, seeming seldom or never to stray from an object that offers a likely holdfast:

however, when compelled by circumstances, it can readily move about, which it does by means of its broad foot. Nova Scotia to South America.

C. convexa. (Fig. 361.) A smaller species than the preceding. Differs in being more convex with the shorter side of the shell almost vertical. The diaphragm is deep-seated and unlike that of the species just described in being convex instead of concave. Color brown. Length 12 mm; breadth 8 mm; height 4 mm. Fairly common. Occurs in shallow water. Nova Scotia to Florida.

Fig. 361— **C. plana.** (Plate XXX.) The flatboat shell. As is indicated C. convexa. by its popular name, the shell in this species is flat; not infrequently it is concave. The diaphragm is about one half the length of the shell, and it is convex. It cannot easily be mistaken for any of the other boat shells, for the pure white, polished, porcelanous interior and its habit of living within the aperture of large dead shells are peculiar to this species alone. Length 40 mm; breadth 25 mm. It is a very common species occurring in shallow and offshore waters from Maine to Texas.

FAMILY LITTORINIDAE

GENUS *Littorina*

L. litorea. (Plate XXX.) The edible periwinkle. The littorinas are characterized by a thick, solid, conical, spiral, shell and a horny operculum. The shell has from 4 to 6 whorls, and is without an umbilicus. There are about 175 species in the single genus *Littorina,* and all of them are alike in having a projecting head bearing conical tentacles which are provided with eyes at their base. The radula is long, being 2 or 3 times the body length. There is a longitudinal groove in the foot, which divides this structure so that each side advances alternately when the animal creeps. Although there is a rudimentary siphonal fold in the mantle, the aperture of the shell is without a siphonal canal or notched lip. In the present species the shell may be diagnosed by its rough exterior, which is dark brown, reddish, or yellowish in color, and marked with dark revolving bands. The interior is white or brownish. The columella is broad and white. Length 25 mm; breadth 16 mm. This species was originally introduced on our Atlantic Coast from

Europe, coming by the way of Iceland and Greenland and thence down
the shores of Labrador to Delaware Bay, which is the southern limit
of its range. Since its introduction it has become one of the commonest
and most abundant of littoral mollusks; for it is strictly a species of the
tidal zone.

L. rudis. (Fig. 362.) This species has the general characters of the
one just described, but the shell differs in having whorls which are
marked usually with revolving grooves and ridges, often with interrupted spiral, whitish bands, and with spots. Color very variable; usually
whitish, reddish, or black. Length 14 mm; breadth 10 mm. Very common. Occurs on rocks near the high-water mark. Arctic Ocean to New
Jersey.

Fig. 362—   Fig. 363—   Fig. 364—   Fig. 365—
L. rudis.   L. irrorata.   L. palliata.   L. vincta.

L. irrorata. (Fig. 363.) A species which can be recognized by the
flattened whorls of the shell, the indistinct sutures, and the relatively
high spire. Surface with numerous revolving ridges spotted with short,
brown lines. The general color is yellowish-white or green. Aperture
white within, with reddish tinge on columella. Length 25 mm. Found
between the tide marks. Occurs from Vineyard Sound to Florida, but
is commoner toward the south.

L. palliata. (Fig. 364.) The shell of this littorina is quite globular
and is with a low spire and a broadly flattened columella. The surface is smooth and shining. Color variable; usually yellow, brown, or
red; often with revolving black bands. The head of the animal is orange-
colored. Length 15 mm. Found usually on *Fucus* in the tidal zone; often
in abundance. Arctic Ocean to New Jersey.

Genus *Lacuna*

L. vincta. (Fig. 365.) A near relative of the littorinas, its shell thin
and conical; there are 5 whorls produced into a pointed spire. The char-

acteristic feature of the shell is the long umbilical groove extending along the columella to the deep umbilicus. Color gray, yellowish, or purplish, usually with 2 dark brown bands on the upper whorl and 4 on the lower one. Length 12 mm; breadth 7 mm. Very abundant on algæ in protected waters and on the holdfasts of *Laminaria* washed ashore. Circumpolar. Arctic Ocean to Long Island Sound.

FAMILY RISSOIDAE

GENUS *Rissoa*

R. minuta. (Fig. 366.) The shell of this little species is spiral and turreted, and with a regularly rounded aperture having a slightly expanded lip. There are 5 whorls, and the shell is thin and polished. The operculum is horny. The animal has eyes at the base of the tentacles, and the forepart of the foot is produced into filaments. Color yellowish, often with a pigmented coating of green. Length 4 mm; breadth 2 mm. Common. Found on seaweeds and under stones. Gulf of St. Lawrence to New Jersey.

Fig. 366—
R. minuta.

Fig. 367—
R. aculeus.

R. aculeus. (Fig. 367.) A species with most of the characters of the one just described, but the shell differs in having a proportionately longer and large spire. There are 6 whorls, the surface of which is marked with very minute revolving lines. Color light yellow. Length 4 mm; breadth 1.7 mm. Common on algæ. Greenland to Long Island Sound.

FAMILY CERITHIIDAE

GENUS *Bittium*

B. alternatum. (Fig. 368.) Shell elongated, with about 8 whorls, and with a rounded aperture having a very small anterior canal. Surface sculptured with a fine network of raised lines. Operculum horny and with 4 spirals. Color bluish or blackish. Length 6 mm; breadth 2 mm. Found in shallow water. Often abundant. Occurs from Massachusetts Bay to New Jersey.

Genus *Triforis*

T. perversa. (Fig. 369.) The shell in this species is long and sinistral; that is to say, it has the whorls turning to the left instead of to the right as is the case with the majority of univalve shells. There are 12 or more whorls, the surface of which is sculptured with 3 revolving series of granules; and the aperture is prolonged to form a closed anterior tube. Color reddish-black. Length 7 mm; breadth 1.7 mm. Fairly common. Found among the marine plants near the low-water mark. Cape Cod to west coast of Florida.

Fig. 368—B. alternatum.   Fig. 369— T. perversa.   Fig. 370— C. greeni.   Fig. 371—C. terebralis.

Genus *Cerithiopsis*

C. greeni. (Fig. 370.) Shell elongated, with 10 to 12 whorls, and the aperture having an anterior and a posterior canal, the posterior one being the less distinct of the two. Surface bearing a series of granules formed by longitudinal ridges and revolving lines. Color reddish-black. Length 5 mm; breadth 1.2 mm. Found in shallow water. Massachusetts Bay to Texas.

C. terebralis. (Fig. 371.) Shell slender and with 10 to 12 flattened whorls, each sculptured with 3 revolving ridges and numerous fine longitudinal lines between the ridges. Color reddish-brown. Length 12 mm; breadth 3 mm. Occurs in shallow water. Cape Cod to Texas.

Family **VERMETIDAE**

Genus *Vermetus*

V. radicula. (Plate XXX.) This singular shell would not readily be taken for that of a mollusk by the inexpert collector, its resemblance being strikingly close to the calcareous tube of some annelid. At the apex, the shell begins as an irregular spiral with about 8 of the whorls in contact; then the whorls become separated and extend aimlessly, the lower portion of the shell assuming a contorted appearance. There is a circular, horny operculum present. A number of smooth, concave partitions occur inside the shell. The visceral mass of the animal is greatly elongated, and, owing to its stationary habit of living, the foot is rudimentary. The head is long and bears 2 tentacles provided with eyes at their base, and there are 2 other tentacles at the side of the mouth. Color of animal light brown with black spots. Color of shell variable; usually white, yellowish, pink, or purplish. Length of spiral portion 12 mm; length of free portion up to 20 cm. Found in shallow water. Often numerous individuals will be found together with their shells closely intertwined. It is common from Cape Cod to Florida, but is much more abundant toward the southern limit of its range.

FAMILY STROMBIDAE

GENUS *Strombus*

S. gigas. (Plate XXX.) The queen conch. The shell of this species is very large and solid; it is with a conical spire, a large body whorl, a long and somewhat narrow aperture, and a wide flaring outer lip. The lip has an anterior canal and a less distinct posterior one, and a deep notch for the head near the lower edge. The surface is uneven and bears numerous protuberances. A striking feature is the polished pink surface of the interior. The animal is provided with a pair of eyes which are at the end of long stalks. A pair of tentacles project from the eye stalks. The foot is long and narrow, the metapodium, or hinder part, bearing the operculum. The operculum is curious in that it has the appearance and function of a claw; for the stromb does not creep as do most other gasteropods, but leaps, using the clawlike operculum to effect a purchase or foothold in the sand. Color brownish. This species is the largest gastropod mollusk on the Atlantic Coast, measuring 25 cm or more in length and weighing up to 5 pounds. It is edible, and its shell is often used as a dinner horn; also it is used as a common garden and household shelf ornament, and for making cameos. Common among coral reefs in shallow water where it lives as a scavenger. Florida Keys and West Indies.

S. pugilis. (Plate XXX.) The fighting conch. The habits and general characters of the species just described will apply to the present form; but here the shell differs in having a less expanded outer lip, which is orange or deep red blending into purple, then pink. The spire of this species is often set with sharp knobs on the upper margin of each whorl. Color light reddish-brown, often with areas or bands of dark brown or purplish. Length 9.5 cm. Found in shallow water. Cape Hatteras to Panama.

FAMILY CYPRAEIDAE

GENUS *Cypraea*

C. exanthema. (Fig. 372.) The measled cowry. This form, like all other cowries, has a shell which is solid, ovate, and with the surface very smooth and highly polished. In nearly all of them the spire is concealed by the last whorl, which is with an aperture as long as the shell and channeled at both ends. There is no operculum. In the animal the mantle is expanded on each side and reflexed, meeting over the top of the shell and completely covering it. The foot is well developed. In the present species the shell is characterized by its aperture, which is toothed on both sides and has the lip turned in; and by its color, which is brown with white spots, the spots often having brown centers. There is a yellowish band lengthwise along the middle of the dorsal region of the shell. The interior of the lip is violet. Length 11 cm. Lives in shallow water. Not uncommon in the south; rarer in the northern part of its range. Cape Hatteras to Panama.

Fig. 372—C. exanthema.

C. spurca. (Plate XXX.) The yellow cowry. A species with many of the characters possessed by the one just described. It may be distinguished by its color, which is yellowish on the upper side and pure white on the base. Examination will show that the yellowish color consists of numerous closely crowded yellow dots on a white background. Length about 12 mm. Found along the shore in southern Florida.

GENUS *Trivia*

T. pediculus. (Fig. 373.) The coffee-bean cowry. The shell of this species is somewhat like that of *Cypræa,* but the surface instead of being smooth is marked with transverse

Fig. 373—T. pediculus.

ridges which start from a median longitudinal groove on the back and extend around into the aperture. Color rose-pink with 6 brownish-black spots on the back. Length 20 mm. Found in shallow water and strewn along the beaches. Occurs in Florida.

FAMILY DOLIIDAE

GENUS *Dolium*

D. galea. (Fig. 374.) The shell of this species is thin and produced into numerous rounded revolving ridges. The spire is very short and with deeply channeled sutures; the body whorl is very large and swollen and with a crenulated outer lip. There is no operculum. The animal has a long proboscis which is curved over the back of the shell. Color light brownish-yellow. Length 25 cm. Lives in shallow water. Cape Hatteras to Texas; also in the West Indies.

Fig. 374—
D. galea.

D. perdix. (Fig. 375.) Similar to *D. galea* but with higher spire and less pronounced revolving ridges. Color brownish with crescentic spots of white. Length about 20 cm. Habitat and occurrence on our coast same as that of preceding species.

Fig. 375—D. perdix.

GENUS *Pyrula*

P. papyratia. (Plate XXX.) The paper fig shell. Shell thin and ornamented with numerous small revolving ridges crossed with distinct longitudinal growth lines. Spire nearly flat; body whorl very large and with flaring aperture which is produced anteriorly into a tapering canal. General outlines of shell fig- or pear-shaped. Color of exterior whitish, yellowish or brownish with brown streaks extending from spire; interior golden or chestnut. Length about 10 cm. Shallow water. Florida.

FAMILY CASSIDIDAE

GENUS *Cassis*

C. tuberosa. (Fig. 376.) The sardonyx helmet shell. This is a thick, solid shell with a short spire and large body whorl having a long, narrow aperture which terminates anteriorly in a canal that is bent backward. The outer lip is broad, much thickened and bears a number of white teeth. There is a horny operculum. The columella is covered with a broad callus and the columellar lip is ribbed.

There are 3 rows of protuberances on the body whorl, and numerous fine revolving and longitudinal grooves cross the surface of the shell. The general outline of the shell is triangular. Color yellow with brown patches. Length 20 cm. Cameos are often made from this shell, its composition being such that a

Fig. 376—C. tuberosa.

white subject can be cut to contrast with a black background. Found in sandy situations, where the animal preys chiefly on bivalve mollusks. North Carolina to West Florida; also in the West Indies.

Fig. 377—C. cameo.

C. cameo. (Fig. 377.) The black or cameo helmet shell. This shell can be recognized by the dark brown color of the teeth and ridges surrounding the aperture. The general color of the shell is yellowish-white tinged with yellowish-brown, the lip and columella being yellowish-cream. In outline, this shell is more swollen than the preceding form. Length 25 cm. Like *C. tuberosa,* this species is much used in cutting cameos. North Carolina to west Florida and in the West Indies.

DIVISION **GYMNOGLOSSA**

The gastropods in this division have no radula or jaws.

FAMILY **PYRAMIDELLIDAE**

GENUS *Turbonilla*

T. elegans. (Fig. 378.) Shell elongate and with rounded whorls. Surface marked with numerous fine revolving grooves and coarse longitudinal ridges. Lip not fully formed. Horny operculum present. Animal with ear-shaped tentacles joined at the base, and sessile eyes. Color light yellow. Length 9 mm; breadth 2.5 mm. Often common in shallow water. Massachusetts Bay to Florida.

Fig. 378— T. elegans.

Genus *Odostomia*

O. seminuda. (Fig. 379.) Shell elongate, with 6 or 7 flattened whorls, pointed apex and ovate aperture. Columella with oblique fold. Surface marked with coarse revolving lines crossed with longitudinal grooves. Animal with tentacles and eyes similar to those of *T. elegans*. Color of shell glossy white. Length 4 mm; breadth 2 mm. Rather common in shallow water. Massachusetts Bay to Florida.

Fig. 379—O. seminuda.

Division **RACHIGLOSSA**

The gastropods in this division are characterized by the possession of a long, narrow radula which is made up of three longitudinal rows of teeth consisting of a central row and two lateral rows. The teeth in the majority of species are composed of a base and a number of cusps. There is a well developed siphon, and the proboscis is long and retractile. The shell is spiral, is usually with an operculum which is invariably horny, and is with an aperture produced anteriorly to form a siphonal canal.

Family **MURICIDAE**

Genus *Murex*

M. pomum. (Plate XXX.) Shell with prominent spire and oblong-ovate aperture which has the siphonal canal nearly closed and somewhat bent back. Surface rough and with revolving ribs crossing varices and nodules. The varices are large and conspicuous, and they indicate the former position of the outer lip during the growth of the shell. Between each two varices occur 1 or more nodes. Color ashy or yellowish with brownish patches. Length 12 cm. Common. Cape Hatteras to Texas and West Indies.

Genus *Urosalpinx*

U. cinereus. (Plate XXXI.) The oyster drill. This form is closely allied to the one just described, but it is much smaller and has a greater range. It can be recognized by the 12 or more longitudinal ridgelike undulations crossing the whorls. The whorls number about 5 or 6 and the

PLATE XXXII

1—*Modiolus demissus*; 2/3x.   2—*Lithophagus lithophagus*, in section of rock; 1x.

3—*Pinna seminuda*; 1/3x.   4—*Ostrea virginica*; 1/3x.

5—*Pecten irradians*; 1/3x.   6—*Ensis directus*; 1/3x.   7—*Loripes edentula*; 1x.

8—*Tellina radiata*; ½x.   9—*Tellina alternata*; 2/3x.

10—*Labiosa canaliculata*; 2/3x.   11—*Venus mercenaria*; 1/3x.

12—*Venus cancellata*; ¾x.   13—*Petricola pholadiformis*; 1¼x.

14—*Cardium magnum*; 1/3x. 15—*Laevicardium mortoni*; 1¼x. 16—*Tagelus gibbus*; ½x.

PLATE XXXIII

1—*Mactra solidissima*; ½x.   2—*Pecten magellanicus*; ½x.
3—*Mytilus edulis*, typical form; 1x.   4—*Spirula peroni*; 4x

surface is also marked with numerous very fine revolving lines. Color brownish or gray; brownish inside. Length 25 mm; breadth 15 mm. Very common, especially on oyster beds, where its predatory habit causes considerable losses to the oystermen. It kills its victims by boring a hole through the shell with the radula and then thrusting in its proboscis to devour the animal. The range of this species on our coast is from Massachusetts Bay to Florida. It also occurs on the Pacific Coast.

Genus *Eupleura*

E. caudata. (Plate XXXI.) This shell has a characteristic flattened appearance owing to the situation and development of the varices on each side. There are 7 angulated whorls, each being with a shoulder below the suture. The aperture is elongated and with thick crenulated lips, the anterior portion being produced into a straight siphonal canal. Color gray or brownish. Length 25 mm; breadth 15 mm. Common. Lives near the low-tide level and in shallow water. Cape Cod to west Florida.

Genus *Purpura*

P. lapillus. (Fig. 380.) The shell of this species is oval and with a large body whorl and aperture. Within certain limits, the relative proportions of the spire and the body whorl are quite variable, those forms occurring in exposed situations being usually distinguished by relatively higher spires. The shell is thick and consists of 5 or 6 whorls, each bearing numerous revolving ridges and deep furrows. The outer lip is arched and ridged within the margin; the columella is nearly flat, smooth, with the lower end slightly deflected;

Fig. 380—P. lapillus.

and the aperture is with a siphonal canal that is so short as to form merely a notch at the anterior end. Color whitish, yellowish, or reddish; occasionally banded. Length 30 mm; breadth 17 mm. This is a predatory species, and is very common among barnacles and bivalve mollusks, these forming its principal food. It occurs near the low-water mark from Greenland to the eastern end of Long Island.

Family COLUMBELLIDAE

Genus *Columbella*

C. lunatia. (Fig. 381.) Shell slender, conical and with an elongated narrow aperture having thick lips and prominent teeth. A notch forms the siphonal canal at the anterior end. The whorls number about 6, and are smooth but marked with a single revolving line and several near the base. The operculum is very small. Color reddish-brown with 2 or 3 rows of crescentic spots of ivory encircling the lower whorl. Length 5 mm; breadth 2.5 mm. Common in shallow water. Massachusetts Bay to Florida.

Fig. 381— C. lunatia.

C. avara. (Fig. 382.) This is another common species, but it is less so than the one just described. It can be recognized by 10 to 15 undulating ridges on the body whorl which are crossed with revolving lines. Color yellowish with brownish patches. Length 12 mm; breadth 5.5 mm. Found near the low-water mark. Cape Cod to Florida.

Fig. 382— C. avara.

Family NASSIDAE

Genus *Nassa*

N. obsoleta. (Plate XXXI. Also see Plate XXIX.) In this species the shell is conical, with a prominent spire and an oval aperture which ends anteriorly with a slight notch. There are 6 whorls, these being with simple sutures, and without the angular shoulders that occur in *N. trivittata,* another member of the same genus with which it may become confused. Also the spire in the present species is blunter than in the species just mentioned. The surface is covered with numerous slightly developed revolving and longitudinal ridges. The posterior portion of the animal's foot is not bifurcated (not terminated with two appendices). Color of exterior brown; interior purple or black. Length 25 mm; breadth 12 mm. This is a very common and abundant species living in the mud flats and the shallow water of bays and inlets. It has the habit of collecting in shallow pools in such numbers that the bottom is often completely covered. Although a predaceous species destroying other mollusks by drilling a hole through their shells, it lives largely as a scavenger. Perfect shells of this species are the exception rather than the rule, as the spire is more often than not disintegrated or broken off at the apex. Gulf of St. Lawrence to Florida.

N. trivittata. (Plate XXXI.) This species and the one just described are nearly alike in general appearance; and as its range and habitat are the same as that of the other, there my be some confusion of identity when either is taken by the inexpert collector. This confusion can be avoided by observing that the shell of the present species differs from that of *N. obsoleta* in having a more prominent and pointed spire, and whorls with more deeply impressed sutures and with angular shoulders. Moreover, the surface is with numerous strongly developed revolving and longitudinal ridges, making a decussated (crossed; intersected) pattern on the exterior of the shell. It should be noted, also, that the foot of the animal is bifurcate behind. Color of exterior yellowish-white or greenish-white, often with brown revolving bands; interior white. Length 18 mm; breadth 8 mm. Very common. With the habits of *N. obsoleta.*

N. vibex. (Fig. 383.) The shell in this species is with a relatively larger spire and smaller body whorl than that in the two preceding forms. There are 6 whorls, these being somewhat flattened and with slightly impressed sutures. Chiefly characterizing the shell, however, is the surface sculpturing of the body whorl, which consists of 12 revolving lines and 12 undulating longitudinal ridges. The aperture is with a rather deep anterior notch. Like that of *N. trivittata,* the foot of this animal is bifurcate behind. Color whitish or brownish, with spots or a revolving band of darker shade. Length 12 mm; breadth 8 mm. Occurs from Cape Cod to the Gulf of Mexico and the West Indies, but is commoner south of Cape Hatteras.

Fig. 383—
N. vibex.

FAMILY BUCCINIDAE

GENUS *Buccinum*

B. undatum. (Plate XXXI.) A species familiar to all collectors along our temperate shores. The shell has a prominent conical spire and a large body whorl, the total number of whorls being 6. There is a large aperture, somewhat oval in shape, with a simple outer lip, and a twisted columellar lip. The aperture ends anteriorly in a wide notch. A laterally situated nucleus characterizes the operculum. The surface of the shell is covered with a yellowish horny periostracum. Its principal diagnostic feature, however, is found in the sculpturing of the whorls; these have 12 prominent oblique undulating ridges which are crossed with numerous elevated revolving ribs. Color grayish or yellowish exteriorly; white or yellowish interiorly. Length 60 mm; breadth 35 mm. This species represents the familiar "whelk" of the seas around Great Britain, where it is used for food. It is very common on our shores, particularly along the coast of Maine, and occurs from the low-tide level to deep water. Greenland to New Jersey.

Genus *Neptunea*

N. decemcostata. (Fig. 384.) Recognized by the 10 costæ, or revolving ribs on the body whorl, the upper whorls having but 2. The ribs are rounded and tend to decrease in size toward the lower, or anterior, end of the shell. There are 6 whorls; the spire is prominent and the body whorl is large. Aperture oval and with short anterior canal. Color of shell variable, usually white with yellowish, horny periostracum; interior pure white. Length 75 mm; breadth 45 mm. This form lives in rather deep water but it is often cast up on the beach by storms. Cape Cod northward.

Fig. 384—N. decemcostata.

N. pygmaea. (Fig. 385.) Shell with 6 whorls. Spire high; aperture rather small and elongated, ending anteriorly in a somewhat long siphonal canal. Surface with numerous revolving grooves and covered with a hairy periostracum which is conspicuously corrugated. Color grayish. Length 20 mm. Found in shallow and offshore waters. Gulf of St. Lawrence to Long Island Sound.

Fig. 385—N. pygmaea.

Family TURBINELLIDAE

Genus *Fulgur*

F. carica. (Plate XXXI.) The knobbed whelk. One of the commonest, as well as one of the largest univalves occurring on our shores. This

huge pear-shaped shell with its low spire and very large body whorl tapering anteriorly into a long twisted siphonal canal is familiar to all seashore
collectors from Cape Cod southward. It may be distinguished at once
from other whelks by the revolving row of knobs, or nodules, on the
shoulder of the body whorl, these being continued on the bases of spiral
whorls. Like that of most gastropods, the shell of this species is dextral,
that is to say, the aperture is on the right when the spire is upward, and
the whorls are with simple sutures, thus giving a flattened surface to the
spire. Color gray or brownish, with bright vermilion aperture. Length
22 cm; breadth 11 cm. The egg capsules of this species are often washed
ashore where their discovery excites the curiosity of the seashore visitor. These capsules are purselike containers attached in a row to a
common chord and may be distinguished apart from those of other
species by their squared, or angular, margins (see Plate XXIX).
Lives on sandy or gravelly bottoms in shallow and offshore waters
where it is predatory on other mollusks. Cape Cod to the Gulf of
Mexico.

F. canaliculatum. (Plate XXXI.) The channeled whelk. Like the
knobbed whelk, this form has a large, dextral, pear-shaped shell terminating anteriorly with a long, tapering, twisted siphonal canal. *F.
canaliculatum,* however, is immediately recognizable by the absence of
knobs on the whorls, and by the presence of a very distinct, deeply impressed, broad canal at the suture. The periostracum is quite thick and
hairy. Color of exterior gray or brownish; interior yellowish. Length
13 cm; breadth 7 cm. This species is usually associated with the knobbed
whelk and it is hardly less common, but it is more abundant locally in
its northern range. Its egg capsules differ from those of *F. carica*
in having the margins thin, or wedge-shaped in cross-section. (See
Plate XXIX.) Cape Cod to the Gulf of Mexico.

F. perversa. (Plate XXXI.) A southern form similar to *F. carica*
but with the shell characteristically sinistral. The spire is more depressed than that of the knobbed whelk, and in younger specimens the
shell is marked with longitudinal brownish bands or streaks. Older
specimens are less conspicuously colored, often being a dull white. The
animal has a black body. Length about 20 cm. Common in shallow water.
Cape Hatteras to Texas; also in the West Indies.

Family OLIVIDAE

Genus *Oliva*

O. literata. (Plate XXXI.) This handsome shell is easy to identify because of its beauty and of its being one of the very few so-called "olive shells" occurring on our coast. It is cylindrical, tapering at both ends, and the surface is smooth, highly polished and porcelanous in appearance. The spire is very short and the body whorl very large, the latter composing the greater part of the shell, while the aperture is long, narrow and with a deep anterior notch. Channeled sutures separate the whorls, and the columella is covered with a pure white, enamel-like, obliquely striated callus deposit. There is no operculum. The animal has a large foot, and the mantle extends laterally around the sides of shell and covers it. Color whitish or yellowish with longitudinal zigzag markings of brown; interior of aperture purplish. Length 7 cm; breadth 2.6 cm. Found along sandy shores, from North Carolina to Key West and the West Indies.

Genus *Olivella*

O. mutica. (Fig. 386.) This shell has a spire somewhat higher than that in the preceding species, and the aperture is shorter, being about half of the total length. The surface is white and shining and decorated with revolving wavy brown bands on the body whorl. Length 12 mm. Found along sandy shores, being particularly common in Florida waters. North Carolina to Key West and the West Indies.

Fig. 386— O. mutica.

Division TOXOGLOSSA

This division comprises certain gasteropods in which the radula is long and narrow and provided with only two rows of teeth, these being the marginals; central and lateral teeth are wanting.

Family CONIDAE

Genus *Conus*

C. floridanus. (Plate XXXI.) The distinctive shape of this shell with large body whorl tapering almost to a point below, thus having nearly the exact outlines of a cone, makes it easily identified. It has a conspicuous spire, pagodalike in appearance, and a long narrow aperture with a simple outer lip. The shoulder of the whorls is flattened and distinctly keeled. Color yellowish with revolving rows of brownish spots. Length 5 cm. Not uncommon in sandy shallow areas. Occurs from Cape Hatteras to Florida, but is less common toward the north.

FAMILY CANCELLARIIDAE

GENUS *Cancellaria*

C. reticulata. (Fig. 387.) Shell somewhat elongate, with conspicuous spire and large body whorl. Whorls with shoulders flattened below the sutures. Aperture elongate, ribbed within, and produced anteriorly to form a small siphonal canal. Columella with 2 distinct plaits. Surface of shell marked with prominent revolving and longitudinal grooves, thus giving a strongly cancellated appearance. Color whitish with bands of yellowish or reddish-brown. Length 5.5 cm. Found in shallow water. Cape Hatteras to Florida.

Fig. 387—C. reticulata.

CLASS PELECYPODA

The pelecypods are symmetrical bivalve mollusks without a distinct head. Clams and oysters may be taken as familiar representatives of the class. (See Fig. 388.) The two valves which compose the shell are hinged together dorsally and are not invariably symmetrical. In certain instances the shell may be rudimentary. Many shells have a thin, horny outer layer called the periostracum; in some species this non-calcareous layer looks like a coat of varnish, whereas in other species it is hairy. The inner, mother-of-pearl layer owes its iridescence to the diffraction of light by the exceedingly thin strata of nacre that make up this region of the shell. Both the outer and the middle layers are secreted by the thick outer edge of the mantle, but the pearly layer is secreted by the whole general surface of the mantle. The oldest or first-formed area of the shell is the *umbo,* or beak, at the dorsal margin, from which point the concentric successive lines of growth may be seen to extend to the ventral and lateral margins. This beak is often prominent and projects toward the anterior end of the shell. In some species there is present in front of the umbo a heart-shaped dorsal depression, the *lunule*. The two halves of the shell are held together, usually behind the umbones, by a ligament made up of the external ligament proper and an internal cartilaginous structure of very elastic properties. The ligament may be external, situated in excavated recesses of the shell margin, or it may be internal, occupying a groove or spoon-shaped hollow. Interlocking teeth called the *cardinals* and the *laterals* complete the hinge, the cardinals being immediately below the umbones, and the laterals, usually consisting of long ridges, being situated in front and behind the umbones, from which they derive the designations anterior and posterior laterals, respectively. The shell of the living animal is opened and closed by the adductor muscles,

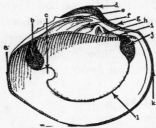

Fig. 388—Pelecypod shell. a, posterior margin; b, posterior muscle attachment area; c, pallial sinus; d, umbo and dorsal side; e, ligament; f, posterior lateral tooth; g, lunule; h, cardinal teeth; i, anterior lateral tooth; j, anterior muscle attachment area; k, anterior margin; l, pallial line; m, ventral margin.

and the areas of their attachment—called "muscle scars"—are often prominently indicated on the inner surface of the dead shell. Besides the muscle scars (some species have one adductor, others have two), there are often present on the inner surface a *pallial line* connecting the anterior and posterior muscle attachment area. This line represents the region of insertion for the retractor muscles of the mantle margin. An indentation occurs in the pallial line which is caused by the insertion of the siphonal retractor muscle, and it is known as the *pallial sinus*.

The mantle (see Fig. 389) which surrounds the visceral mass, is, like the shell, in two parts, and certain regions of the margin are modified to control the incoming and outgoing currents of

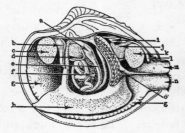

Fig. 389—Diagram of pelecypod, showing soft parts. a, stomach; b, palps; c, anterior muscle; d, liver; e, intestine; f, reproductive organs; g, mantle; h, foot; i, heart; j, kidney; k, posterior muscle; l, rectum; m, excurrent siphon; n, incurrent siphon; o, gills.

water, this modification very often being in the form of two tubes, or siphons, through the lower one of which the water flows into the mantle cavity bringing oxygen and the minute organisms upon which the mollusk lives. Water bearing waste products is carried out through the upper siphon. In certain species the siphons are long and well developed, in others they are little more than coalescence of the margin of the mantle, while in still others, such as the oyster, they are wanting altogether, the water currents merely flowing into the mantle cavity at the hinder and lower margins and escaping in the region of the anus, this circulation being aided by the constant beating of microscopic whiplike processes, or cilia.

The gills, lying in the mantle cavity where they are washed by the water currents, are sometimes called *ctenidia* because of their comblike structure. These are attached along their dorsal margins, there generally being two gills on each side. Gill types differ among the various groups of pelecypods, and it is upon these types that the orders of the class have been founded. (Fig. 390.) The fact that the gills are commonly arranged in layers or plates has suggested the older class name *Lamellibranchiata*. Osphradia, presumably olfactory organs, occur as sensory epithelial patches lying near the gills. As a rule, however, sense organs are not so highly organized as in other groups; ocelli, or eyes, particularly are often absent; when ocelli are present, however, these often may be well developed.

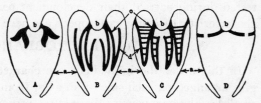

Fig. 390—Diagrams of pelecypod gills. A, *Protobranchiata;* B, *Filibranchiata;* C, *Eulamellibranchiata;* D, *Septibranchiata.* a, mantle; b, visceral mass; c, descending lamella of gill; d, ascending lamella of gill.

The visceral mass has at its forward end the mouth, which is without the radula, or rasping organ, characteristic of the class *Gasteropoda.* There is no distinct head; but the mouth has two leaflike lips, or *labial palps.* A short œsophagus connects the mouth with the stomach, which in turn is connected with the anus at the posterior end by a long tubular intestine winding about through the visceral mass. In certain forms, such as *Mya,* the soft-shelled clam, a crystalline rod of unknown function is present in a tubular pouch usually opening from the forward end of the intestine into the stomach. Surrounding the stomach is a very large liver.

The blood vascular system consists of a heart made up of a muscular ventricle, which sends the blood through the arteries and a pair of auricles, which receive the blood from the gills. The arteries convey the blood to hollow spaces in the body tissues, and veins transmit it from these spaces to the kidneys and the gills, and from these organs to the heart. There are two kidneys, each opening into the tissue surrounding the heart and into the mantle cavity.

The nervous system is composed of three principal paired ganglia; these are widely separated and are characteristic of most bivalve mollusks. They consist of the cerebral ganglia, situated immediately above the mouth; the pedal ganglia, located in the foot; and the visceral ganglia, which lie at the hinder part of the visceral mass. A pair of connective nerves joins the cerebral ganglia with the other pairs of ganglia.

Most pelecypods are unisexual, but hermaphrodites are not infrequent. Fertilization takes place usually in the water, but in some species the eggs are fertilized within the mantle cavity, where they remain protected during part of their development period. The special egg cases characteristic of many gasteropods are never formed by pelecypods.

ORDER **PROTOBRANCHIATA**

In this order the individuals are characterized by their featherlike gills arranged on each side of an axis, from which depend two rows of small leaflike expansions; these latter project freely backward into the mantle cavity. (See Fig. 390.) Two adductor muscles are present; the foot is more or less flattened and forms a sort of sole wherewith the animal creeps.

FAMILY NUCULIDAE

GENUS *Nucula*

Fig. 391—P. proxima.

P. proxima. (Fig. 391.) The shell is obliquely shaped in this species, thick and with conspicuous umbones. The hinge has 18 anterior and 12 posterior teeth, all of them being comparatively large, and the two series form almost a right angle. A central pit for the ligament occurs between the two rows of teeth. No siphons are present. Length 10 mm; height

9 mm. A species often common in muddy or gravelly situations near the shore. Gulf of St. Lawrence to South Carolina.

Genus *Yoldia*

Y. limutala. (Fig. 392.) The shell of this species is rounded in front and somewhat tapering behind, and the surface is covered with a smooth glossy periostracum. Umbones nearly central. The teeth are prominent, 23 being anterior and 18 posterior with a central pit reserved for the ligament between the two rows. Small retractile siphons present. Color of shell dark greenish outside, light bluish and nacreous inside. Length 48 mm; height 23 mm. This is an active species, being able to crawl or leap with considerable rapidity, which it does by using its muscular foot. Common in shallow water. Labrador to North Carolina.

Fig. 392—Y. limutala.

Fig. 393—Y. sapotilla.

Y. sapotilla. (Fig. 393.) A species closely resembling the preceding except that it has from 16 to 18 teeth on each side of the central space occupied by the ligament. The hinder part of the shell is broader than that of *Y. limutala,* but its total size is much less, measuring 22 mm long and 11 mm high. The interior is pearly-white. Often common in comparatively shallow and in offshore waters. Arctic Ocean to Long Island.

Y. thraciaeformis. (Fig. 394.) This species has many of the characters of the preceding described forms belonging to the genus. It may easily be distinguished, however, by its larger size and by the riblike undulation extending obliquely from the umbones of the shell to the posterior ventral margin. The shell also differs quite markedly from that of the others in being broadest behind. There are 12 teeth on each side of the space for the ligament. Color dark greenish. Length 5 cm; height 4 cm. Not uncommon. Occurs in offshore waters. Arctic Ocean to Long Island.

Fig. 394—Y. thraciæformis.

Genus *Leda*

L. tenuisulcata. (Plate XXXI.) In this species the shell is elongated with the front end rounded and the hinder part produced into a rather narrow angle, blunt and slightly gaping at the tip. There are 12 anterior and 16 posterior teeth with a central pit for the ligament between the two rows. The umbones are smooth and a number of concentric grooves mark the outer surface of the shell. Small, united retractile siphons are present. Color greenish externally; nacreous within. Length 25 mm; height 12 mm. Common. Found usually on muddy bottoms in shallow water. New England.

FAMILY SOLEMYIDAE

GENUS *Solemya*

S. velum. (Fig. 395.) The shell in this species is elongate and conspicuously marked with about 15 impressed lines radiating from the

Fig. 395—S. velum.

umbones to the lateral and ventral margins. The shell is delicately thin and very fragile, but it is protected by a strong horny periostracum which extends considerably beyond the margins. The mantle of this animal is fused below with an opening only for the foot and small siphon. Color of periostracum chestnut-brown; internal color of shell bluish-white. Length 25 mm; height 12 mm. Sometimes abundant. Lives usually buried in the sand in shallow water. Nova Scotia to North Carolina.

S. borealis is a larger and very similar species, but it is darker and much rarer. It is 5 cm long and 22 mm high. Its habitat is much the same as that of *S. velum,* but its range does not extend quite so far southward.

ORDER FILIBRANCHIATA

The gills characterizing the animals of this order are filamentous. They depend freely into the mantle cavity, each row of gills having the filaments recurved, or folded back. (See Fig. 390.) The foot is provided with a gland for the secretion of a hornlike fibrous substance used by the animal to secure itself firmly to surrounding objects. This fibrous holdfast is known as the *byssus,* and it usually occurs in the form of numerous fine, silklike threads cemented at their distal ends to the object of attachment. (See Fig. 399.) In some forms the byssus is a single, thick, calcified projection. The feeble development of the foot is characteristic of mollusks having a strong byssus. There are two adductor muscles present in members of this order.

Family ANOMIIDAE

Genus *Anomia*

A. ephippium. (Plate XXXI.) The common jingle shell. In this species the shell is variable in outline, but it is most commonly circular or oval, the shape being usually modified by the character of the surface to which the animal is attached. The right, or lower, valve is flattened and has a large elongated hole near the hinge through which a short, thick, calcified byssus passes, by means of which the mollusk anchors itself to some object, and with the entire surface of the valve closely approximated. The left, or upper, valve is convex. Both valves are thin and fragile, with the outer surface often scaly and dark in the living animal, but in the dead shells this covering is worn away and the light greenish or copperish or golden nacre is exposed. There are large gills, the mantle is fringed, and the foot, though not well developed (a feature which is correlated with a stationary habit), is cylindrical with the end expanded and grooved. Siphons, as well as oral palps, are wanting. Diameter of shell 25 mm or more. Very abundant on stones, oysters, dead shells and other solid objects in shallow water. All ranges.

A. aculeata. (Fig. 396.) A smaller species somewhat similar to the preceding, but with the surface of the upper valve covered with radiating prickly scales. The lower valve is comparatively thin and the hole for the passage of the byssus is nearly

Fig. 396—A. aculeata. A, upper valve; B, lower valve.

circular in outline. It is found in shallow water, attached to stones and other solid objects, often about the holdfasts of rockweeds. Arctic Ocean to Long Island; commoner toward the north.

Family ARCIDAE

Genus *Arca*

A. pexata. (Plate XXXI.) The bloody clam. The shell is oblong with the umbones prominent and directed very obliquely forwards.

Extending from the umbo to the margin of each valve are from 32 to 36 radiating ribs, causing the inner margin to appear deeply scalloped. The periostracum is thick and bristly with a dark brown color. The hinge is with a row of comblike teeth posterior to the umbones, while just below the umbones are a number of irregular ligament pits. The mantle is open below and the edges bear a row of compound eyes. The gills are obliquely arranged in position, and have the filaments completely free. There is a strong, well developed pointed foot with a byssus, but no siphons are present. Both the gills and the circulatory fluid of this species are red, from which fact it gets its popular name. Length 56 mm; height 53 mm. Often common. Found in shallow water on the bottom or attached to stones, etc. All ranges.

A. transversa. (Fig. 397.) Shell rhomboidal in outline and with 32 to 36 deeply cut ribs marked with fine scaly striæ and crossed with a few deep, concentric lines. Color brown. Length 37 mm; height 25 mm. In general this species resembles *A. noae,* but can be recognized by its rhomboidal shape and the umbones which do not point obliquely forwards. Not uncommon. Found in shallow water. Cape Cod to Florida.

Fig. 397—A. transversa.

A. ponderosa. (Plate XXXI.) The ponderous ark shell. A species in which the shell is short and thick, the umbones being prominent, directed somewhat forward, and rather widely separated, thus giving the shell from an end view a symmetrically heart-shaped outline. There are from 25 to 28 ribs radiating from the umbo of each valve to the margin. The surface of the living shell is covered with a very dark furry periostracum. The hinge is with a row of comblike teeth, those in the middle being somewhat smaller. The soft parts are like those of *A. pexata.* Length about 6.5 cm; height 5 cm. Common. Found in shallow and offshore waters. Cape Cod to Florida. The fossil valves of this shell are said also to occur along the New Jersey coast.

A. noae. (Plate XXXI.) Noah's Ark shell. The shell of this species is in the form of an irregularly shaped quadrangular box. The umbones are prominent, situated very far forward, and separated by a wide dorsal repression above the hinge margin, which latter is long and very straight. From the umbone to the abruptly truncated posterior margin of each valve extends a gradually widening groove, while the general surface is strongly ribbed over all. The ventral margin is sinuous with a gap at the middle for the extension of the byssus. This species secretes a byssus of numerous fine threads, which, however, hardens into a horny cone consisting of thin plates. There are about 50 comblike teeth in the hinge. The soft parts have the same anatomical features as those of *A. pexata*. Length 10 cm; height 6 cm. Common. Found in usually among stones and rock crevices in shallow water. Cape Hatteras to Florida; West Indies.

FAMILY MYTILIDAE

GENUS *Mytilus*

M. edulus. (Plate XXXIII.) The edible mussel. The shell is wedge-shaped, tapering to a point in front, and rounded behind. It is characterized by having the umbones at the extreme anterior end. Hinge teeth are wanting. The mantle margins are fringed, and are joined at a certain point to form an excurrent anal siphon. No distinct branchial siphon is present. The foot is cylindrical and grooved, and it is provided with a byssal gland that secretes very tough, fibrous byssal threads. The labial palps are large. Color of periostracum, black or dark brown; color of shell interior, pearly with purple margins; color of flesh, white. Length 10 cm; height 3.5 cm. This mussel, in common with other members of its family, can move about by extending the foot and attaching itself successively by newly formed byssal threads. It is associated in enormous numbers in shallow water or between the tide marks, often forming vast tracts of closely crowded individuals, and it is generally anchored to solid objects or its fellows by its byssus in such a way that the umbones or pointed ends of the shell are directed downwards. Very common. Arctic Ocean to North Carolina.

Fig. 398—M. hamatus.

M. edulus, variety *pellucidus*. (Plate XXXI.) Found associated with the typical form and differs only in the color of the periostracum, which is of a greenish or yellowish cast, through which are visible the shell markings consisting of radiating lines of deep purple.

M. hamatus. (Fig. 398.) The hooked mussel. This species has the general characters of the two preceding, but is unlike them in having the shell conspicuously bent below the umbonal region, and in having the surface covered with fine, irregular, radiating ridges. Color deep grayish-brown. Length 5 cm; height 3 cm. Common and very abundant. Chesapeake Bay to Florida.

Genus *Modiolus*

M. modiolus. (Fig. 399.) The bearded mussel. The common horse mussel. Because of its wedge-shaped outline, this form has a superficial

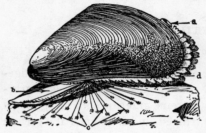

resemblance to the edible mussel, but careful examination will show several striking points of difference. The umbones are almost, but not quite, at the front end of the shell, and this region is considerably swollen. The periostracum is thick and coarse in texture and is produced into a shaggy growth, particularly toward the ventral posterior margin of the shell. Another distinguishing feature is found in the fleshy parts, which are orange-colored; the foot is red. The soft parts are similar, anatomically, to those of the edible mussel. Color of periostracum dark brown. Length usually about 4 cm; height about 2 cm; however, individuals measuring up to 15 cm are not uncommonly found. Found occasionally in rock crevices in the tidal zone; common in offshore waters. Arctic Ocean to New Jersey.

Fig. 399—M. modiolus. a, siphon; b, foot; c, byssal threads; d, mantle margin.

M. demissus. (Plate XXXII.) The ribbed mussel. This form, often bearing the species name *plicatula,* can be distinguished at once by the elongate, wedge-shaped shell marked with numerous radiating ribs, those toward the front end being much finer than the others. The umbones, like those of the preceding species, are not quite at the front end of the shell. A gap, through which the byssal threads pass, occurs in the ventral margin. The surface is dingy, the periostracum being a drab yellowish-green in color. Internally, the shell is pearly, tinted with purple. Length 7.5 cm; height 3 cm. This is one of the commonest pelecypods on our coast, occurring in great abundance in salt marshes and mud flats. All ranges.

Genus *Modiolaria*

M. nigra. (Fig. 400.) The black mussel. A form having a somewhat elliptical shell, the surface of which is marked with concentric growth lines and bears a small area of fine radial ribs near the anterior end and a much larger ribbed area toward the posterior end; between the two areas occurs a considerable space without ribs. The soft parts are somewhat similar in structure to those of the edible mussel, but the anal siphon is long, and the foot may be used as a prehensile organ for clinging to seaweeds and other objects as well as for affixing the byssus. Although popularly known as the "black mussel," the color is variable, ranging from purplish or brownish to black. Length 5 cm; height 2.8 cm. Not uncommon. Found in shallow water. Arctic Ocean to Long Island.

Fig. 400—M. nigra.

Fig. 401—
M. discors.

Fig. 402—C.
glandula.

M. discors. (Fig. 401.) A species much resembling *M. nigra,* but about one half its size and marked with 16 ribs in the area near the front. Found in shallow and offshore waters. Common locally. Arctic Ocean to Long Island.

Genus *Crenella*

C. glandula. (Fig. 402.) Shell elliptical or rhomboidal in outline, and surface marked with numerous fine radiating ribs usually crossed with a few concentric growth lines. Umbones on middle dorsal margin. A tooth occurs in the hinge of each valve. This species represents a genus in which the byssus consists of a single thread. The foot has a disklike expansion at the end. Color brownish-yellow. Length 12 mm; height 9 mm. Common in shallow water. Gulf of St. Lawrence to New Jersey.

Genus *Lithophagus*

L. lithophagus. (Plate XXXII.) Shell cylindrical, rounded in front and somewhat tapering behind. Surface with numerous fine longitudinal and transverse lines. Color chestnut-brown. Length up to 8 cm. This is a rock-boring species and when young attaches itself by the byssus; later it bores or dissolves out a burrow in coral rocks or dead coral, within which it lives for the remainder of its life, the diameter of the opening usually remaining smaller than that of the cavity enlarged by the growing animal. Common. Florida.

ORDER **PSEUDOLAMELLIBRANCHIATA**

The gill structure characterizing the members of this order consists of a series of more or less connected filamentous leaflets in each row, with both the descending and the ascending limb of each leaflet joined by interfilamentary partitions. (See Fig. 390.) The foot is poorly developed or wanting. Usually but one adductor muscle is present; if there are two, the anterior one is small and feeble. There are no siphons, the mantle being entirely open.

FAMILY **AVICULIDAE**

Genus *Pinna*

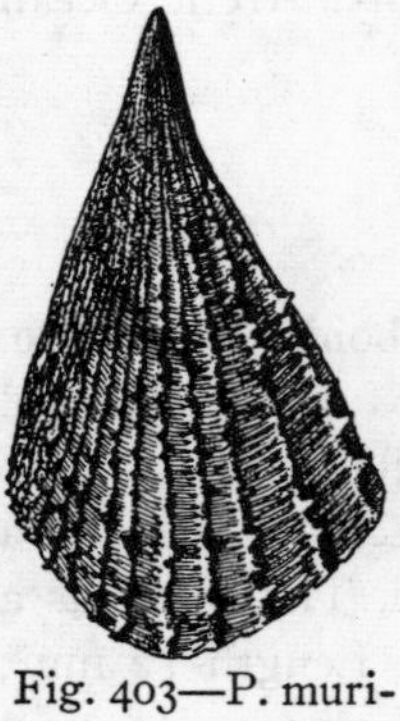

Fig. 403—P. muri-
cata.

P. muricata. (Fig. 403.) The fan shell. The razor shell; so called because of its sharp edges. In outline the shell is somewhat triangular or wedge-shaped. At the pointed end are the umbones. The shell has a translucent whitish crystalline structure, and the surface is produced into a number of radiating ribs bearing sharp, spinelike scales. Length 17 cm. This animal has the habit of attaching itself with its byssus so that the pointed end is directed downward, leaving the razor-sharp edges of the broad end exposed just above the surface of the mud or sand in which it lives, thus endangering the unshod or incautious collector who may chance to tread upon it. Common. Found in shallow water. North Carolina to Texas.

P. seminuda. (Plate XXXII.) A fan shell often found associated with the preceding form. It may be distinguished from the other by the far more numerous ribs and scales. These are more prominent toward the broad end of the shell, but toward the umbones they become so minute and indistinct that they usually disappear entirely, leaving the apical half of the shell bare—hence the scientific name *seminuda*. As a rule, the margin at the broad end of the shell in this species forms a straight line.

FAMILY OSTREIDAE

GENUS *Ostrea*

O. virginica. (Plate XXXII.) The American, or common, oyster. In this well known species, the shell is made up of two unequal valves, the upper valve being rather flat while the left, or lower, valve is convex and is the one that is attached to the objects on which the animal finds an anchorage when young. The shell is exceedingly variable in outline, and the structure is thick and frequently in folded layers with a rough, shingly surface. There is but 1 adductor muscle; this is toward the middle, with the gills curving around it and with the heart immediately in front of it. Color whitish. Length up to 45 cm, but more commonly about 12 cm. Very common and abundant. Lives in shallow and offshore waters; often occurs in water that is brackish. Gulf of St. Lawrence to Gulf of Mexico.

FAMILY PECTINIDAE

GENUS *Pecten*

P. irradians. (Plate XXXII.) The common scallop. In this form, like other true scallops, the valves of the shell are with "ears," or "wings," which are projections from each side of the umbones and which form the hinged dorsal margin. The present species has these wings about equal in size, and about 20 radiating ribs and numerous growth lines marking the surface of the shell. The adductor muscle is near the middle of the body and is encircled by the filamentous gills, but these are attached by only one lamella. The mantle is completely open and has a row of brilliant blue eyes along each margin. Color of shell variable, the upper valve being usually the darker; color of mantle margins scarlet or orange. Length 75 cm; breadth 73 cm. Scallops are an important article of food, the adductor muscle being the part that is commonly eaten. They are active mollusks, being capable of leaping or swimming about by snapping together the valves of their shells, and make very interesting occupants of the aquarium. *P. irradians* is found in shallow and offshore waters, often in abundance over mud flats and among eelgrass. Cape Cod to Texas.

P. islandicus. (Fig. 404.) This species has the wings unequal in size, and there are more than 50 (sometimes as many as 100) radiating ribs marking the surface of the shell. The lower valve is generally flatter than the upper. Color variable, but usually light orange to reddish-brown. Length 9 cm; breadth 7.5 cm. The range of this species is more northerly than that of *P. irradians,* extending from Cape Cod to the Arctic Ocean.

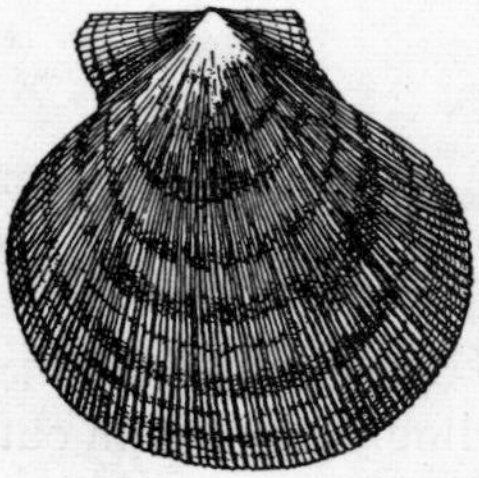

Fig. 404—P. islandicus.

P. magellanicus. (Plate XXXIII.) The giant scallop. Wings about equal in size. Surface of shell marked with fine radiating striations instead of the prominent rounded ribs characteristic of the preceding described scallops. Color of upper valve brown; lower valve white. Length 17 cm; breadth slightly less. Not uncommon. Found in the deeper bays and harbors and other offshore waters. Labrador to New Jersey.

ORDER **EULAMELLIBRANCHIATA**

In this order the gills are placed two on each side, each gill consisting of filaments and having a descending and an ascending limb, the two limbs being joined by processes, or interlamellar partitions. The adjacent filaments are likewise joined with each other, but the interfilamentary connections are vascular channels, or *trabeculae.* Thus the entire gill structure has a reticulated appearance. (See Fig. 309.) The two limbs each form a continuous lamella, the margin of the outer lamella of the outer gill uniting with the mantle. Two adductor muscles are commonly present. The mantle margins are more or less fused, and the presence of siphons, either in the form of closed tubes or not, is an invariable characteristic. The majority of pelecypods are included in this order.

Suborder SUBMYTILACEA

Shell with well developed cardinal and lateral teeth. Two adductor muscles present. The siphons are closed, and the pallial sinus is generally wanting. Mantle margins open below.

Family CARDITIDAE

Genus *Venericardia*

V. borealis. (Fig. 405.) The shell of this species is obliquely heart-shaped with the umbones elevated and directed forward. About 20 rounded, radiating ribs mark the surface, which latter is covered with a rough rusty brown periostracum. The hinge is with 2 conspicuous oblique cardinal teeth and 1 lateral tooth in each valve. The ligament of the hinge is external. Length 25 mm; height 17 mm. Occurs in off-shore waters, but the dead shells are commonly cast upon the beach. Labrador to New Jersey.

Fig. 405—
V.borealis.

Family ASTARTIDAE

Genus *Astarte*

A. castenea. (Fig. 406.) Shell thick and heavy. Surface slightly undulated and marked with concentric striations. Periostracum thick

Fig. 406—A. castenea.

and colored chestnut-brown. Umbones prominent, directed forward, and situated nearly in the middle of dorsal margin. Hinge with 2 well developed cardinal teeth in each valve. Lateral teeth wanting. Ligament external. Distinct lunule present. There is a short anal siphon. The foot of this mollusk is red. Length 25 mm; height about the same. Not uncommon. Found in shallow and offshore waters. Nova Scotia to New Jersey.

A. undata. (Fig. 407.) A species somewhat similar but larger than the preceding and with 10 to 20 prominent concentric ridges. Color brownish-olive. Length 25 mm; height 27 mm. Common. Found in shallow water. Gulf of St. Lawrence to Long Island Sound.

Fig. 407—A.
undata.

Fig. 408—C. islandica.

Fig. 409—L.
filosa.

FAMILY CYPRINIDAE
GENUS *Cyprina*

C. islandica. (Fig. 408.) The shell of this species has the general shape of that of an *Astarte,* but it is very much larger. It is recognized by its coarse, wrinkled periostracum and the hinge, which has 3 diverging cardinal teeth and 1 posterior lateral tooth in each valve. Color dark brown or black. Length 80 mm; breadth 75 mm. Occurs in water of considerable depth, but the valves occasionally are found on the shore after a storm. Arctic Ocean to Long Island.

FAMILY LUCINIDAE
GENUS *Lucina*

L. filosa. (Fig. 409.) A species with a thick shell having rounded outlines and prominent growth lines on the surface. The hinge margin is straight, while the hinge is with 1 cardinal tooth in the left valve and 2 teeth in the right valve. Lateral teeth are not present. Ligament semi-external. There is a distinct lunule. Pallial sinus wanting. Mantle open below. Foot very long and wormlike. Color whitish. Length and height 37 mm. Occurs in water of considerable depth, but the dead shells are not uncommonly cast up on the beach. Arctic Ocean to Patagonia.

L. tigrina. (Fig. 410.) Shell flatly convex with circular outline. Surface marked by numerous radiating ribs which are crossed by many concentric ridges. Otherwise, except in size, this form is distinguished by the same general characters as the preceding. Length and height about 75 mm. Florida to Texas.

Fig. 410—L. tigrina.

L. dentata. (Fig. 411.) A form having the general characters of *L. filosa,* but differing in the surface markings of the shell, these consisting of a

Fig. 411—L. dentata.

number of concentric growth lines between which traverse numerous deep oblique narrow grooves crossed by well marked concentric lines. The margins bear teeth formed by the grooved character of the surface. The valves are rather thin and white. Length and breadth 25 mm. Occurs in offshore waters, but the dead shells are not uncommon on the beach. Cape Cod to Brazil.

GENUS *Loripes*

L. edentula. (Plate XXXII.) A species with the circular outlines of *Lucina,* but with more inflated valves. It may be recognized by the finely striated growth lines on a white exterior, whereas the interior is colored a bright yellow or orange, particularly in the region of the pallial line and the areas where the adductor muscles are attached to the shell. The hinge teeth are not well developed, and the ligament is very weak. Length and breadth about 6 cm. This is a deep-water species; its dead shells, however, are often abundant along the shore. Cape Hatteras to the Gulf of Mexico.

SUBORDER TELLINACEA

The shells of the pelecypods in this order are elongate or somewhat triangular with a very large pallial sinus, the latter not infrequently extending forward nearly to the anterior adductor muscle. The valves are almost or exactly alike in shape. There are two adductor muscles. The mantle margins are not fused below, while the siphons are very long and most often are separate. The foot is large; likewise the labial palps.

FAMILY **TELLINIDAE**

GENUS *Tellina*

T. tenera. (Fig. 412.) Shell porcelanous, with a somewhat ovate outline, rounded in front and obliquely produced behind. Umbones just posterior to the middle. There are a posterior and an anterior lateral tooth in each valve, and 1 of the 2 cardinal teeth present is rudimentary. Ligament external. The surface is polished and marked with delicate concentric growth lines. The shell is delicate with a whitish or pinkish iridescent cast. In the genus to which this species belongs the siphons are remarkably long and slender. Length 14 mm; height 9 mm. Lives in offshore waters, but the dead shells often are numerous on sandy beaches. Gulf of St. Lawrence to west coast of Florida.

Fig. 412—T. tenera.

T. radiata. (Plate XXXII.) The sunset shell. This handsome species has the shell plainly marked by 3 broad pink rays on a shining white ground, the rays extending from the umbo of each valve to the ventral margin, a striking feature from which it derives its popular name. In outline it is somewhat narrowly elliptical, broad in front and narrowing behind with the umbones a little back of the middle. The surface is smooth, highly polished, and with faintly impressed concentric growth lines. Texture porcelanous. A yellowish tinge usually suffuses the umbonal region, while in some varieties the shell is without the rays and is cream-colored throughout and with pink umbones. Length about 8 cm; height 3 cm. Not uncommon. Occurs in shallow, sandy areas. Cape Hatteras to the Gulf of Mexico.

T. alternata. (Plate XXXII.) A species somewhat resembling the preceding and having the same habitat and range. It may be distinguised from *T. radiata* by the more pointed hinder end and the slight deflection to one side in this region of the shell; also, the entire surface is marked with distinctly impressed, fine, concentric, close-set grooves, every other one usually being obscured toward the posterior end, which is produced into an angular ridge extending diagonally from the umbones to the hinder margin. Color and variation as in *T. radiata*.

GENUS *Macoma*

M. tenta. (Fig. 413.) Shell elliptical, narrowed and slightly warped posteriorly, thin and with pointed umbones. Exterior marked with fine,

concentric growth lines; interior polished and shining. Right valve with
2 cardinals; left valve with 1; no laterals. Color white. Length 15 mm;
height 10 mm. Occurs in muddy bays. Common. Cape Cod to Florida
and West Indies.

M. baltica. (Fig. 414.) Shell almost circular, narrowing somewhat
toward the hinder end. Species variable in choice of habitat and in appearance: when living in a clean shallow sandy areas, the shell is pure
white, pinkish or yellowish and thin; when inhabiting muddy bays or
estuaries, the shell is bluish or rusty, is covered with a dingy periostracum, and is rather thick throughout. Hinge with 2 cardinals in each
valve; no lateral teeth present. Length 23 mm; height 18 mm. Very
common. Arctic Ocean to Georgia.

Fig. 413— M. tenta.   Fig. 414—M. bal- tica.   Fig. 415—D. fossor.

FAMILY DONACIDAE

GENUS *Donax*

D. fossor. (Fig. 415.) The shell of this species is roughly triangular
with the forward region elongated and obliquely rounded at the end and
with the hinder part abruptly truncated so as to be nearly square at the
end. Surface marked with radiating striations. Each valve with 2 cardinal teeth and 1 lateral. Hinge ligament external. Inner margin of shell
crenulated. The foot of this animal is very long, while the gills are very
unequal in size. Color white. Length 12 mm; height 7 mm. A shallow-
water form. Common. South shore of Long Island to Texas.

FAMILY MACTRIDAE

GENUS *Mactra*

M. solidissima. (Plate XXXIII.) The surf or giant clam. This form
has a shell that is somewhat triangular, massive, and slightly gaping
at the ends. It can be easily recognized by the characters of the hinge:

there is a large, conspicuous, heart-shaped internal ligament fossette, or pit, from the apex of which extend 2 long bladelike lateral teeth bearing cross-striations. A V-shaped cardinal tooth is also present, but this is commonly broken off in separated valves. There are 2 ligaments, an external and an internal. The mantle is open below, and the siphons are long, fused, and with fringed openings. Foot strong, bent, and tongue-shaped. The calcareous part of the shell is white, but the valves are protected by a brownish or yellowish periostracum, which, however, is often much worn away in the adult animals. This is the largest bivalve mollusk on our coast; it measures 15 cm in length and 12 cm in height. Very common. Inhabits sandy bottoms in shallow water. Labrador to Cape Hatteras.

Fig. 416—M. lateralis.

M. lateralis. (Fig. 416.) This is a much smaller but very common *Mactra,* and has a shell with many of the characters that distinguish the preceding form. It is, however, with more pointed umbones and with a strongly marked taper toward the anterior end. Length 15 mm; height 12 mm. Often very abundant, locally, in muddy areas, salt marshes and shallow estuaries. Nova Scotia to Texas and West Indies.

Genus *Labiosa*

L. canaliculata. (Plate XXXII.) Shell rounded and swollen in front, tapering rather abruptly and compressed behind, the latter region with a gap between the valves. Structurally thin and fragile. The surface is characteristically and strongly marked with fairly evenly spaced concentric ribs and grooves which become increasingly heavier as they approach the ventral margin. A distinct ridge is present below the posterior dorsal margin. The ligament pit in this species resembles that of *Mactra,* but is proportionately smaller. Hinge teeth not well developed. Color white. Length 5 cm; height 4 cm. Not uncommon. Occurs in sandy shallow regions. New Jersey to Mexico.

Suborder **VENERACEA**

In this suborder of bivalves the shells reach their highest degree of development. In outline they range from roundish to oval or elongate. A pallial sinus is present; and there are two adductor muscles. There is usually no fusion of the mantle edges below, but there are siphons, which in most cases are short and more or less united.

FAMILY VENERIDAE

GENUS *Venus*

V. mercenaria. (Plate XXXII.) The hard-shell clam or quahog. The shell of this species is obliquely rounded or somewhat heart-shaped with prominent umbones and a distinct lunule. The umbones are directed forward. The surface is marked with a number of concentric growth ridges. Structurally the shell is thick, solid, and well developed. The hinge is with an external ligament, and with 3 stout cardinal teeth in each valve, but without laterals. A minute crenulation occurs along the inner lower margin of each valve. Externally the shell is a dingy white, while the inner surface is a pure dull white with violet areas on the lower margin and about the muscle scars. Length 11 cm; height 9 cm. This is the common edible, or "little-neck," clam extensively served in hotels and restaurants. It is very abundant in shallow, muddy, or sandy areas, particularly south of Cape Cod. Gulf of St. Lawrence to Texas.

V. cancellata. (Plate XXXII.) Shell somewhat heart-shaped. Surface strongly marked with radiating ribs crossed with less numerous concentric ridges. Hinge with 3 cardinal teeth in each valve; no laterals; ligament external. Color white without, and usually with brownish spots; interior white with violet areas. Length 4 cm; height 2.7 cm. Very abundant on sandy bottoms in shallow water. Cape Hatteras to west coast of Florida; West Indies.

GENUS *Callocardia*

C. convexa. (Fig. 417.) This species may easily be mistaken for *Venus mercenaria,* which it very closely resembles. If, however, the collector will note that the present species has no purple on the inner surfaces of the shell, and that the inner margins are not crenulated, there is likely to be no confusion. Moreover, the hinge is with lateral teeth as well as cardinals, and the siphons are very long. Length 43 mm; height 35 mm. Nova Scotia to Florida.

Fig. 417—C. convexa.

Family **PETRICOLIDAE**

Genus *Petricola*

P. pholadiformis. (Plate XXXII.) This elongated, somewhat elliptical shell with its widely gaping posterior end and sculptured surfaces is a familiar object to Atlantic shore collectors, being cast up on the beach in considerable numbers. The umbones are at the anterior end, and from these extend radiating grooves and ribs, those on the front of the valves being coarse and prominent, whereas those on the long produced hinder part are fine and appear as striations. The surface is also marked with numerous concentric growth lines, and where these cross the large anterior ribs, the latter are produced into a series of arched scales. In texture, the shell is chalky, and the valves are thin and fragile. There are 2 delicate cardinals in each valve, and no laterals. There is a deep pallial sinus. The mantle is closed for the most part, leaving an opening only for the small foot to protrude. The siphons are long and united about one third of their length at the base. Color dull chalky white. Length 32 mm; height 17 mm. This is a boring species inhabiting burrows the diameter of its shell in clay, limestone, and coral situations in the tidal zone and shallow water. It is especially common south of Cape Cod, boring in clay. Gulf of St. Lawrence to Texas.

Suborder **CARDIACEA**

The forms contained in this suborder are generally distinguished by a heart-shaped shell. The ligament of the hinge is external. Except for an opening for the extrusion of the foot, the mantle is closed below. Short siphons are present. Many of the species are known as "cockles."

Family **CARDIIDAE**

Genus *Cardium*

C. islandicum. (Fig. 418.) The large rounded shell of this species is characterized by 36 to 38 three-sided, radiating ribs which are separated by deep rounded grooves. Crossing the ribs and grooves are a number of concentric growth lines. In the living animal the shell is protected by a yellowish-brown periostracum, which projects in the form of a fringe

along the ribs. Umbones near the middle of dorsal margin. Hinge in
each valve with 2 unequal-sized cardinals and with an anterior and a
posterior lateral placed some distance from the cardinals. The border of
the inner surface has a row of prominent denticu-
lations. Mantle with papillæ along the margin.
Siphons separate. Foot large and recurved.
Length 5 cm; height a trifle less. Lives in offshore waters. Long Island to Arctic Ocean, and
circumpolar; very rare south of Cape Cod,
however.

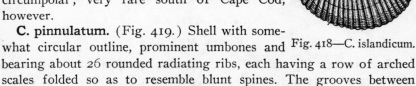

Fig. 418—C. islandicum.

C. pinnulatum. (Fig. 419.) Shell with somewhat circular outline, prominent umbones and
bearing about 26 rounded radiating ribs, each having a row of arched
scales folded so as to resemble blunt spines. The grooves between

Fig. 419—C.
pinnulatum.

the ribs are rounded and deeply indented. Hinge and
soft parts with the characters of the preceding species.
Color of outer surface of shell white; inner surface white
or pink. Length 11 mm; height 10 mm. Occurs in offshore waters. Very abundant. Labrador to Long Island
Sound.

C. magnum. (Plate XXXII.) Shell typically heart-shaped and characterized by 33 to 37 broad, radiating ribs. It can be easily distinguished
from other cockles by its striking and attractive color, this being a soft
yellowish-brown decorated with spots and transverse, interrupted rows
of chestnut or purple. Margins crenulated. Hinge and soft parts as in
C. islandicum. This is the largest, as well as the most beautiful, of our
cardiums, measuring 10 cm in length and 11 cm in height. Very abundant in the tidal zone on open beaches; shallow water. Virginia to Texas.

Genus *Laevicardium*

L. mortoni. (Plate XXXII.) In this little form the shell is obliquely
rounded and inflated. The surface of the exterior is smooth, being without ribs or grooves. Color of exterior whitish or fawn with minute,
scattered spots of brown; interior bright yellow, usually with a purplish
blotch on the posterior margin. A characteristic feature of the living
animal is the presence of long conspicuous cirri around the orifices of
the short separated siphons. Length 25 mm; height 17 mm. This is a

common shallow-water species, occurring especially plentifully along sandy beaches in Long Island Sound, where it not infrequently is found living within the tidal zone. Nova Scotia to the west coast of Florida.

Suborder MYACEA

The shells of the mollusks contained in this suborder are more or less chalky in structure, never nacreous, or pearly. The mantle is closed below and is with an opening only for the extension of the foot. All the species are burrowing forms inhabiting sand, mud, or rock.

Family MYIDAE

Genus *Mya*

M. arenaria. (Fig. 420.) The long or soft-shell clam. This species has an elongated ovate shell of rather fragile structure, which can be recog-

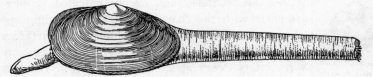

Fig. 420—M. arenaria.

nized by the gaping posterior end and the strikingly long, stout siphons covered with a strong brownish periostracum. The siphons are almost but not wholly retractile. There are no hinge teeth, but a prominent shelflike process occurs in the left valve near the hinge, which is the base of attachment for the internal ligament. The umbones are situated a little forward of the middle. When the animal is young, it forms a byssus, although it does not do so as an adult. Color of shell white with brownish periostracum. Length 10 cm; height 6.5 cm. This is the common edible species found buried in the sand, mud, or gravel between the tide marks. Its presence is indicated at low tide by a small hole in the ground through which it projects its siphons during high water. Arctic Ocean to Cape Hatteras.

Family SOLENIDAE

Genus *Ensis*

E. directus. (Plate XXXII.) The common razor clam. This form with its shell resembling the handle of the so-called "straight razor" should not be hard to identify. The very elongated shell is slightly curved and is nearly square at the ends, which have rounded corners. The umbones and the hinge are at the anterior end. The hinge is with 2 teeth in the right valve and 3 in the left, the teeth in the right valve consisting of 1 projecting tooth and an elongated ridgelike tooth back of it, while the teeth in the left valve are made up of 2 projecting teeth and a double ridge. Siphons short and united for about half their length. This form is quite active in its movements; it is capable of burrowing or swimming with considerable rapidity by means of its long protrusible and protean foot which can be alternately expanded and constricted at the end to form a wedge or resistance whereby it is enabled to draw close its shell. Color white, with glossy yellowish or greenish periostracum. Length 15 cm; height 25 mm. Common in sand in the tidal zone and shallow water. Labrador to the west coast of Florida.

E. viridis is a smaller form that might easily be mistaken for a young specimen of the preceding. It differs somewhat in its proportions and in its outlines, being very nearly straight with the forward end abruptly truncated and the hinder end rounded. The presence of a single tooth in each valve will serve at once to distinguish it from the other. Color glossy greenish. Length 5 cm; height 10 mm. Rhode Island to the west coast of Florida.

Genus *Tagelus*

T. gibbus. (Plate XXXII.) Shell elongated, gaping at both ends; ends rounded; dorsal and ventral margins nearly parallel. Umbones just back of the middle. Hinge with 2 small cardinal teeth in each valve. Foot large and muscular. Mantle edge having the appearance of being continuous with periostracum extending over shell margin. Siphons very long, separated throughout, and provided orange-colored ocelli. Color of shell white with yellowish periostracum. Length 9 cm; height 3 cm. Lives buried in sand or mud in the tidal zone and shallow water. Corresponding to its separate siphons, double holes in the ground mark the site of its burrow. Common. Cape Cod to Texas.

Suborder **PHOLADACEA**

The mollusks in this suborder have boring habits, and use their shells to rasp out long chambers in clay, rock, or wood. A characteristic of the shell is the absence of a hinge and ligament. There are three adductor muscles. The mantle usually is closed, consequently the foot is short and very often poorly developed. Very long siphons are present which are united nearly their entire length.

Family PHOLADIDAE

Genus *Pholas*

P. costata. (Fig. 421.) The angel-wing shell. A clay- or mud-boring species. This form has a thin, fragile shell, gaping at both ends, the

anterior dorsal margin of which is reflected back over the umbones. An interior spoon-shaped process for the attachment of muscles curves out from within the umbonal cavity nearly to the center of the shell. The dorsal mar-

Fig. 421—P. costata.

gin is generally protected by 2 horny accessory valves external to the original shell. This latter is elongated and quite convex with the outer surface sculptured into about 30 radiating, serrated ribs; the inner surface bears corresponding depressions. Color white. Length 15 cm; height 5 cm. This species inhabits sand as well as less resistant material. Common in the tidal zone in southern States; infrequent north of Cape Hatteras. Cape Cod to South America.

P. truncata. (Fig. 422.) A form with the habits and most of the characters distinguishing the preceding described species. Differs in structure of accessory valve, which is calcareous instead of horny; also differs in size, proportion, and outline, being 73 mm long and 37 mm high with anterior end tapering somewhat abruptly into a sharp point and the posterior end broad and roughly rectangular. Radiating ridges occur only on the anterior triangular region. Common, particularly in southern range. Massachusetts Bay to the west coast of Florida.

Fig. 422—P. truncata.

Genus *Zirphaea*

Z. crispata. (Fig. 423.) In this species the anterior dorsal margin of the shell is only slightly reflected. There are no accessory valves as in *Pholas,* but the umbones are protected by a membranous covering. The anterior part is marked with radiating ribs, whereas the posterior part is smooth except for growth lines. The chief distinguishing feature of this shell, however, and the one that also characterizes the genus, is the division of the surface of each valve into equal halves by a broad groove extending somewhat obliquely from the umbones to the ventral margins. Length 60 mm; height 35 mm. Common in northern range. Circumpolar. Arctic Ocean to Cape Hatteras.

Fig. 423—Z. crispata.

FAMILY TEREDIDAE

GENUS *Teredo*

T. navalis. (Fig. 424.) The common shipworm. This very widely known mollusk is a wood-boring species, wormlike in shape, and with the valves of the shell small, 3-lobed, and carried on the forward end of the body. At the hinder end of the body are long siphons which have placed at their sides 2 calcareous structures known as the pallets, structures used to close the orifice of the animal's burrow against enemies, at the same time allowing the free circulation of water required by the animal—that is to say, the water for respiration, which also holds the minute organic food particles on which the shipworm subsists. The entrance to the burrow is very small, although the animal within may be of a considerable diameter. This is so because the burrow is started when the animal is young; as the latter grows it enlarges the cavity by a rasping motion of the valves at the posterior end, at first working across the grain of the wood and later boring with the grain, meanwhile maintaining a small opening at the entrance for the extension of its slender siphons guarded by the pallets. The burrow is lined with a calcareous coating, and usually does not run into its neighbor. Shipworms are classified chiefly by the characters of the shell and the pallets. The present species has a shell in which the anterior lobes are considerably smaller than the posterior lobes. The pallets are paddlelike in shape, with the ends concave and the sides produced backwards into 2 sharp points.

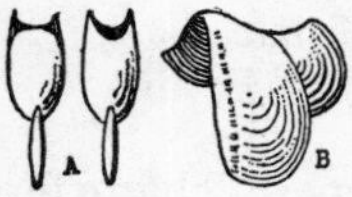

Fig. 424—T. navalis. A, pallets. B, shell.

Blade and stalk are of equal length. Length of burrow variable, but usually about 25 cm; diameter 6 to 10 mm. Length of shell 6 mm; height 2 mm. Very common in submerged timber of every description. Arctic Ocean to Texas.

T. megotara. (Fig. 425.) With the general characters of *T. navalis*. Shell with posterior lobes large and separated from the umbones by a narrow, deep, rounded notch in each valve. Pallets with paddlelike blades, broad toward the ends. Blade twice as long as stalk. Length of burrow up to 45 cm. Length and height of shell 6 to 12 mm. Length of pallet 6 mm; width 3 mm. Common in floating driftwood. Arctic Ocean to South Carolina.

Fig. 425—T.
megotara.　A,
pallets. B, shell.

Fig. 426—T.
chloritica.　A,
pallets. B, shell.

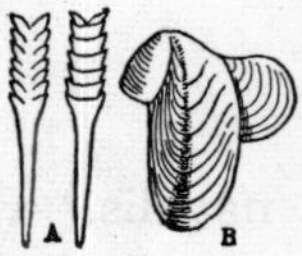

Fig. 427—X.
fimbriata.　A,
pallets. B, shell.

T. chloritica. (Fig. 426. Also see Plate XXIX.) General characters similar to those of *T. navalis;* burrows, however, are close together, very small, and directed across the grain of the wood, causing the wood to have a honey-comb appearance. Shell 3 mm in length and height. Pallets paddle-shaped and 5 mm in length. Massachusetts Bay to Florida.

Genus *Xylotrya*

X. fimbriata. (Fig 427.) A species with the general appearance and characters of *T. navalis,* but can be distinguished from that form by the pallets, which are long and feather-shaped. Its burrow is 30 cm or more long. Not uncommon. Massachusetts Bay to Florida.

Suborder **ANATINACEA**

The pelecypods contained in this suborder are characterized by the possession of a thin, nacreous shell with an external ligament. Many of them are degenerate; hence these are without certain well developed organs; particularly the foot, adductor muscles, labial palps, and gills are often wanting.

Family **PANDORIDAE**

Genus *Clidiophora*

C. trilineata. (Fig. 428.) Shell closed and with valves close together. Right valve flat; left valve somewhat convex. There is a diverging ridge on the right valve, while the left valve has 2 diverging grooves near the hinge. In outline the shell is rounded in front, and tapering behind, being curved upward into a gaping tip through which the short siphons project. Color whitish. Length 32 mm; height 17 mm. Common in shallow water. Nova Scotia to Texas.

Fig. 428—C. trilineata.

CLASS **CEPHALOPODA**

The cephalopods are the most intelligent and highly organized of the mollusks. Of the various forms that are included in the group, perhaps the best known are the squids, devilfishes, cuttlefishes and the chambered, or pearly, nautili, the two last-named, however, not being represented on the Atlantic coast of North America. To most persons other than naturalists, there would seem to be little in common between the cephalopods and the snails or clams; yet there are homologous structures possessed by the first-named organisms and the latter that reveal a very close kinship.

Most cephalopods are readily recognized by the large bodies bearing conspicuous, well developed eyes and the circle of arms, often provided with sucker disks or hooks, arising from the head. The body is bilaterally symmetrical and is modified in such a manner that what in other mollusks is usually the normal, or dorsal, side of the back is in the cephalopods the posterior end of the body; the anterior end, which comprises the arms and siphon, represents the ventral side, or, more specifically, the foot. The visceral mass is compact and covered with a thin wall, the whole being enclosed by a thick muscular mantle. There is a large posterior mantle cavity in which are contained the gills and the vents of the intestines, kidneys, and organs of reproduction. By expansion of the mantle, water for both locomotion and respiration is taken into the mantle cavity between the free edge, or collar, of the mantle and the body near the head ; it may be expelled through the short tube, or siphon, within the collar space by the closing of the mantle edge and the contraction of the mantle, the force of the water through the siphon propelling the animal in a direction opposite to that in which the siphon is pointed ; or, if the animal is at rest, the water can pass out of the mantle cavity through the same opening by which it entered. The usual direction in which a cephalopod swims is backward, the numerous arms trailing after so as to offer the least resistance to the water ; but this direction can be easily reversed, merely by changing the direction of the siphon. All cephalopods are able and active swimmers.

The head is provided with a pair of strong, sharp, horny jaws, which are situated within the circle of arms at the base and which lie behind a circular lip surrounding the mouth. These jaws have a close resemblance to the beak of a parrot, and are used in killing and rending prey. There is a radula present in the pharynx, but it is relatively small. The alimentary tract receives digestive fluids from salivary glands, a liver, and a pancreas. The stomach is large and empties into an intestine which passes forward to the anal opening in the mantle cavity. All cephalopods except nautili have a singular protective provision in the form of a glandular pocket, called the ink sac, just within the anus, which secretes a brown or black fluid. When the animal wishes to escape from or confuse its enemies it ejects this fluid into the water, thus creating an inky screen which obscures it from view and permits its retreat.

Respiration is carried on through the pair of gills, or ctenidia, lying within the mantle cavity. The blood is received from these by a single arterial heart, which in turn sends it throughout the body by way of the arteries to return by the way of large veins and thence through two branchial hearts at the base of the gills. From this it will be observed that there are three hearts.

The special senses are far better developed than in other groups of mollusks. There is a large brain which is protected by a sheath of cartilage. Under this cartilage lie a pair of lithocysts, or organs of equilibrium. The eyes in all cephalopods except the pearly nautili are with a lens ; in some forms there is a transparent cornea in addition to, and in front of, the lens, but this similarity of the cephalopod eye to that possessed by vertebrates is merely accidental, for the retina lining in the eyes of cephalopods is on the inner surface of the eye and receives the light directly. In *Nautilus* the eye is merely a chamber with a small opening through which the light reaches the retina lining.

Cephalopods are unisexual, the male as a rule being smaller than the female. The male fertilizes the eggs by using one of his arms to place a capsule, or spermatophore, filled with sperm cells within the mantle cavity of the female; in certain forms there is a specialized arm for this purpose, which is called the *hectocotylus* arm. In *Argonauta,* the paper nautilus, and certain other octopods the tip of the hectocotylus arm becomes separated and is left within the mantle cavity of the female.

Except among most octopods, which have no shell at all, there is a shell present in cephalopods, but it is usually internal. In *Nautilus* and *Spirula* the shell is external and is a calcareous chambered structure coiled in a single plane. What is termed the shell in *Argonauta* is a calcareous egg case secreted by glands near the tips of two of the arms, and it has no relation to the shells in other forms. Cuttlefishes have a straight, calcareous, internal shell, while the squids have a slender chitinous shell within the mantle, which from its characteristic appearance is called the "pen." No cuttlefishes, however, occur on our coast. Cephalopods are often called "chameleons of the sea" owing to the fact that they can produce rapid and striking color changes over the body.

The *Cephalopoda* usually consists of two orders, the *Tetrabranchiata,* characterized by the possession of four gills and a coiled external shell, and the *Dibranchiata,* characterized by the possession of two gills and a shell which may be coiled and external, or straight and internal, or absent. Only certain forms belonging to the last-named order occur on our coast.

ORDER **DIBRANCHIATA**

(With the characters outlined in the preceding paragraph.)

Suborder **DECAPODA**

Cephalopods having ten arms. Eight of the arms, called sessile arms, are provided with suckers situated along their entire length on the inner side: two of the arms are longer than the others, and are known as tentacular arms; these latter are usually retractile and have the suckers restricted to an expanded portion at the end. In all cases the suckers are with a horny supporting ring and are borne on short stalks. The body has a pair of fins, usually at the hinder end, which are lateral projections of the mantle. A shell is invariably present.

Division PHRAGMOPHORA

This is a division of decapods in which the shell is usually spiral and coiled in one plane, and consists of numerous chambers the partitions of which are perforated by a small tube, or siphuncle, extending around the inner side of the whorl.

Family SPIRULIDAE

Genus *Spirula*

S. peroni. (Fig. 429. Also see Plate XXXIII.) While the living animal of this form will seldom or never be taken by the average shore collec-

tor, its calcareous shell will not infrequently be found cast up on the beach. It is a loosely coiled spiral consisting of about 3 whorls composed of numerous chambers such as were described above in reference to the characters of the division. There is a paucity of information regarding the living animal, since it has been rarely seen, but it is

Fig. 429—S. peroni.

known that the shell is born partly internally and partly externally. The shell is nacreous and white in color. Length 5 mm; gross diameter 25 mm. It is found locally from Nantucket to the Caribbean Sea.

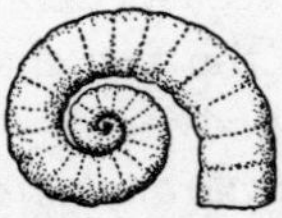

Division CHONDROPHORA

The squids. In this division of decapods the shell when present is a horny, penlike structure bearing a groove on its upper side.

Family SEPIOLIDAE

Genus *Rossia*

R. sublevis. (Fig. 430.) The body of this animal is cylindrical but so short as to be globose. There is a relatively large head. The fins are

rather expansive, rounded in outline, and constricted at the base; they project from near the
middle of the mantle. The eyes are with a cornea.
The mantle in this species is not united with the
head, and the pen is comparatively small. Each
sessile arm bears 2 rows of suckers along its entire length. Length of body 46 mm; breadth 22
mm; length of longest sessile arms 22 mm; length
of tentacular arms 25 mm. Color pinkish with purplish-brown spottings above. Common. Occurs in
offshore waters from Cape Cod northward.

Fig. 430—R. sub-levis.

FAMILY LOLIGINIDAE
GENUS *Loligo*

L. pealei. (Fig. 431.) The common squid. In this species the body is
elongate and cylindrical but somewhat flattened and tapering to a point

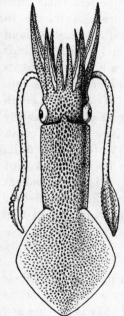

at the hinder end. The fins are broad, rhombic and
more than half the length of the body, and are
terminal. The head is a trifle narrower than the
mantle, and the eyes are with a cornea. There are
2 rows of suckers on the sessile arms, while the
tentacular arms bear 4 rows. The tentacular arms
are partially retractile. In this species the hectocotylus arm of the male is the fourth one on the
left, the tip of this arm being modified for the
transference of the spermatophore. The shell is
chitinous, long and characteristically pen-shaped.
Color whitish or a very pale neutral tint with spottings of brownish-red on the upper side. Length
20 cm; diameter 6 cm. Very abundant in both inshore and offshore waters. Occurs from Maine to
South Carolina.

L. pallida. (Plate XXIX.) A species very
closely resembling the one just described. Some
naturalists regard it merely as a variety of *L. pealei*
and not as a distinct species. Its principal difference

Fig. 431—L. pealei.

is in its proportions and color, being somewhat

thicker in body than the other and with fewer spottings. It is very common in Long Island Sound where its egg clusters are often washed ashore in great numbers. The eggs are enclosed in long, cylindrical jelly-like capsules, each capsule containing many eggs and being attached at one end with numerous others by a common stem to seaweeds or other objects.

FAMILY OMMASTREPHIDAE

GENUS *Ommastrephes*

O. illecebrosus. (Fig. 432.) The sea arrow, or flying squid. The elongate and nearly cylindrical body of this species bears the fins at the

hinder end. These are broad and about half as long as the mantle, and have their forward margins rounded, while the hinder margins form nearly a right angle. The head is as broad as the mantle. The sessile arms are comparatively short and are provided with 2 rows of suckers; the tentacular arms are with 4 rows of suckers and are not retractile. In this species the eyes are bare, that is to say, without a cornea; the siphon is valved and united by bands with the head; and the pen has a hollow cone at its hinder end. Color pale or deep blue, blending into orange or reddish with spottings of dark red. Length 30 cm; diameter 3 cm. This form is used extensively as bait in the cod fishery of the New England coast. It is a very swift swimmer and often projects its body clear out of the water in its fleet pursuit of fishes which form its chief prey. It is very abundant north of Cape Cod, and inhabits chiefly the offshore waters from the Bay of Fundy to the eastern

Fig. 432—O. illecebrosus.

end of Long Island Sound.

SUBORDER OCTOPODA

The cephalopods in this group are characterized principally by their possession of eight arms. The suckers are generally without the stalks that are present in the decapods. In most cases the body is globose and usually without fins.

CEPHALOPODA 417

Family ARGONAUTIDAE

Genus *Argonauta*

A. argo. (Fig. 433.) The paper nautilus. It is only the female in this species which possesses the remarkable shell that is used by her as an egg case. This shell is thin, white, and paperlike, and while in the water it is soft and flexible. There are 2 rows of pointed projections decorating the periphery, and a number of undulating or rounded ribs and grooves radiate from the inner side

Fig. 433—A. argo.

of the whorl. It is without partitions or chambers. The animal is not in physical attachment with the shell, and may even desert it, since it merely serves to hold the eggs during the breeding season. Its likeness to a typical spiral molluscan shell is only accidental. The argonaut rests in her shell and during the breeding season is found swimming at the surface, but at other times appears to inhabit the depths. The body of the animal in general resembles that of an octopus, except that the topmost pair of arms bear at the ends large leaflike expansions. It is these expansions that secrete the delicate shell. Length of shell 20 cm; length of female 20 cm. Color whitish with red spottings. The male argonaut is only about one tenth as large as the female. This is a tropical form, but it is occasionally washed ashore as far north as Long Island where it is carried by the Gulf Stream; it also occurs not infrequently along the coast of the southern states.

Family OCTOPODIDAE

Genus *Octopus*

O. americanus. (See the frontispiece.) The American devilfish. The body of this form is rounded and with a large head from which arise 8 similar sucker arms united at the base by an umbrellalike web. There are 2 rows of suckers; these suckers are sessile and occur along the entire length of each arm. The dorsal side of the mantle is continuous with that of the head, the mantle and head being united by a broad band of connecting tissue. The third arm on the right is the hectocotylus arm. When full grown this animal is about 1 m, or a yard, long. In color it is pale or white, which renders it all but invisible as it moves over the white sandy bottom; but it is capable of making startling and bizarre color changes when excited or irritated. It has the habit of secreting itself within the rocky crevices of coral reefs, where it lies in wait for prey, such as shelled mollusks and crustacea upon which it largely feeds. It is not an uncommon species, and occurs along the Florida coast and in the West Indies.

O. bairdi. (Fig. 434.) A northern species with a short, plump body covered with small tubercles. The arms are comparatively short, webbed

Fig. 434—O. bairdi.

at the base for a third of their length, and bear 2 rows of sessile suckers. The third arm on the right is hectocotylized. Mantle united to head with a broad dorsal connective tissue. Length 7.5 cm. This is a common species, but it is found in offshore waters where it lives usually at considerable depths. It occurs north of Cape Cod.

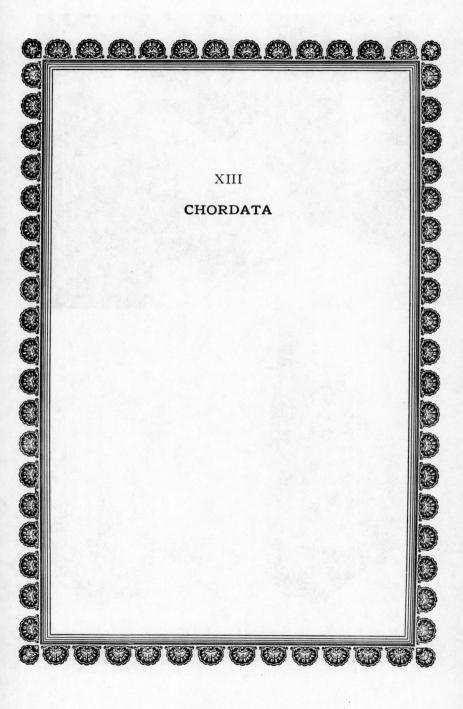

XIII

CHORDATA

PLATE XXXIV

1—*Molgula pellucida*; 2½x.
2—*Botryllus schlosseri*; attached to shell of living *Mytilus*; natural size.
3—*Gorgonia flabellum*; skeleton.

CHAPTER XIII

PHYLUM **CHORDATA**

(BALANOGLOSSIDS, SALPAS, SEA SQUIRTS, ETC.)

THE chordates comprise a considerable and diversified group ranging from the wormlike balanoglossids to man. One of the chief structural features characterizing the group is the *notochord,* an elastic, rodlike support for the body, which lies above the digestive tube. The notochord in certain low chordates remains throughout the life of the animal; in higher forms it becomes surrounded with cartilage or bone which nearly replaces it and constitutes a better support. The notochord is represented in all chordates, with the possible exception of the balanoglossids; in the higher vertebrates it becomes the bony spinal column.

Another feature characteristic of chordates is found in the pharyngeal respiratory apparatus which communicates with the exterior of the body through gill slits. In most of the lower chordates, the gill slits are present and function throughout life; in the higher and air-breathing vertebrates, however, these structures disappear after the early embryonic stages are passed, and are replaced by the lungs.

Then, again, a feature that most chordates have in common is a tubular, central nervous system lying dorsal to the notochord. In vertebrates a comparatively large brain develops from the tubular nervous tissue in the head region.

There are four subphyla of the *Chordata,* and the group is commonly classified as follows:

PHYLUM CHORDATA

SUBPHYLUM **Enteropneusta**

ORDERS ....{ **Balanoglossida**
Cephalodiscida (Not included in this book.)

Subphylum **Tunicata**

Classes ... { **Larvacea**
Thaliacea
Ascidiacea

Subphylum **Leptocardia**   Family **Branchiostomidae**

Subphylum **Vertebrata** (Not within the scope of this work.)

Subphylum ENTEROPNEUSTA

Chordates in which the notochord is represented by a hollow, dorsal projection of the forward part of the digestive canal.

ORDER BALANOGLOSSIDA

Balanoglossids are soft-bodied, elongate, wormlike animals having a proboscis which is separated from the rest of the body by a narrow neck. The proboscis is cylindrical or ovoid in shape. and it encloses a portion of the coelom. The main part of the body, called the trunk, has a collarlike ring encircling the forward end at the neck, and usually tapers to a point at the hinder end. The mouth is at the front of the collar, and the anal opening is at the posterior end of the trunk. There are five body cavities and these connect with the outside through pores in the proboscis and the collar. At the forward end of the trunk is a paired row of gill slits through which the whole forward portion of the digestive tract communicates with the exterior. Back of the gill slits, the paired liver sacs in certain forms can be plainly seen through the integument of the body. The nervous system is not highly organized, but there do occur definite aggregations of nerve cells and fibers, and there is a ring nerve behind the collar; also, in the collar there is a portion of the dorsal strand that forms a neural canal, like the central nervous system in vertebrate animals. The sexes are separate.

The balanoglossids on our coast live usually in tide flats and shallow water where they inhabit burrows. They burrow with the aid of their proboscis, and, as the animal is unable to close the mouth, the sand or mud passes through the digestive tract and in certain species leaves the body in a continuous column. The presence of these continuous, coiled castings on tide flats will often indicate the location of most forms.

FAMILY **HARRIMANIIDAE**

GENUS *Dolichoglossus*

D. kowalevskyi. (Fig. 435.) With the characters of the order. There is a relatively long proboscis, and it has but one pore. Liver sacs are not present in this species. Length about 15 cm. Color of body orange-yellow; proboscis and collar pinkish yellow, but collar is of a darker shade. Often common. Lives in clean, sandy tide flats. Massachusetts Bay to Cape Hatteras.

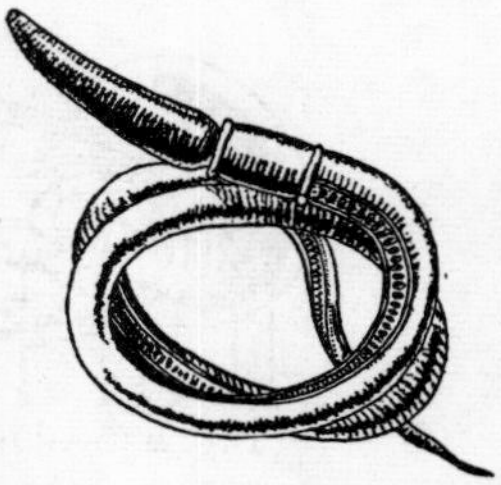

SUBPHYLUM **TUNICATA**

Tunicates are degenerate chordates. They are called tunicates from the fact that they are covered with a characteristic cuticular or cellu-
lose tunic. Under the tunic is another distinct covering called the mantle, which is composed of connective tissue fibers and encloses muscle tissue and blood vessels. A striking feature of tunicates is the large pharynx, usually known as the branchial sac, with its *endostyle,* the latter being a ciliated groove in the mid-ventral region of the pharynx, which is provided with glandular secretions for the entanglement of minute food organisms swept into the mouth. The mouth opens into the pharynx, which is the anterior part of the digestive tract. The pharynx also serves as a respiratory organ; its walls are well supplied with blood vessels, and the water enters it by way of the mouth and either leaves by the way of perforated slits in the walls or passes into a large peribranchial sac and thence out through an *atrial,* or excurrent, siphon. The water that oxygenates the blood of the branchial sac flows in a steady stream propelled by the cilia of the endostyle. There is a tubular heart and it has the singular power to reverse the direction of the blood stream. The nervous system starts out as a dorsal structure very much like that in higher vertebrates, but it usually degenerates into a small ganglion in the adult animal. Most tunicates are hermaphroditic, the sexual organs emptying their products into the cloacal chamber; however, asexual

Fig. 435—D. kowalevskyi.

reproduction is common, often leading to the formation of colonial, or so-called *compound,* forms. There are about 1300 species of tunicates and all of them are marine.

CLASS **LARVACEA**

The animals contained in this class are usually known among naturalists as *appendicularians,* a term derived from their tadpolelike shape and the long tail by which they are propelled through the water. All of them are free-swimming, transparent, and very small, averaging about

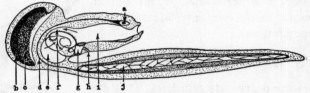

Fig. 436—Diagram of an appendicularian. a, otocyst; b, testis; c, ovary; d, nerve; e, stomach; f, œsophagus; g, atrial canal; h, anus; i, pharynx; j, noto chord.

10 mm long. The body consists of two principal regions, the trunk and the long tail. The trunk contains the pharynx and the viscera, and the tail contains the notochord and dorsal nerve chord. (Fig. 436.) The organization of appendicularians is much like that of larval sea squirts. There is no peribranchial cavity present, and the tunic is without cellulose. The tunic is a large transparent envelope, often many times larger than the body and within which the animal can move about freely. Appendicularians are pelagic, living at the surface usually a considerable distance from land. There are fewer than 50 recognized species of these animals throughout the world, and since there is some uncertainty regarding species definitely known to occur along our coasts, no more than the foregoing brief reference will be given, and this merely to acquaint the reader with existence of the group.

CLASS **THALIACEA**

This class contains the salpas, which are free-swimming pelagic animals found usually at the surface far out at sea. The body is transparent, somewhat cylindrical or barrel-shaped with an opening at both ends; and it is by taking water in at the oral, or anterior, end and forcing it out through the cloacal, or posterior, end that they propel themselves through the water. They are invested with a cellulose tunic and a mantle, the mantle containing a number of contractile muscle bands by means of which the water is forced out of the posterior opening to drive the animal forward. As the water goes through the body it passes the respiratory organ, which usually forms a pierced, or slotted, partition dividing the body cavity into two chambers: the forward, or pharyngeal, and the hinder, or cloacal, chamber. In some forms, such as *Salpa,* this partition is represented merely by a ciliated rodlike structure. The visceral organs are generally small and brightly colored, the principal ones being imbedded in the mantle on the ventral side of the body at the posterior end.

Reproduction in this group is rather peculiar and often complex. While an individual is like its grandparents, it is not like its parents; a generation of solitary individuals alternates with a generation of aggregated individuals. This scheme of reproduction is usually called alternation of generations, but in certain forms reproduction does not take place in exactly the way that is indicated by the strict meaning of that term. Again, in some cases four different types of individuals are included in the life history.

ORDER MULTISTIGMATEA

The animals in this order are characterized by a barrel-shaped body in which the muscle bands form complete rings. The tunic is thin; the respiratory partition is with two rows of stigmata, or pierced orifices; and the openings at the opposite ends of the body are lobed.

GENUS *Doliolum*

D. denticulatum. (Fig. 437.) This form may not infrequently be taken with the tow-net in the warmer offshore waters near our coast. It is with the characters of the order and has a bent respiratory partition containing about 40 stigmata on each side. The digestive tube is straight, a feature which distinguishes it from *Salpa.* The embryo of *Doliolum* develops into a solitary asexual animal which has a process, called the *stolon,* arising from the posterior dorsal end of the body upon which a chain of buds is produced. The buds in this chain eventually mi-

Fig. 437—O. den-ticulatum.

grate to the dorsal outgrowth of the body and become arranged in 5 rows. This is accompanied by a change in the appearance of the animal, its muscle bands becoming wider and many of its internal organs disappearing. Thus, it now serves as little more than a living perambulator for the developing offspring, and is, in fact, in this stage called the "nurse." Three kinds of individuals are produced among the buds: the nutritive animal, the foster animal and the aggregated, or sexual, animals. The nutritive animals arise from the side rows of buds and remain attached to the nurse; they have poorly developed muscles and function merely as aids in obtaining food and oxygen for the colony. The foster animals arise from the middle rows of buds and in time become free. However, before the foster animals are set free, the sexual individuals become attached to them as very young buds and remain with the others until maturity when separation takes place. The full-grown foster animal and the sexual animal are somewhat alike, but the former has no organs of reproduction, whereas the latter has them. The germ cells of the sexual animal develop into tailed larvæ which grow into asexual, or solitary, animals that produce buds, thus completing the life-history cycle. The length of the solitary, or sexual, animal is 3 mm, but old nurses with their chains may measure 5 or 6 times this length.

ORDER **ASTIGMATEA**

In this order the animals are characterized by bodies which are somewhat flattened dorsoventrally, and in which the muscle bands often form incomplete rings. The tunic is thick; the respiratory partition is a ciliated rodlike structure extending diagonally from the dorsal body wall to the ventral œsophagus.

GENUS *Salpa*

S. democratica. (Fig. 438.) This is a very abundant species off the coast of New England. The life-history includes an alternation of 2 types of individuals, a solitary type and an aggregated type, a form of reproduction that is somewhat different from that of the species just described. In the solitary animal a mass of germ cells is borne near the base of the budding stolon. The chain of buds is double; as each bud arises it receives one of the germs. The buds gradually develop and increase in size, the distal ones of the chain being the oldest and largest. The chain increases in length until it may contain from 20 to 30 pairs of salpas. This chain of aggregated individuals often separates from the parent and remains together for some time be-

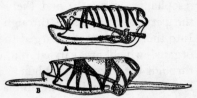

Fig. 438—S. democratica. A, solitary generation; B, aggregate generation.

fore portions break away and swim off; however, before the salpa chain has parted from the parent, the individuals have begun the muscular contractions by means of which water is passed through their bodies. The germ cell contained in each individual of the chain—the individual now functioning merely as a nursery—in time develops into a solitary animal which again, like its grandparent, reproduces by budding. The solitary form of this species is 24 mm long and may be recognized by the 2 long projections at the hinder end of the body; the aggregated individuals are each 15 mm long and are without these projections. In both forms the visceral parts are blue and contain the digestive tube, which is bent on itself.

CLASS **ASCIDIACEA**

The ascidians. These are solitary animals or aggregated forms living a colonial life. With the exception of *Pyrosoma,* all of them are sessile, usually attaching themselves to rocks, wharf piles, shells, seaweed, etc. The body is sac-shaped with two openings; one of them being the oral, or incurrent, opening, and the other being the cloacal, or excurrent, opening. The two openings are usually near each other, each is at the end of a contractile projection of the body wall called the siphon, and may be closed by sphincter muscles. There is a cellulose tunic which is variable in structure, in some forms being gelatinous or transparent, in other forms being thick, firm and opaque, often with a rough, warty exterior. There is a very large branchial sac with a series of slits, or stigmata, opening through its entire wall, giving it a latticelike appearance; and its upper margin, just back of the mouth, is surrounded with a circlet of tentacles. Near the branchial sac are the œsophagus, stomach, and the short intestine, while on each side is the peribranchial space through which passes the respiratory water. Into this space are discharged the products of the genital and digestive organs, which are carried out of the body with the water flowing through the excurrent siphon. There are present an endostyle, a peribranchial band, and dorsal lamina. The principal ganglion of the nervous system is contained in the mantle between the siphons. These tunicates are hermaphridotic but self-fertilization does not often take place. The egg hatches into a larva which is very similar to an adult appendicularian. The tail is long and the body has three projections, or papillæ, at the forward end, by means of which the larva attaches itself to some object to undergo metamorphosis. This metamorphosis is complex, but a simple statement of its essentials would describe it as including the absorption and disappearance of the tail, the appearance and development of the peribranchial sac, and the acquisition of the characteristic sac-shaped form of the full-grown animal.

ORDER ASCIDIAE SIMPLICES

The simple ascidians, or sea squirts. Solitary, but often loosely associated forms. Some are not fixed, but none is free-swimming.

FAMILY MOLGULIDAE

GENUS *Molgula*

M. manhattensis. (Fig. 439.) The body of this ascidian is some what globular or ovoid with the siphons touching each other at their

Fig. 439—M. manhattensis.

bases. The siphons are relatively long and are very contractile. The tunic is rather thin, translucent, and with a corrugated outside which is usually covered with sand, dirt, or organic detritus. Color greenish. Length 25 mm. Often very abundant in shallow water near the low-tide level where it frequently is found in large crowded clusters. Casco Bay to North Carolina.

M. arenata. (Plate IV.) A form similar to the preceding but with laterally compressed body and shorter siphons which are rather widely separated. Body usually encrusted with closely adhering sand particles.

Length 20 mm. Found on shelly, gravelly, or sandy bottoms in shallow water. Rather common in Long Island and Vineyard Sounds.

M. pellucida. (Plate XXXIII.) Body nearly globular, and with a smooth translucent tunic. In this species there is no foreign matter adhering to the surface. The siphons are fairly large, broad at their bases which are close to each other, and diverging at the ends. The viscera can be seen through the thin tunic. Length 25 mm. Often numerous. Found fixed or free on shelly or sandy bottoms in shallow water. Massachusetts Bay to North Carolina.

GENUS *Eugyra*

E. pilularis. (Fig. 440.) Body nearly globular and with long, slender and somewhat closely approximated siphons. Siphons contractile; when extended are as long as the body mass. Diameter up to 8 mm. This form is able to move about. Often abundant in the mud or fine sand of shallow and offshore waters. Vineyard Sound northward.

Fig. 440—E. pilularis.

FAMILY CYNTHIIDAE

GENUS *Styela*

S. partita. (Fig. 441.) This ascidian has a low, flattened body with a broad base and squarish, somewhat elevated siphons, the openings

Fig. 441—S. partita.

of which are figured with alternating triangles of white and of purple. The surface of the animal is hard, leathery, and wrinkled. Color dark brown. Length 25 mm; breadth 12 mm. Common in shallow water. Found associated in groups; each individual always attached, never free. Massachusetts Bay to North Carolina.

GENUS *Cynthia*

C. carnea. (Fig. 442.) Body low, flattened and attached by a broad oval base. Siphons small, separated and with squarish openings. Color deep red. Length 12 mm. Not uncommon. Found in shallow and offshore waters. Labrador to Martha's Vineyard.

Fig. 442—C. carnea.

C. pyriformis. (Fig. 443.) The sea peach. This form is often attached by a narrow stalklike base. The body is globular or elliptical

in outline and has large cylindrical siphons at the top. The surface is of a velvety texture and is tinted a pleasing pink or yellow. Length up to 75 mm; breadth 30 mm. Common locally in clear shallow water. Greenland to Massachusetts Bay.

Fig. 443—C. pyriformis.

Genus *Boltenia*

B. rubra. (Fig. 444.) The sea potato. Body somewhat elliptical and surmounting a long stalk with which it is continuous, the stalk tapering from the hinder under side. The siphons are rather inconspicuous, and are not at the top but at the side of the body. The openings are each with 4 lobes. The surface is rough and leathery. Color red. Length of body mass 4 cm; length of stalk 26 cm. This is an abundant species, but it lives attached to the bottom in offshore waters; it is, however, not infrequently washed ashore by storms. Cape Cod northward.

Family **ASCIDIIDAE**

Genus *Ciona*

C. intestinalis. (Plate I.) This attractive species, often bearing the scientific name *C. tenella* in textbooks, should soon or late reward the diligent collector who seeks it within its range. The body is very elongate, somewhat cylindrical, and invested with a gelatinous, more or less transparent tunic through which the internal organs are fairly visible. The siphons are at the upper end of the body and are with lobed openings, the incurrent openings being with 8 lobes and the excurrent openings being with 6. Usually colorless, but occasionally yellowish. Length 10 cm; breadth 2 cm. Found attached to wharf piles, stones, and shells

Fig. 444—B. rubra.

in shallow water near the low-tide level. Labrador to Long Island Sound, but is commoner toward the north.

ORDER **ASCIDIAE COMPOSITAE**

The compound ascidians. These are colonial aggregated forms, the individuals multiplying by a process of budding from a single parent, and usually being grouped together in star- or flower-shaped clusters which form the common mass. Each individual is joined to the others by a common tunic but is without any internal connection. In many cases there is a common cloaca.

FAMILY **BOTRYLLIDAE**

GENUS *Botryllus*

B. schlosseri. (Plate XXXIV.) This species forms colonies which occur as fleshy incrustations on 'seaweeds, eelgrass, stones, etc. The zooids form groups of circular or elliptical or typical stellate patterns, each group consisting of from 5 to 10 radially arranged individuals, and the whole colony being embedded in a common gelatinous tunic. There is a common cloaca in the center of each group, while near the periphery each individual bears an incurrent siphonal aperture. Color very variable, black or purplish being common. Diameter of colony 10 cm or more; diameter of radial group 3 to 6 mm. This is a very common form in shallow water. It occurs abundantly in Long Island Sound and northward.

FAMILY **POLYCLINIDAE**

GENUS *Amaroucium*

A. pellucidum. (Fig. 445.) A species forming massive, translucent, gelatinous colonies, usually hemispherical or irregularly rounded in shape, and often covered with a thick, closely adhering incrustation of sand. The colony is made up of closely compacted elongate lobules arising from a common base by narrow stalks,

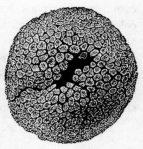

Fig. 445—A. pellucidum. (Top view of colony.)

each lobule increasing in diameter toward the surface of the colony. Contained in each lobule are a number of very elongate zooids arranged in circular or elliptical groups around a common cloaca. The long bodies of the zooids are perpendicular to the surface of the colony, and only the oral and cloacal regions are exposed. Color neutral, with bright orange or red stomach. Diameter of colony up to 15 cm; diameter of lobule 5 to 10 mm; length of zooid up to 25 mm. Very common in shallow and offshore waters. Vineyard Sound to North Carolina.

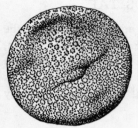

Fig. 446—A. constellatum. (Top view of colony.)

A. constellatum. (Fig. 446.) This species is considered by some naturalists to be merely a form of *A. pellucidum*. It forms thick, hemispherical crusts on piles, rocks, and coarse sand, etc. The general characters are like those of the preceding species, but the surface of the colony is smooth instead of being marked with lobular interstices. The zooid groups are irregularly stellate or elliptical. Not uncommon. Found in shallow and offshore waters. Long Island Sound and northward.

A. stellatum. (Fig. 447.) Sea pork. The colonies formed by this species are large, smooth, crestlike lobes of often thick, irregular, vertical plates. The zooids are arranged in stellate groups having from 6 to 20 individuals in each, and the whole colony is enclosed by a common gelatinous tunic. The mass formed by the colony may be 60 cm long, 15 cm high, and 2.5 thick, and because of its size and texture, which make it appear not unlike slices of salt pork, the fishermen have given the species its popular name. General color bluish or pinkish; branchial sac and intestine orange. Very common in shallow and offshore waters. Occurs northward of Cape Cod to North Carolina.

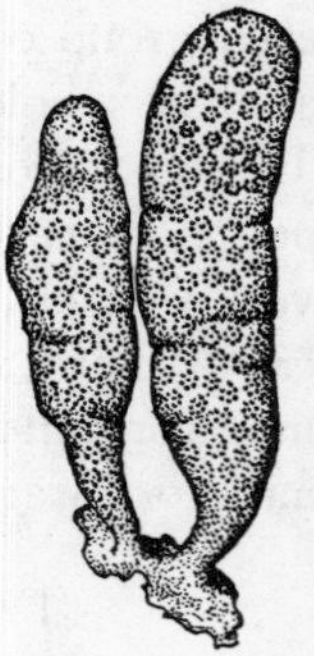

Fig. 447—A. stellatum.

Family **DIDEMNIDAE**
Genus *Didemnum*

D. lutarium. (Fig. 448.) This species forms thin incrusting colonies on stones, shells, etc. It differs from any of the compound ascidians previously described in that the tunic contains calcareous spicules. These spicules are stellate in form and thickly distributed throughout the tunic. The surface is somewhat wrinkled and roughly indicates the arrangement of the zooids, the latter being very minute in size. Color whitish, yellowish, or reddish. The colonies may be as large as 30 cm in diameter with a thickness of from 2 to 4 mm or more. Common. Found in shallow water. Labrador to Long Island Sound.

Fig. 448—D. lutarium. A, colony; B, spicules.

ORDER **ASCIDIAE LUCIAE**

Luminiscent, colonial, free-swimming ascidians. The colony is enclosed in a common tunic and is variously shaped but forms essentially an elongated hollow (usually cylindrical) chamber open at only one end. The individual animals are arranged at right angles to the outer surface with their incurrent openings on the outer side of the cylindrical chamber and their excurrent openings on the inner side. Each zooid has a pair of light-producing organs, and all colonies are brilliantly and beautifully phosphorescent. The colony swims by suddenly contracting and expelling the water from the open end of the chamber, thus propelling it along with the closed end foremost.

Genus *Pyrosoma*

P. atlanticum. (Fig. 449.) The form of this colony is somewhat like that of an elongated thimble. From the outer surface there arise

from the tunic numerous fingerlike projections. It is beautifully translucent with a greenish, pinkish, or yellowish cast. Length of colony 25 cm or more; diameter 3 or 4 cm. This is an inhabitant of tropical waters and is often taken in

Fig. 449—P. atlanticum.

offshore towings in southern ranges, but sometimes it may be washed ashore.

Subphylum LEPTOCARDIA

In this subphylum are contained certain elongate, fish-shaped chordates which have the notochord or cylindrical rod extending the whole length of the body. The head in these forms is wanting, but there is a mouth which is surrounded by long cirri, or bristlelike appendages, and which leads into a large pharynx that is pierced by paired gill slits. This mouth is ventrally situated, just back of the pointed snout at the anterior end, but is not in exact alignment with the body, being deflected somewhat to the left. The anal opening is at the hinder end and is also inclined to the left of the median line. The pharynx collects food, which passes on to a straight digestive tract, and it serves as a respiratory organ besides. Respiratory water passing through the gill slits streams into the peribranchial chamber and leaves the body through a median ventral opening anterior to the anus, an opening known as the *atriopore*. A series of prominent muscle bands, or myotomes, occurs along each side of the body. The chordate character of these animals is also represented in the position of the nerve chord which is dorsal to the notochord, the latter in turn lying immediately above the digestive tube. The sexes are separate.

Family BRANCHIOSTOMIDAE

Genus *Branchiostoma*

B. lanceolatum. (Fig. 450.) A lancelet. This form is commonly known by the older scientific name of *Amphioxis*. The body is symmetrical, laterally compressed, and in general typically fish-shaped. The tail fin is long and pointed. It has about 26 pairs of

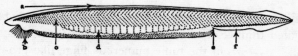

Fig. 450—B. lanceolatum. a, dorsal fin; b, oral cirri; c, muscle segments; d, gonads; e, ventral pore of peribranchial chamber; f, ventral fin.

gonads; these lie along each side of the body near the pharynx and project into the peribranchial chamber. Into the peribranchial chamber are discharged the products of the gonads as well as those of the excretory system. Length about 48 mm. Lancelets are found in tropical and temperate regions living in the sand in shallow, sheltered waters. They bury themselves in the bottom in an upright position with the mouth projecting above the sand. The present species occurs in Chesapeake Bay and ranges southerly.

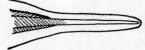

Genus *Asymmetron*

Fig. 451—A. macricaudatum. (Left side of tail.)

A. macricaudatum. (Fig. 451.) A species of lancelet in which the body is asymmetrical. Ventral fin poorly developed and without fin rays. Tail long and slender. About 26 gonads present extending from the 11th to the 37th myotome inclusive. Found in sheltered sandy bottoms on the Florida coast.

BIBLIOGRAPHY

The following references are recommended as supplementary sources of information for the student of marine invertebrate life. These selections have been made from a voluminous literature; all the selections are in the English language, but they do not include generally inaccessible books or the scarcer papers of scientific societies. Most of the titles listed should be found in any well stocked public library.

Such papers, journals and other technical publications as are included here, usually have their names abbreviated. The customary method of abbreviating is used, and is as follows :—

Bull. Bur. Fish.—Bulletin of the Bureau of Fisheries.

Bull. Mus. Comp. Zool. Harvard.—Bulletin of the Museum of Comparative Zoology of Harvard College.

Bull. U. S. Fish. Com.—Bulletin of the United States Fisheries Commission.

Bull. U, S. Nat. Mus.—Bulletin of the United States National Museum.

Ill. Cat. Mus. Comp. Zool. Harvard.—Illustrated Catalog of the Museum of Comparative Anatomy of Harvard College.

Mem. Boston Soc. Nat. Hist.—Memoirs of the Boston Society of Natural History.

Mem. Mus. Comp. Zool. Harvard.—Memoirs of the Museum of Comparative Zoology of Harvard College.

Mem. Nat. Acad. Sci.—Memoirs of the National Academy of Sciences.

Popular and General Works

Arnold, Augusta Foote
The Sea-Beach at Ebb Tide. The Century Co., New York. 1901. (A guide to the study of the seaweeds and the lower animals found between the tide-marks.)

Crowder, William
Dwellers of the Sea and Shore. The Macmillan Company, New York. 1923. (A descriptive account of the plant and animal life of the seashore, the tide pool and the open sea.)
A Naturalist at the Seashore. The Century Co., New York. 1928. (A narrative work detailing the lives and habits of marine animals.)

Flattely, F. W. and Walton, C. L.
The Biology of the Seashore. The Macmillan Company, New York. 1922. (A survey of physical and biological factors in the habits of seashore animals.)

Mayer, A. G.
Seashore Life. New York Zoological Society. 1911. (A guide for the identification of the commoner marine invertebrates occurring along the coast of New York State.)

Parker, T. J. and Haswell, W. A.
A Text-book of Zoology, Vol. I. The Macmillan Company, New York. 1897. (A work on the embryology and anatomy of invertebrate forms.)

Pratt, H. S.
A manual of the Common Invertebrate Animals. A. C. McClurg and Co. Chicago. 1916. (A standard reference work on the classification and description of the common invertebrate animals—exclusive of insects—occurring in the United States.)

Verrill, A. E. and Smith, S. I.
Report on the Invertebrate Animals of Vineyard Sound. Report of the United States Fisheries Commission, 1871–2. 1874. (A survey of the marine invertebrates and their habitats occurring in the New England region.)

Specific References

MICROSCOPY

Cross, M. I. and Coles, M. J.
Modern Microscopy. Chicago Medical Book Company. 1922. (A handbook for beginners.)

Giger, Michael F.
Animal Micrology. University of Chicago Press, Chicago. 1906 and recent. (Practical exercises in microscopical methods.)

Kingsbury, B. F. and Johannsen, O. A.
Histological Technique. John Wiley and Sons. New York. 1927. (Treats of special methods for various animal forms including invertebrates, plankton, etc.)

Lee, Arthur Bolles
The Microtomist's Vade-mecum. P. Blackiston's Sons. Philadelphia. (Designed for the advanced worker, but of value to the beginner in some problems.)

PROTOZOA

Calkins, G. N.
Marine Protozoa of Woods Hole. Bull. U. S. Fish. Comm., 1901.

Cushman, J. A.
Foraminifers of the Atlantic Ocean. U. S. National Museum, Washington, D. C. 1918–24.

BIBLIOGRAPHY 439

Lebour, M. V.

The Dinoflagellates of Northern Seas. Marine Biological Association of the United Kingdom, Plymouth, England. 1925.

SPONGES

George, W. C. and Wilson H. V.

Sponges of Beaufort and Vicinity. Bull. Bur. Fish., Vol. 37, 1917–18.

Hyatt, A.

Revision of North American Poriferae. Mem. Boston Soc. Nat. Hist., Vol. 2, 1875–77.

Moore, H. F.

The Commercial Sponges and Sponge Fisheries. Bull. Bur. Fish., Vol. 28, 1910.

Smith, H. M.

The Commercial Sponges of Florida. Bull. Bur. Fish., Vol. 17, 1897.

COELENTERATES

Agassiz, A.

North American Acalephae. Ill. Cat. Comp. Zool. Harvard, No. 2. 1865.

Fraser, C. M.

Some Hydroids of Beaufort, N. C., Bull. Bur. Fish., Vol. 30, 1912.

Mayer, A. G.

Medusae of the World. Vol. I—The Hydromedusae; Vol. II—The Hydromedusae; Vol. III—the Scyphomedusae. Carnegie Institution Bulletin 109. 1910.

Ctenophores of the Atlantic Coast of North America. Carnegie Institution Bulletin 162. 1911.

Nutting, C. C.

American Hydroids. I. Plumularidae. Bull. U. S. Nat. Mus. 1900.

American Hydroids. II. Sertularidae. Bull. U. S. Nat. Mus. 1904.

American Hydroids. III. Campanularidae and Bonneviellidae. Bull. U. S. Nat. Mus. 1915.

LOWER WORMS

Harmer, S. F. and Shipley, A. E.

Cambridge Natural History, Vol. 2. See "Flatworms and Metazoa, Nemertines, Threadworms, Sagitta, Rotifers, etc." The Macmillan Company, New York. 1896.

MOLLUSCOIDA

Bassler, R. S.

The Bryozoa or Moss Animals. Smithsonian Report for 1920, Publication 2633. Smithsonian Institution, Washington, D. C. 1922.

Morse, E. W.
Observations on Living Brachiopods. Mem. Boston Soc. Nat. Hist., Vol. 2, also Vol. 5. 1901.
Osburn, R. C.
The Bryozoa of the Woods Hole Region. Bull. Bur. Fish., Vol. 30, 1912.

ANNELIDS

Verrill, A. E.
New England Annelids. Transactions of the Connecticut Academy, Vol. 4. 1881.
Andrews, E. A.
The Annelida Chætopoda of Beaufort, N. C. Proceedings of the U. S. Nat. Mus., Vol. 14. 1891.

ECHINODERMS

Clark H. L.
Echinoderms of the Woods Hole Region. Bull. U. S. Fish. Com., Vol. 22, 1902.
Coe, W. R.
Echinoderms of Connecticut. State Geological and Natural History Survey, Bull. No. 19. Hartford. 1912.
Mead, A. D.
Natural History of the Starfish. Bull. U. S. Fish. Com., 1899.

ARTHROPODS

Callman, W. T.
The Crustacea of New Jersey. Report of the New Jersey State Museum, Trenton. 1911.
Hay, W. P. and Shore, C. A.
The Decapod Crustaceans of Beaufort, N. C. and the Surrounding Region. Bull. Bur. Fish., Vol. 35, 1915–16.
Holmes, S. J.
The Amphipoda of New England. Bull. Bur. Fish., Vol. 24, 1904.
Kunkel, B. W.
The arthrostraca of Connecticut. State Geological and Natural History Survey, Bulletin 26. Hartford. 1918.
Pilsbry, H. A.
The Sessile Barnacles. Bull. U. S. Nat. Mus., No. 93. 1916.
Paulmier, F. C.
Higher Crustacea of New York City. N. Y. State Museum Bulletin, No. 91, Zoology 12. 1905.

MOLLUSKS

Rogers, Julia E.
  The Shell Book. Doubleday, Page and Company. 1909.
Tryon, G. W. and Pilsbry, H. A.
  Manual of Conchology. Philadelphia. 1878 and recent.

CHORDATES

Castle, W. E.
  The Sea Squirt (*Ciona intestinalis*). Bull. Mus. Comp. Zool. Harvard, Vol. 27. 1896.
Van Name, W.
  Simple Ascidians of the Coast of New England. Proceedings of the Boston Soc. Nat. Hist., Vol. 34. 1912.
  Compound Ascidians of the Coast of New England. Proceedings of the Boston Soc. Nat. Hist., Vol. 34. 1910.

GLOSSARY

Abdomen. The principal posterior body-division in many invertebrates.

Aboral. The side or region of the body opposite to the mouth.

Acontium. A threadlike organ provided with nettle cells.

Actinal. Pertaining to the region in which the tube feet of echinoderms are located.

Actinule. A larval form of certain hydroids.

Adaptive gills. Structures that function as gills, but which in mollusks are not true ctenidia.

Adductor muscle. A muscle which in contracting pulls the parts of the body together.

Adoral. Near the Mouth.

Algae. Simple plants which include the seaweeds.

Alimentary tract. The digestive canal.

Alternation of generations. A reproductive cycle in which a generation of sexual and a generation of asexual forms alternate.

Ambulacral feet. The tube feet provided with sucker disks in echinoderms.

Ambulacral groove. In starfish, the furrow sheltering the tube feet; it occupies a position on the under side of each ray.

Ambulacral pores. Minute orifices perforating the body wall in starfishes and sea urchins.

Ampulla. A bulblike expansion of the ambulacral feet in echinoderms.

Ancestrula. The first-formed individual of a bryozoan colony.

Antenna. One of the "feelers." An appendage of the head; it is usually segmented and has a sensory function.

Antennal sinus. An indentation in the shell near the antennal region of certain ostracods.

Antennule. A first antenna. One of the anterior pairs of antennae when two pairs are present.

Anterior. At or near the forward end of the body.

Anus. The vent. The posterior opening of the digestive tract.

Apex. The tip of the spire on the shell of a univalve mollusk.

Artery. A blood vessel conducting the blood from the heart.

Asexual. Reproduction without the agency of sex.

Auricle. A chamber of the heart into which the blood is conducted by the veins. A ciliated organ in ctenophores.

Autozooid. A feeding polyp in an alcyonarian colony.

Avicularium. An appendage usually resembling a bird's head in bryozoans.

Axial. Along a central line of the body or an appendage.

443

Bilateral symmetry. Having a form which may be divided lengthwise into similar halves.

Biramous. Two-branched.

Bivalve. A mollusk having a shell composed of two similar parts, or valves.

Blastostyle. The reproductive individual of a campanularian hydroid colony.

Branchia. A gill.

Byssus. Threads secreted by the foot of certain bivalve mollusks for attachment to the substratum.

Caecum. A blind gut or pouchlike extension occurring on the alimentary canal.

Calyx. The cuplike body of a crinoid.

Carapace. In crustaceans, the hard exterior or covering of the cephalothorax.

Cardinal teeth. In bivalve shells, the hinged teeth below the umbones.

Carina. The median dorsal plate in the shell of a barnacle.

Cartilage. The internal part of the ligament in bivalve shells.

Cephalic. Pertaining to the head.

Cephalothorax. In arthropods, the body-division formed by the fusion of the head and the thorax.

Cerata. Dorsal extensions which function as gills.

Chelate. Provided with pincerlike claws.

Cheliped. A leg provided with a pincerlike claw.

Chitin. The hard, horny substance which composes the outer covering of many arthropods.

Chromatophores. Colored pigment cells.

Cilia. Minute, hairlike, vibratory projections of the cells forming the inner and the outer surface of certain animals.

Cirrus. A filamentous appendage, often having a sensory function. In flatworms, an organ of copulation.

Cloaca. The space in the intestinal tract near the anus which receives the discharged products of the body.

Cnidoblast. In coelenterates, a stinging cell which contains the nematocyst.

Cnidocil. A minute spine projecting from a nettle cell.

Coenosarc. The soft parts of a coral.

Coenenchym. The soft parts of an alcyonarian colony.

Collar cell. A cell provided with a flagellum around the base of which is a collar.

Columella. The axis or central pillar of a spiral shell or of a corallite.

Corallite. An individual coral animal.

Corallum. An assemblage, or colony, of coral animals.

Corbula. A podlike structure on the branches of a plumularian hydroids.

Costae. In hydroids, the plates composing the corbula.

Ctenidium. A type of gill resembling a comb in mollusks.

Cuticle. The protective or outer skin.

Distal. Farthest removed.

Dorsal. On or toward the back, or dorsum.

Ectoderm. In sponges and coelenterates, the outermost layer of cells composing the body.

Ectosarc. In protozoans, the outermost layer of non-granular protoplasm.

Embryo. In animals, the young individual undergoing development usually within egg membranes.

Endoderm. The innermost layer of cells in sponges and coelenterates.

Endosarc. In protozoans, the interior granular protoplasm.

Extensor muscle. A muscle that extends the parts of the body.

Exumbrella. The upper, or aboral, side of a jellyfish.

Flagellum. A vibratory threadlike extension of a cell. In crustaceans, the terminal joints of the antennae.

Flame cell. A primitive type of excretory cell in flatworms.

Food vacuole. In protozoans, a globule in which food particles are contained.

Furca. A pair of forklike projections at the hinder end of the body.

Ganglion. A small cluster of nerve cells.

Gastric. Pertaining to the stomach.

Gastrovascular space. The central cavity in coelenterates.

Gastrozooid. A feeding polyp in a siphonophore colony.

Gemmule. A capsulated reproductive cell in sponges.

Genital. Pertaining to reproduction.

Gill. A breathing organ adapted to obtain the air contained in water.

Gill filaments. In pelecypods, the ciliated vertical ridges on the gills.

Girdle. In chitons, a fold of the integument commonly at the margin of the shell.

Gizzard. A tough, muscular portion of the alimentary tract which is used in grinding or reducing the food.

Gnathopod. A grasping claw near the mouth in amphipods.

Gonad. An organ of sex which produces reproductive cells.

Gonangium. The gonotheca. A cuticular covering of the blastostyle in campanularian hydroids.

Gonosome. The medusoid form of a hydromedusan.

Gonotheca. In hydroids, the cuticular transparent covering protecting the reproductive polyp.

Hectocotylus arm. In cephalopods, a modified arm of the male used to transfer the sperm to the female.

Hermaphroditic. Having the reproductive organs of both sexes in one individual.

Hydranth. The nutritive zooid, or individual feeding polyp in hydroid colony.

Hydrocaulus. The stem of a hydroid colony.

Hydrorhiza. The rootlike holdfast of a hydroid.

Hydrotheca. The protective cuplike structure of a polyp in a hydroid colony.

Hypostome. In a hydroid, the projection of the polyp's body on which the mouth is borne.

Infundibulum. A longitudinal fold in the wall of the pharynx in ascidians.

Interradius. In starfishes, the interspace between two rays.

Introvert. The anterior part of a sipunclulid's body, which can be invaginated, or retracted.

Labium. An under lip.

Lamella. A structure composed of thin, leaflike plates.

Larva. A young, free, self-sustaining animal that has not completed its development or achieved its adult form.

Lateral. To the right or left of a median line.

Lip. In gastropods, the margin of the shell aperture.

Lithocyst. A calcareous sense-organ in medusae and ctenophores.

Lophophore. The circular or horseshoe-shaped tentacle-bearing ridge in molluscoids.

Lorica. The shell investing the body of certain rotifers.

Lunule. In pelecypods, a depressed area, usually heart-shaped, in front of and close to the umbones.

Macronucleus. The larger nucleus in infusorians.

Madreporite. (Madreporic plate.)—The porous plate or sieve through which the water enters the water-vascular system in echinoderms.

Mandibles. In arthropods, the first, or anterior, pair of mouth-parts.

Mantle. In brachiopods and mollusks, the fold of the integument that secretes the shell. In ascidians, the body wall underneath the tunic.

Manubrium. The central downward projecting portion of the body of a jellyfish, which bears the mouth.

Maxillae. In arthropods, the paired mouth-parts immediately behind the mandibles.

Maxillipeds. In arthropods, the appendages immediately posterior to the maxillae.

Metric system. All measurements given in this work are made according to the metric system. The unit of measurement is the millimeter (mm.) which is equivalent to .0394, or a trifle more than $\frac{1}{25}$ of an inch. For all practical purposes the following equivalents will serve: 2.5 mm = $\frac{1}{10}$ inch; 12.5 = $\frac{1}{2}$ inch;

·25 mm = 1 inch. A reproduction of a metric scale is given in the accompanying figure.

The smaller divisions of the scale indicate the length of a millimeter; the divisions shown by the longer lines, 1, 2, 3, etc., represent centimeters.

Medusa. A jellyfish.

Medusoid. Like a medusa. The sexual generation in hydromedusans and in scyphomedusans.

Megalops. An advanced larval form of certain crustaceans.

Megascleres. The larger spicules in sponges.

Membranelles. Diminutive membranes in some protozoans.

Mesentery. A supporting tissue of one of the viscera. In coelenterates, a partition extending inward from the body wall.

Mesogloea. The tissue lying between the ectoderm and the endoderm in coelenterates.

Micronucleus. The smaller nucleus in protozoans.

Microscleres. The minute spicules in sponges.

Nauplius. An early larval form of some crustaceans.

Nectophore. A swimming individual in siphonophores.

Nematocyst. In coelenterates, the stinging structure within the cnidoblast.

Nematophore. A defensive polyp armed with nematocysts in certain coelenterates.

Nephridium. A urinary tubule.

Nerve commisure. A nerve connecting two ganglia consisting of a pair.

Nerve connective. A nerve connecting two separate pairs of ganglia.

Neuropodium. The lower part of the parapodium of an annelid.

Notochord. A cylindrical, elastic cord lying between the nerve cord and the alimentary canal of young chordate embryos.

Notopodium. The upper part of the parapodium of annelids.

Nucleus. A structure, usually spherical and usually central, within a cell, and which is the dominant factor in its development and activities.

Ocellus. A minute, primitive eye.

Oesophagus. The gullet. The region of the alimentary canal leading from the pharynx to the stomach.

Operculum. A structure, usually platelike, closing an aperture or forming a covering.

Oral. Pertaining to the mouth or mouth region.

Osculum. The excurrent, or cloacal, aperture in sponges.

Ossicle. In echinoderms, a calcareous plate or rod.

Otocyst. A primitive organ presumed to be sensitive to sound.

Ovary. The organ producing the eggs.

Ovicell. In bryozoans, the chamber containing the developing embryo.

Pallets. Calcareous opercular plates in the siphons of *Teredo*.

Pallial line. A line formed by the attachment of the mantle margin in the shell of pelecypods.

Pallial sinus. An indentation in the pallial line marking the attachment area of the siphonal retractor muscle in pelecypods.

Palp. A sensory organ in the mouth region.

Papulae. In echinoderms, delicate projections of the body wall functioning as gills.

Parapodium. In annelids, the jointless, paired appendages of the segments; in certain mollusks, the epipodium, or lateral prolongation of the foot.

Parenchyma. The connective tissue filling the body cavity of flatworms.

Paxillae. Calcareous rods supporting minute spines, occurring on the aboral surface of certain starfishes.

Pedicel. A stalk or stem.

Pedecellariae. Minute pincerlike structures on the surface of certain echinoderms.

Peduncle. A stalk or stemlike structure forming an organ of attachment in sessile animals.

Pericardium. The membrane surrounding the heart.

Periopod. In crustaceans, one of the appendages of the thorax which are posterior to the maxillipeds.

Periostracum. The covering, or skin, investing the outer surface of a mollusk's shell.

Perisarc. The outer covering of a hydroid.

Periproct. The region surrounding the anus.

Peristome. The region surrounding the mouth, in echinoderms.

Petaloid area. In echinoids, the petallike figure on the aboral side indicating the ambulacral pores.

Pharynx. The region of the alimentary canal posterior to the mouth.

Plankton. Minute forms found free in the water. These may either be free-floating or free-swimming.

Planula. In coelenterates, a free-swimming larval form bearing cilia.

Pleopod. In crustaceans, one of the abdominal appendages.

Pluteus. In echinoderms, a free-swimming larval form.

Pneumatophore. The organ of flotation in siphonophores.

Polypide. The soft animal structure of a bryozoan.

Polyp. A sessile individual coelenterate or bryozoan.

Proboscis. In worms, the extensible organ bearing the mouth-parts; it is usually part of the pharynx.

Pseudopodium. In sarcodinas, a temporary extension of the body functioning as a foot.

Rachis. The upper, or polyp-bearing part of a sea pennatulid.

Radula. The lingual ribbon bearing the minute calcareous teeth in mollusks.

Rectum. The posterior portion of the alimentary canal.

Retractor muscle. A muscle serving to draw an organ towards its point of attachment.

Rhinophores. In nudibranchs, the posterior pair of tentacles.

Rhopalia. The sense organs of scyphomedusans.

Ring canal. In echinoderms, the central circular canal of the water-vascular system.

Rostrum. In crustaceans, the forward-projecting extension of the carapace.

Scutum. In barnacles, a shieldlike structure forming one of the two pairs of hinged plates of the shell.

Scyphistoma. A hydroid generation in the development of scyphomedusans.

Segment. In annelids, one of the annular divisions of the body.

Septum. A partition.

Sessile. Fixed to one place without the power of locomotion.

Seta. A bristle.

Siphon. In mollusks and ascidians, an organ through which water passes to or from the branchial organs.

Siphonal canal. A groove in the lip of the shell of some gastropods to enclose the siphon.

Siphuncle. In nautilids, a tubular passage through the chambers of the shell.

Spermatophore. A capsulated mass of spermatozoa.

Spermatozoa. The male reproductive sexual cells.

Sphaeridia. Minute modified spines on the outer surface of some echinoderms.

Spicule. Minute calcareous or silicious structures forming part or all of the framework of certain sponges. In echinoderms, these structures serve to stiffen the body.

Spire. In mollusks, that part of the shell which is first formed and which is above the larger lower whorl.

Statocyst. A sense organ presumed to be related to the function of maintaining equilibrium.

Stolon. The holdfast of a hydroid or a bryozoan; the portion of the stem which is attached to the substratum.

Stone canal. The calcareous tube leading from the madreporite to the ring canal in echinoderms.

Strobila. The fixed compound form of a scyphomedusan.

Subchela. A pinching claw in which the terminal segment works on the principle of a folding knife-blade.

Subumbrella. The oral, or under surface, of a medusa.

Tactile. Pertaining to the sense of touch.

Telson. The terminal segment, or tail, of a crustacean or arachnoid.

Tentacle. An elongated organ, tactile in function and often used for grasping.

Tentaculocyst. A sense organ in scyphozoans.

Test. In echinoids, the hard shell, or skeleton; in ascidians, the tunic; in foraminifers, the rigid calcareous framework.

Testis. The organ producing the spermatozoa.

Thorax. The portion of the body between the head and the abdomen.

Tooth papillae. Projections from the jaw in ophiurans.

Tunic. In tunicates, the outer covering.

Umbilicus. In gastropods, a depression in the shell at the base of the axis.

Umbo. (plural, *umbones.*) In pelecypods, a protuberance of the shell near the hinge indicating the primary area of growth; in barnacles, the projection above the hinge.

Unisexual. Not hermaphroditic. Having but one kind of sex organs, either male or female.

Univalve. A gastropod. Having a shell that consists of but one part.

Uropod. Modified swimming appendage at the side of the telson in crustacea.

Viviparous. Live-bearing. Bringing forth active young which have been formed within the parent's body.

Veliger larva. The free-swimming, ciliated larva of certain gastropods.

Velum. In hydromedusans, the circular, locomotor membrane projecting inwards from the umbrella, or bell.

Ventral. The under side of an animal; opposite to the dorsal.

Vibraculum. The whiplike defensive appendage extending from the zooecium of certain bryozoans.

Viscera. The organs within the body cavities.

Zooea. An intermediate larval form of certain crustaceans.

Zooecium. The chamber or sac containing the polypide of a bryozoan; it may be of calcareous, semicalcareous or chitinous composition.

Zooid. The individual member of a hydroid or bryozoan colony.

INDEX

A CATALOGUE OF SELECTED DOVER BOOKS
IN ALL FIELDS OF INTEREST

A CATALOGUE OF SELECTED DOVER BOOKS
IN ALL FIELDS OF INTEREST

AMERICA'S OLD MASTERS, James T. Flexner. Four men emerged unexpectedly from provincial 18th century America to leadership in European art: Benjamin West, J. S. Copley, C. R. Peale, Gilbert Stuart. Brilliant coverage of lives and contributions. Revised, 1967 edition. 69 plates. 365pp. of text.
21806-6 Paperbound $3.00

FIRST FLOWERS OF OUR WILDERNESS: AMERICAN PAINTING, THE COLONIAL PERIOD, James T. Flexner. Painters, and regional painting traditions from earliest Colonial times up to the emergence of Copley, West and Peale Sr., Foster, Gustavus Hesselius, Feke, John Smibert and many anonymous painters in the primitive manner. Engaging presentation, with 162 illustrations. xxii + 368pp.
22180-6 Paperbound $3.50

THE LIGHT OF DISTANT SKIES: AMERICAN PAINTING, 1760-1835, James T. Flexner. The great generation of early American painters goes to Europe to learn and to teach: West, Copley, Gilbert Stuart and others. Allston, Trumbull, Morse; also contemporary American painters—primitives, derivatives, academics—who remained in America. 102 illustrations. xiii + 306pp.
22179-2 Paperbound $3.50

A HISTORY OF THE RISE AND PROGRESS OF THE ARTS OF DESIGN IN THE UNITED STATES, William Dunlap. Much the richest mine of information on early American painters, sculptors, architects, engravers, miniaturists, etc. The only source of information for scores of artists, the major primary source for many others. Unabridged reprint of rare original 1834 edition, with new introduction by James T. Flexner, and 394 new illustrations. Edited by Rita Weiss. 6⅝ x 9⅝.
21695-0, 21696-9, 21697-7 Three volumes, Paperbound $15.00

EPOCHS OF CHINESE AND JAPANESE ART, Ernest F. Fenollosa. From primitive Chinese art to the 20th century, thorough history, explanation of every important art period and form, including Japanese woodcuts; main stress on China and Japan, but Tibet, Korea also included. Still unexcelled for its detailed, rich coverage of cultural background, aesthetic elements, diffusion studies, particularly of the historical period. 2nd, 1913 edition. 242 illustrations. lii + 439pp. of text.
20364-6, 20365-4 Two volumes, Paperbound $6.00

THE GENTLE ART OF MAKING ENEMIES, James A. M. Whistler. Greatest wit of his day deflates Oscar Wilde, Ruskin, Swinburne; strikes back at inane critics, exhibitions, art journalism; aesthetics of impressionist revolution in most striking form. Highly readable classic by great painter. Reproduction of edition designed by Whistler. Introduction by Alfred Werner. xxxvi + 334pp.
21875-9 Paperbound $3.00

"ESSENTIAL GRAMMAR" SERIES

All you really need to know about modern, colloquial grammar. Many educational shortcuts help you learn faster, understand better. Detailed cognate lists teach you to recognize similarities between English and foreign words and roots—make learning vocabulary easy and interesting. Excellent for independent study or as a supplement to record courses.

ESSENTIAL FRENCH GRAMMAR, Seymour Resnick. 2500-item cognate list. 159pp.
(EBE) 20419-7 Paperbound $1.50

ESSENTIAL GERMAN GRAMMAR, Guy Stern and Everett F. Bleiler. Unusual short-cuts on noun declension, word order, compound verbs. 124pp.
(EBE) 20422-7 Paperbound $1.25

ESSENTIAL ITALIAN GRAMMAR, Olga Ragusa. 111pp.
(EBE) 20779-X Paperbound $1.25

ESSENTIAL JAPANESE GRAMMAR, Everett F. Bleiler. In Romaji transcription; no characters needed. Japanese grammar is regular and simple. 156pp.
21027-8 Paperbound $1.50

ESSENTIAL PORTUGUESE GRAMMAR, Alexander da R. Prista. vi + 114pp.
21650-0 Paperbound $1.35

ESSENTIAL SPANISH GRAMMAR, Seymour Resnick. 2500 word cognate list. 115pp.
(EBE) 20780-3 Paperbound $1.25

ESSENTIAL ENGLISH GRAMMAR, Philip Gucker. Combines best features of modern, functional and traditional approaches. For refresher, class use, home study. x + 177pp.
21649-7 Paperbound $1.75

A PHRASE AND SENTENCE DICTIONARY OF SPOKEN SPANISH. Prepared for U. S. War Department by U. S. linguists. As above, unit is idiom, phrase or sentence rather than word. English-Spanish and Spanish-English sections contain modern equivalents of over 18,000 sentences. Introduction and appendix as above. iv + 513pp.
20495-2 Paperbound $3.50

A PHRASE AND SENTENCE DICTIONARY OF SPOKEN RUSSIAN. Dictionary prepared for U. S. War Department by U. S. linguists. Basic unit is not the word, but the idiom, phrase or sentence. English-Russian and Russian-English sections contain modern equivalents for over 30,000 phrases. Grammatical introduction covers phonetics, writing, syntax. Appendix of word lists for food, numbers, geographical names, etc. vi + 573 pp. 6⅛ x 9¼.
20496-0 Paperbound $5.50

CONVERSATIONAL CHINESE FOR BEGINNERS, Morris Swadesh. Phonetic system, beginner's course in Pai Hua Mandarin Chinese covering most important, most useful speech patterns. Emphasis on modern colloquial usage. Formerly *Chinese in Your Pocket*. xvi + 158pp.
21123-1 Paperbound $1.75

Two Little Savages; Being the Adventures of Two Boys Who Lived as Indians and What They Learned, Ernest Thompson Seton. Great classic of nature and boyhood provides a vast range of woodlore in most palatable form, a genuinely entertaining story. Two farm boys build a teepee in woods and live in it for a month, working out Indian solutions to living problems, star lore, birds and animals, plants, etc. 293 illustrations. vii + 286pp.

20985-7 Paperbound $2.50

Peter Piper's Practical Principles of Plain & Perfect Pronunciation. Alliterative jingles and tongue-twisters of surprising charm, that made their first appearance in America about 1830. Republished in full with the spirited woodcut illustrations from this earliest American edition. 32pp. $4\frac{1}{2}$ x $6\frac{3}{8}$.

22560-7 Paperbound $1.00

Science Experiments and Amusements for Children, Charles Vivian. 73 easy experiments, requiring only materials found at home or easily available, such as candles, coins, steel wool, etc.; illustrate basic phenomena like vacuum, simple chemical reaction, etc. All safe. Modern, well-planned. Formerly *Science Games for Children*. 102 photos, numerous drawings. 96pp. $6\frac{1}{8}$ x $9\frac{1}{4}$.

21856-2 Paperbound $1.25

An Introduction to Chess Moves and Tactics Simply Explained, Leonard Barden. Informal intermediate introduction, quite strong in explaining reasons for moves. Covers basic material, tactics, important openings, traps, positional play in middle game, end game. Attempts to isolate patterns and recurrent configurations. Formerly *Chess*. 58 figures. 102pp.   (USO) 21210-6 Paperbound $1.25

Lasker's Manual of Chess, Dr. Emanuel Lasker. Lasker was not only one of the five great World Champions, he was also one of the ablest expositors, theorists, and analysts. In many ways, his Manual, permeated with his philosophy of battle, filled with keen insights, is one of the greatest works ever written on chess. Filled with analyzed games by the great players. A single-volume library that will profit almost any chess player, beginner or master. 308 diagrams. xli x 349pp.

20640-8 Paperbound $2.75

The Master Book of Mathematical Recreations, Fred Schuh. In opinion of many the finest work ever prepared on mathematical puzzles, stunts, recreations; exhaustively thorough explanations of mathematics involved, analysis of effects, citation of puzzles and games. Mathematics involved is elementary. Translated by F. Göbel. 194 figures. xxiv + 430pp.   22134-2 Paperbound $4.00

Mathematics, Magic and Mystery, Martin Gardner. Puzzle editor for Scientific American explains mathematics behind various mystifying tricks: card tricks, stage "mind reading," coin and match tricks, counting out games, geometric dissections, etc. Probability sets, theory of numbers clearly explained. Also provides more than 400 tricks, guaranteed to work, that you can do. 135 illustrations. xii + 176pp.

20335-2 Paperbound $2.00

EAST O' THE SUN AND WEST O' THE MOON, George W. Dasent. Considered the best of all translations of these Norwegian folk tales, this collection has been enjoyed by generations of children (and folklorists too). Includes True and Untrue, Why the Sea is Salt, East O' the Sun and West O' the Moon, Why the Bear is Stumpy-Tailed, Boots and the Troll, The Cock and the Hen, Rich Peter the Pedlar, and 52 more. The only edition with all 59 tales. 77 illustrations by Erik Werenskiold and Theodor Kittelsen. xv + 418pp. 22521-6 Paperbound $3.50

GOOPS AND HOW TO BE THEM, Gelett Burgess. Classic of tongue-in-cheek humor, masquerading as etiquette book. 87 verses, twice as many cartoons, show mischievous Goops as they demonstrate to children virtues of table manners, neatness, courtesy, etc. Favorite for generations. viii + 88pp. 6½ x 9¼.
22233-0 Paperbound $1.50

ALICE'S ADVENTURES UNDER GROUND, Lewis Carroll. The first version, quite different from the final Alice in Wonderland, printed out by Carroll himself with his own illustrations. Complete facsimile of the "million dollar" manuscript Carroll gave to Alice Liddell in 1864. Introduction by Martin Gardner. viii + 96pp. Title and dedication pages in color. 21482-6 Paperbound $1.25

THE BROWNIES, THEIR BOOK, Palmer Cox. Small as mice, cunning as foxes, exuberant and full of mischief, the Brownies go to the zoo, toy shop, seashore, circus, etc., in 24 verse adventures and 266 illustrations. Long a favorite, since their first appearance in St. Nicholas Magazine. xi + 144pp. 6⅝ x 9¼.
21265-3 Paperbound $1.75

SONGS OF CHILDHOOD, Walter De La Mare. Published (under the pseudonym Walter Ramal) when De La Mare was only 29, this charming collection has long been a favorite children's book. A facsimile of the first edition in paper, the 47 poems capture the simplicity of the nursery rhyme and the ballad, including such lyrics as I Met Eve, Tartary, The Silver Penny. vii + 106pp. (USO) 21972-0 Paperbound $1.25

THE COMPLETE NONSENSE OF EDWARD LEAR, Edward Lear. The finest 19th-century humorist-cartoonist in full: all nonsense limericks, zany alphabets, Owl and Pussycat, songs, nonsense botany, and more than 500 illustrations by Lear himself. Edited by Holbrook Jackson. xxix + 287pp. (USO) 20167-8 Paperbound $2.00

BILLY WHISKERS: THE AUTOBIOGRAPHY OF A GOAT, Frances Trego Montgomery. A favorite of children since the early 20th century, here are the escapades of that rambunctious, irresistible and mischievous goat—Billy Whiskers. Much in the spirit of Peck's Bad Boy, this is a book that children never tire of reading or hearing. All the original familiar illustrations by W. H. Fry are included: 6 color plates, 18 black and white drawings. 159pp. 22345-0 Paperbound $2.00

MOTHER GOOSE MELODIES. Faithful republication of the fabulously rare Munroe and Francis "copyright 1833" Boston edition—the most important Mother Goose collection, usually referred to as the "original." Familiar rhymes plus many rare ones, with wonderful old woodcut illustrations. Edited by E. F. Bleiler. 128pp. 4½ x 6⅜. 22577-1 Paperbound $1.00

THE RED FAIRY BOOK, Andrew Lang. Lang's color fairy books have long been children's favorites. This volume includes Rapunzel, Jack and the Bean-stalk and 35 other stories, familiar and unfamiliar. 4 plates, 93 illustrations x + 367pp.
21673-X Paperbound $2.50

THE BLUE FAIRY BOOK, Andrew Lang. Lang's tales come from all countries and all times. Here are 37 tales from Grimm, the Arabian Nights, Greek Mythology, and other fascinating sources. 8 plates, 130 illustrations. xi + 390pp.
21437-0 Paperbound $2.75

HOUSEHOLD STORIES BY THE BROTHERS GRIMM. Classic English-language edition of the well-known tales — Rumpelstiltskin, Snow White, Hansel and Gretel, The Twelve Brothers, Faithful John, Rapunzel, Tom Thumb (52 stories in all). Translated into simple, straightforward English by Lucy Crane. Ornamented with headpieces, vignettes, elaborate decorative initials and a dozen full-page illustrations by Walter Crane. x + 269pp.
21080-4 Paperbound **$2.00**

THE MERRY ADVENTURES OF ROBIN HOOD, Howard Pyle. The finest modern versions of the traditional ballads and tales about the great English outlaw. Howard Pyle's complete prose version, with every word, every illustration of the first edition. Do not confuse this facsimile of the original (1883) with modern editions that change text or illustrations. 23 plates plus many page decorations. xxii + 296pp.
22043-5 Paperbound $2.75

THE STORY OF KING ARTHUR AND HIS KNIGHTS, Howard Pyle. The finest children's version of the life of King Arthur; brilliantly retold by Pyle, with 48 of his most imaginative illustrations. xviii + 313pp. 6⅛ x 9¼.
21445-1 Paperbound $2.50

THE WONDERFUL WIZARD OF OZ, L. Frank Baum. America's finest children's book in facsimile of first edition with all Denslow illustrations in full color. The edition a child should have. Introduction by Martin Gardner. 23 color plates, scores of drawings. iv + 267pp.
20691-2 Paperbound $3.50

THE MARVELOUS LAND OF OZ, L. Frank Baum. The second Oz book, every bit as imaginative as the Wizard. The hero is a boy named Tip, but the Scarecrow and the Tin Woodman are back, as is the Oz magic. 16 color plates, 120 drawings by John R. Neill. 287pp.
20692-0 Paperbound $2.50

THE MAGICAL MONARCH OF MO, L. Frank Baum. Remarkable adventures in a land even stranger than Oz. The best of Baum's books not in the Oz series. 15 color plates and dozens of drawings by Frank Verbeck. xviii + 237pp.
21892-9 Paperbound $2.25

THE BAD CHILD'S BOOK OF BEASTS, MORE BEASTS FOR WORSE CHILDREN, A MORAL ALPHABET, Hilaire Belloc. Three complete humor classics in one volume. Be kind to the frog, and do not call him names . . . and 28 other whimsical animals. Familiar favorites and some not so well known. Illustrated by Basil Blackwell. 156pp.
(USO) 20749-8 Paperbound $1.50

THE ARCHITECTURE OF COUNTRY HOUSES, Andrew J. Downing. Together with Vaux's *Villas and Cottages* this is the basic book for Hudson River Gothic architecture of the middle Victorian period. Full, sound discussions of general aspects of housing, architecture, style, decoration, furnishing, together with scores of detailed house plans, illustrations of specific buildings, accompanied by full text. Perhaps the most influential single American architectural book. 1850 edition. Introduction by J. Stewart Johnson. 321 figures, 34 architectural designs. xvi + 560pp.
22003-6 Paperbound $5.00

LOST EXAMPLES OF COLONIAL ARCHITECTURE, John Mead Howells. Full-page photographs of buildings that have disappeared or been so altered as to be denatured, including many designed by major early American architects. 245 plates. xvii + 248pp. 7⅞ x 10¾. 21143-6 Paperbound $3.50

DOMESTIC ARCHITECTURE OF THE AMERICAN COLONIES AND OF THE EARLY REPUBLIC, Fiske Kimball. Foremost architect and restorer of Williamsburg and Monticello covers nearly 200 homes between 1620-1825. Architectural details, construction, style features, special fixtures, floor plans, etc. Generally considered finest work in its area. 219 illustrations of houses, doorways, windows, capital mantels. xx + 314pp. 7⅞ x 10¾. 21743-4 Paperbound $4.00

EARLY AMERICAN ROOMS: 1650-1858, edited by Russell Hawes Kettell. Tour of 12 rooms, each representative of a different era in American history and each furnished, decorated, designed and occupied in the style of the era. 72 plans and elevations, 8-page color section, etc., show fabrics, wall papers, arrangements, etc. Full descriptive text. xvii + 200pp. of text. 8⅜ x 11¼.
21633-0 Paperbound $5.00

THE FITZWILLIAM VIRGINAL BOOK, edited by J. Fuller Maitland and W. B. Squire. Full modern printing of famous early 17th-century ms. volume of 300 works by Morley, Byrd, Bull, Gibbons, etc. For piano or other modern keyboard instrument; easy to read format. xxxvi + 938pp. 8⅜ x 11.
21068-5, 21069-3 Two volumes, Paperbound $12.00

KEYBOARD MUSIC, Johann Sebastian Bach. Bach Gesellschaft edition. A rich selection of Bach's masterpieces for the harpsichord: the six English Suites, six French Suites, the six Partitas (Clavierübung part I), the Goldberg Variations (Clavierübung part IV), the fifteen Two-Part Inventions and the fifteen Three-Part Sinfonias. Clearly reproduced on large sheets with ample margins; eminently playable. vi + 312pp. 8⅛ x 11. 22360-4 Paperbound $5.00

THE MUSIC OF BACH: AN INTRODUCTION, Charles Sanford Terry. A fine, nontechnical introduction to Bach's music, both instrumental and vocal. Covers organ music, chamber music, passion music, other types. Analyzes themes, developments, innovations. x + 114pp. 21075-8 Paperbound $1.95

BEETHOVEN AND HIS NINE SYMPHONIES, Sir George Grove. Noted British musicologist provides best history, analysis, commentary on symphonies. Very thorough, rigorously accurate; necessary to both advanced student and amateur music lover. 436 musical passages. vii + 407 pp. 20334-4 Paperbound $4.00

DESIGN BY ACCIDENT; A BOOK OF "ACCIDENTAL EFFECTS" FOR ARTISTS AND DESIGNERS, James F. O'Brien. Create your own unique, striking, imaginative effects by "controlled accident" interaction of materials: paints and lacquers, oil and water based paints, splatter, crackling materials, shatter, similar items. Everything you do will be different; first book on this limitless art, so useful to both fine artist and commercial artist. Full instructions. 192 plates showing "accidents," 8 in color. viii + 215pp. 8⅜ x 11¼.   21942-9 Paperbound $3.75

THE BOOK OF SIGNS, Rudolf Koch. Famed German type designer draws 493 beautiful symbols: religious, mystical, alchemical, imperial, property marks, runes, etc. Remarkable fusion of traditional and modern. Good for suggestions of timelessness, smartness, modernity. Text. vi + 104pp. 6⅛ x 9¼.
20162-7 Paperbound $1.50

HISTORY OF INDIAN AND INDONESIAN ART, Ananda K. Coomaraswamy. An unabridged republication of one of the finest books by a great scholar in Eastern art. Rich in descriptive material, history, social backgrounds; Sunga reliefs, Rajput paintings, Gupta temples, Burmese frescoes, textiles, jewelry, sculpture, etc. 400 photos. viii + 423pp. 6⅜ x 9¾.   21436-2 Paperbound $5.00

PRIMITIVE ART, Franz Boas. America's foremost anthropologist surveys textiles, ceramics, woodcarving, basketry, metalwork, etc.; patterns, technology, creation of symbols, style origins. All areas of world, but very full on Northwest Coast Indians. More than 350 illustrations of baskets, boxes, totem poles, weapons, etc. 378 pp.
20025-6 Paperbound $3.00

THE GENTLEMAN AND CABINET MAKER'S DIRECTOR, Thomas Chippendale. Full reprint (third edition, 1762) of most influential furniture book of all time, by master cabinetmaker. 200 plates, illustrating chairs, sofas, mirrors, tables, cabinets, plus 24 photographs of surviving pieces. Biographical introduction by N. Bienenstock. vi + 249pp. 9⅞ x 12¾.   21601-2 Paperbound $5.00

AMERICAN ANTIQUE FURNITURE, Edgar G. Miller, Jr. The basic coverage of all American furniture before 1840. Individual chapters cover type of furniture—clocks, tables, sideboards, etc.—chronologically, with inexhaustible wealth of data. More than 2100 photographs, all identified, commented on. Essential to all early American collectors. Introduction by H. E. Keyes. vi + 1106pp. 7⅞ x 10¾.
21599-7, 21600-4 Two volumes, Paperbound $11.00

PENNSYLVANIA DUTCH AMERICAN FOLK ART, Henry J. Kauffman. 279 photos, 28 drawings of tulipware, Fraktur script, painted tinware, toys, flowered furniture, quilts, samplers, hex signs, house interiors, etc. Full descriptive text. Excellent for tourist, rewarding for designer, collector. Map. 146pp. 7⅞ x 10¾.
21205-X Paperbound $3.00

EARLY NEW ENGLAND GRAVESTONE RUBBINGS, Edmund V. Gillon, Jr. 43 photographs, 226 carefully reproduced rubbings show heavily symbolic, sometimes macabre early gravestones, up to early 19th century. Remarkable early American primitive art, occasionally strikingly beautiful; always powerful. Text. xxvi + 207pp. 8⅜ x 11¼.   21380-3 Paperbound $4.00

ALPHABETS AND ORNAMENTS, Ernst Lehner. Well-known pictorial source for decorative alphabets, script examples, cartouches, frames, decorative title pages, calligraphic initials, borders, similar material. 14th to 19th century, mostly European. Useful in almost any graphic arts designing, varied styles. 750 illustrations. 256pp. 7 x 10. 21905-4 Paperbound $4.00

PAINTING: A CREATIVE APPROACH, Norman Colquhoun. For the beginner simple guide provides an instructive approach to painting: major stumbling blocks for beginner; overcoming them, technical points; paints and pigments; oil painting; watercolor and other media and color. New section on "plastic" paints. Glossary. Formerly *Paint Your Own Pictures.* 221pp. 22000-1 Paperbound $1.75

THE ENJOYMENT AND USE OF COLOR, Walter Sargent. Explanation of the relations between colors themselves and between colors in nature and art, including hundreds of little-known facts about color values, intensities, effects of high and low illumination, complementary colors. Many practical hints for painters, references to great masters. 7 color plates, 29 illustrations. x + 274pp. 20944-X Paperbound $3.00

THE NOTEBOOKS OF LEONARDO DA VINCI, compiled and edited by Jean Paul Richter. 1566 extracts from original manuscripts reveal the full range of Leonardo's versatile genius: all his writings on painting, sculpture, architecture, anatomy, astronomy, geography, topography, physiology, mining, music, etc., in both Italian and English, with 186 plates of manuscript pages and more than 500 additional drawings. Includes studies for the Last Supper, the lost Sforza monument, and other works. Total of xlvii + 866pp. 7⅞ x 10¾. 22572-0, 22573-9 Two volumes, Paperbound $12.00

MONTGOMERY WARD CATALOGUE OF 1895. Tea gowns, yards of flannel and pillow-case lace, stereoscopes, books of gospel hymns, the New Improved Singer Sewing Machine, side saddles, milk skimmers, straight-edged razors, high-button shoes, spittoons, and on and on . . . listing some 25,000 items, practically all illustrated. Essential to the shoppers of the 1890's, it is our truest record of the spirit of the period. Unaltered reprint of Issue No. 57, Spring and Summer 1895. Introduction by Boris Emmet. Innumerable illustrations. xiii + 624pp. 8½ x 11⅝. 22377-9 Paperbound $8.50

THE CRYSTAL PALACE EXHIBITION ILLUSTRATED CATALOGUE (LONDON, 1851). One of the wonders of the modern world—the Crystal Palace Exhibition in which all the nations of the civilized world exhibited their achievements in the arts and sciences—presented in an equally important illustrated catalogue. More than 1700 items pictured with accompanying text—ceramics, textiles, cast-iron work, carpets, pianos, sleds, razors, wall-papers, billiard tables, beehives, silverware and hundreds of other artifacts—represent the focal point of Victorian culture in the Western World. Probably the largest collection of Victorian decorative art ever assembled—indispensable for antiquarians and designers. Unabridged republication of the Art-Journal Catalogue of the Great Exhibition of 1851, with all terminal essays. New introduction by John Gloag, F.S.A. xxxiv + 426pp. 9 x 12. 22503-8 Paperbound $5.00

A HISTORY OF COSTUME, Carl Köhler. Definitive history, based on surviving pieces of clothing primarily, and paintings, statues, etc. secondarily. Highly readable text, supplemented by 594 illustrations of costumes of the ancient Mediterranean peoples, Greece and Rome, the Teutonic prehistoric period; costumes of the Middle Ages, Renaissance, Baroque, 18th and 19th centuries. Clear, measured patterns are provided for many clothing articles. Approach is practical throughout. Enlarged by Emma von Sichart. 464pp.   21030-8 Paperbound $3.50

ORIENTAL RUGS, ANTIQUE AND MODERN, Walter A. Hawley. A complete and authoritative treatise on the Oriental rug—where they are made, by whom and how, designs and symbols, characteristics in detail of the six major groups, how to distinguish them and how to buy them. Detailed technical data is provided on periods, weaves, warps, wefts, textures, sides, ends and knots, although no technical background is required for an understanding. 11 color plates, 80 halftones, 4 maps. vi + 320pp. 6⅛ x 9⅛.   22366-3 Paperbound $5.00

TEN BOOKS ON ARCHITECTURE, Vitruvius. By any standards the most important book on architecture ever written. Early Roman discussion of aesthetics of building, construction methods, orders, sites, and every other aspect of architecture has inspired, instructed architecture for about 2,000 years. Stands behind Palladio, Michelangelo, Bramante, Wren, countless others. Definitive Morris H. Morgan translation. 68 illustrations. xii + 331pp.   20645-9 Paperbound $3.00

THE FOUR BOOKS OF ARCHITECTURE, Andrea Palladio. Translated into every major Western European language in the two centuries following its publication in 1570, this has been one of the most influential books in the history of architecture. Complete reprint of the 1738 Isaac Ware edition. New introduction by Adolf Placzek, Columbia Univ. 216 plates. xxii + 110pp. of text. 9½ x 12¾.
                          21308-0 Clothbound $12.50

STICKS AND STONES: A STUDY OF AMERICAN ARCHITECTURE AND CIVILIZATION, Lewis Mumford.One of the great classics of American cultural history. American architecture from the medieval-inspired earliest forms to the early 20th century; evolution of structure and style, and reciprocal influences on environment. 21 photographic illustrations. 238pp.   20202-X Paperbound $2.00

THE AMERICAN BUILDER'S COMPANION, Asher Benjamin. The most widely used early 19th century architectural style and source book, for colonial up into Greek Revival periods. Extensive development of geometry of carpentering, construction of sashes, frames, doors, stairs; plans and elevations of domestic and other buildings. Hundreds of thousands of houses were built according to this book, now invaluable to historians, architects, restorers, etc. 1827 edition. 59 plates. 114pp. 7⅞ x 10¾.
                          22236-5 Paperbound $4.00

DUTCH HOUSES IN THE HUDSON VALLEY BEFORE 1776, Helen Wilkinson Reynolds. The standard survey of the Dutch colonial house and outbuildings, with constructional features, decoration, and local history associated with individual homesteads. Introduction by Franklin D. Roosevelt. Map. 150 illustrations. 469pp. 6⅝ x 9¼.   21469-9 Paperbound $5.00

VISUAL ILLUSIONS: THEIR CAUSES, CHARACTERISTICS, AND APPLICATIONS, Matthew Luckiesh. Thorough description and discussion of optical illusion, geometric and perspective, particularly; size and shape distortions, illusions of color, of motion; natural illusions; use of illusion in art and magic, industry, etc. Most useful today with op art, also for classical art. Scores of effects illustrated. Introduction by William H. Ittleson. 100 illustrations. xxi + 252pp.

21530-X Paperbound $2.00

A HANDBOOK OF ANATOMY FOR ART STUDENTS, Arthur Thomson. Thorough, virtually exhaustive coverage of skeletal structure, musculature, etc. Full text, supplemented by anatomical diagrams and drawings and by photographs of undraped figures. Unique in its comparison of male and female forms, pointing out differences of contour, texture, form. 211 figures, 40 drawings, 86 photographs. xx + 459pp. 5⅜ x 8⅜.

21163-0 Paperbound $3.50

150 MASTERPIECES OF DRAWING, Selected by Anthony Toney. Full page reproductions of drawings from the early 16th to the end of the 18th century, all beautifully reproduced: Rembrandt, Michelangelo, Dürer, Fragonard, Urs, Graf, Wouwerman, many others. First-rate browsing book, model book for artists. xviii + 150pp. 8⅜ x 11¼.

21032-4 Paperbound¹ $2.50

THE LATER WORK OF AUBREY BEARDSLEY, Aubrey Beardsley. Exotic, erotic, ironic masterpieces in full maturity: Comedy Ballet, Venus and Tannhauser, Pierrot, Lysistrata, Rape of the Lock, Savoy material, Ali Baba, Volpone, etc. This material revolutionized the art world, and is still powerful, fresh, brilliant. With *The Early Work*, all Beardsley's finest work. 174 plates, 2 in color. xiv + 176pp. 8⅛ x 11.

21817-1 Paperbound $3.75

DRAWINGS OF REMBRANDT, Rembrandt van Rijn. Complete reproduction of fabulously rare edition by Lippmann and Hofstede de Groot, completely reedited, updated, improved by Prof. Seymour Slive, Fogg Museum. Portraits, Biblical sketches, landscapes, Oriental types, nudes, episodes from classical mythology—All Rembrandt's fertile genius. Also selection of drawings by his pupils and followers. "Stunning volumes," *Saturday Review*. 550 illustrations. lxxviii + 552pp. 9⅛ x 12¼.

21485-0, 21486-9 Two volumes, Paperbound $10.00

THE DISASTERS OF WAR, Francisco Goya. One of the masterpieces of Western civilization—83 etchings that record Goya's shattering, bitter reaction to the Napoleonic war that swept through Spain after the insurrection of 1808 and to war in general. Reprint of the first edition, with three additional plates from Boston's Museum of Fine Arts. All plates facsimile size. Introduction by Philip Hofer, Fogg Museum. v + 97pp. 9⅜ x 8¼.

21872-4 Paperbound $2.50

GRAPHIC WORKS OF ODILON REDON. Largest collection of Redon's graphic works ever assembled: 172 lithographs, 28 etchings and engravings, 9 drawings. These include some of his most famous works. All the plates from *Odilon Redon: oeuvre graphique complet,* plus additional plates. New introduction and caption translations by Alfred Werner. 209 illustrations. xxvii + 209pp. 9⅛ x 12¼.

21966-8 Paperbound $4.50

THE PHILOSOPHY OF THE UPANISHADS, Paul Deussen. Clear, detailed statement of upanishadic system of thought, generally considered among best available. History of these works, full exposition of system emergent from them, parallel concepts in the West. Translated by A. S. Geden. xiv + 429pp.

21616-0 Paperbound $3.50

LANGUAGE, TRUTH AND LOGIC, Alfred J. Ayer. Famous, remarkably clear introduction to the Vienna and Cambridge schools of Logical Positivism; function of philosophy, elimination of metaphysical thought, nature of analysis, similar topics. "Wish I had written it myself," Bertrand Russell. 2nd, 1946 edition. 160pp.

20010-8 Paperbound $1.50

THE GUIDE FOR THE PERPLEXED, Moses Maimonides. Great classic of medieval Judaism, major attempt to reconcile revealed religion (Pentateuch, commentaries) and Aristotelian philosophy. Enormously important in all Western thought. Unabridged Friedländer translation. 50-page introduction. lix + 414pp.

(USO) 20351-4 Paperbound $4.50

OCCULT AND SUPERNATURAL PHENOMENA, D. H. Rawcliffe. Full, serious study of the most persistent delusions of mankind: crystal gazing, mediumistic trance, stigmata, lycanthropy, fire walking, dowsing, telepathy, ghosts, ESP, etc., and their relation to common forms of abnormal psychology. Formerly *Illusions and Delusions of the Supernatural and the Occult.* iii + 551pp. 20503-7 Paperbound $4.00

THE EGYPTIAN BOOK OF THE DEAD: THE PAPYRUS OF ANI, E. A. Wallis Budge. Full hieroglyphic text, interlinear transliteration of sounds, word for word translation, then smooth, connected translation; Theban recension. Basic work in Ancient Egyptian civilization; now even more significant than ever for historical importance, dilation of consciousness, etc. clvi + 377pp. 6½ x 9¼.

21866-X Paperbound $4.95

PSYCHOLOGY OF MUSIC, Carl E. Seashore. Basic, thorough survey of everything known about psychology of music up to 1940's; essential reading for psychologists, musicologists. Physical acoustics; auditory apparatus; relationship of physical sound to perceived sound; role of the mind in sorting, altering, suppressing, creating sound sensations; musical learning, testing for ability, absolute pitch, other topics. Records of Caruso, Menuhin analyzed. 88 figures. xix + 408pp.

21851-1 Paperbound $3.50

THE I CHING (THE BOOK OF CHANGES), translated by James Legge. Complete translated text plus appendices by Confucius, of perhaps the most penetrating divination book ever compiled. Indispensable to all study of early Oriental civilizations. 3 plates. xxiii + 448pp. 21062-6 Paperbound $3.50

THE UPANISHADS, translated by Max Müller. Twelve classical upanishads: Chandogya, Kena, Aitareya, Kaushitaki, Isa, Katha, Mundaka, Taittiriyaka, Brhadaranyaka, Svetasvatara, Prasna, Maitriyana. 160-page introduction, analysis by Prof. Müller. Total of 670pp. 20992-X, 20993-8 Two volumes, Paperbound $7.50

PLANETS, STARS AND GALAXIES: DESCRIPTIVE ASTRONOMY FOR BEGINNERS, A. E. Fanning. Comprehensive introductory survey of astronomy: the sun, solar system, stars, galaxies, universe, cosmology; up-to-date, including quasars, radio stars, etc. Preface by Prof. Donald Menzel. 24pp. of photographs. 189pp. 5¼ x 8¼.
21680-2 Paperbound $2.50

TEACH YOURSELF CALCULUS, P. Abbott. With a good background in algebra and trig, you can teach yourself calculus with this book. Simple, straightforward introduction to functions of all kinds, integration, differentiation, series, etc. "Students who are beginning to study calculus method will derive great help from this book." Faraday House Journal. 308pp.
20683-1 Clothbound $2.50

TEACH YOURSELF TRIGONOMETRY, P. Abbott. Geometrical foundations, indices and logarithms, ratios, angles, circular measure, etc. are presented in this sound, easy-to-use text. Excellent for the beginner or as a brush up, this text carries the student through the solution of triangles. 204pp.
20682-3 Clothbound $2.00

BASIC MACHINES AND HOW THEY WORK, U. S. Bureau of Naval Personnel. Originally used in U.S. Naval training schools, this book clearly explains the operation of a progression of machines, from the simplest—lever, wheel and axle, inclined plane, wedge, screw—to the most complex—typewriter, internal combustion engine, computer mechanism. Utilizing an approach that requires only an elementary understanding of mathematics, these explanations build logically upon each other and are assisted by over 200 drawings and diagrams. Perfect as a technical school manual or as a self-teaching aid to the layman. 204 figures. Preface. Index. vii + 161pp. 6½ x 9¼.
21709-4 Paperbound $2.50

THE FRIENDLY STARS, Martha Evans Martin. Classic has taught naked-eye observation of stars, planets to hundreds of thousands, still not surpassed for charm, lucidity, adequacy. Completely updated by Professor Donald H. Menzel, Harvard Observatory. 25 illustrations. 16 x 30 chart. x + 147pp.
21099-5 Paperbound $2.00

MUSIC OF THE SPHERES: THE MATERIAL UNIVERSE FROM ATOM TO QUASAR, SIMPLY EXPLAINED, Guy Murchie. Extremely broad, brilliantly written popular account begins with the solar system and reaches to dividing line between matter and nonmatter; latest understandings presented with exceptional clarity. Volume One: Planets, stars, galaxies, cosmology, geology, celestial mechanics, latest astronomical discoveries; Volume Two: Matter, atoms, waves, radiation, relativity, chemical action, heat, nuclear energy, quantum theory, music, light, color, probability, antimatter, antigravity, and similar topics. 319 figures. 1967 (second) edition. Total of xx + 644pp.
21809-0, 21810-4 Two volumes, Paperbound $5.75

OLD-TIME SCHOOLS AND SCHOOL BOOKS, Clifton Johnson. Illustrations and rhymes from early primers, abundant quotations from early textbooks, many anecdotes of school life enliven this study of elementary schools from Puritans to middle 19th century. Introduction by Carl Withers. 234 illustrations. xxxiii + 381pp.
21031-6 Paperbound $4.00

THE PRINCIPLES OF PSYCHOLOGY, William James. The famous long course, complete and unabridged. Stream of thought, time perception, memory, experimental methods—these are only some of the concerns of a work that was years ahead of its time and still valid, interesting, useful. 94 figures. Total of xviii + 1391pp.
20381-6, 20382-4 Two volumes, Paperbound $9.00

THE STRANGE STORY OF THE QUANTUM, Banesh Hoffmann. Non-mathematical but thorough explanation of work of Planck, Einstein, Bohr, Pauli, de Broglie, Schrödinger, Heisenberg, Dirac, Feynman, etc. No technical background needed. "Of books attempting such an account, this is the best," Henry Margenau, Yale. 40-page "Postscript 1959." xii + 285pp. 20518-5 Paperbound $3.00

THE RISE OF THE NEW PHYSICS, A. d'Abro. Most thorough explanation in print of central core of mathematical physics, both classical and modern; from Newton to Dirac and Heisenberg. Both history and exposition; philosophy of science, causality, explanations of higher mathematics, analytical mechanics, electromagnetism, thermodynamics, phase rule, special and general relativity, matrices. No higher mathematics needed to follow exposition, though treatment is elementary to intermediate in level. Recommended to serious student who wishes verbal understanding. 97 illustrations. xvii + 982pp. 20003-5, 20004-3 Two volumes, Paperbound $10.00

GREAT IDEAS OF OPERATIONS RESEARCH, Jagjit Singh. Easily followed non-technical explanation of mathematical tools, aims, results: statistics, linear programming, game theory, queueing theory, Monte Carlo simulation, etc. Uses only elementary mathematics. Many case studies, several analyzed in detail. Clarity, breadth make this excellent for specialist in another field who wishes background. 41 figures. x + 228pp. 21886-4 Paperbound $2.50

GREAT IDEAS OF MODERN MATHEMATICS: THEIR NATURE AND USE, Jagjit Singh. Internationally famous expositor, winner of Unesco's Kalinga Award for science popularization explains verbally such topics as differential equations, matrices, groups, sets, transformations, mathematical logic and other important modern mathematics, as well as use in physics, astrophysics, and similar fields. Superb exposition for layman, scientist in other areas. viii + 312pp.
20587-8 Paperbound $2.75

GREAT IDEAS IN INFORMATION THEORY, LANGUAGE AND CYBERNETICS, Jagjit Singh. The analog and digital computers, how they work, how they are like and unlike the human brain, the men who developed them, their future applications, computer terminology. An essential book for today, even for readers with little math. Some mathematical demonstrations included for more advanced readers. 118 figures. Tables. ix + 338pp. 21694-2 Paperbound $2.50

CHANCE, LUCK AND STATISTICS, Horace C. Levinson. Non-mathematical presentation of fundamentals of probability theory and science of statistics and their applications. Games of chance, betting odds, misuse of statistics, normal and skew distributions, birth rates, stock speculation, insurance. Enlarged edition. Formerly "The Science of Chance." xiii + 357pp. 21007-3 Paperbound $2.50

THE PRINCIPLES OF PSYCHOLOGY, William James. The famous long course, complete and unabridged. Stream of thought, time perception, memory, experimental methods—these are only some of the concerns of a work that was years ahead of its time and still valid, interesting, useful. 94 figures. Total of xviii + 1391pp.
20381-6, 20382-4 Two volumes, Paperbound $9.00

THE STRANGE STORY OF THE QUANTUM, Banesh Hoffmann. Non-mathematical but thorough explanation of work of Planck, Einstein, Bohr, Pauli, de Broglie, Schrödinger, Heisenberg, Dirac, Feynman, etc. No technical background needed. "Of books attempting such an account, this is the best," Henry Margenau, Yale. 40-page "Postscript 1959." xii + 285pp.
20518-5 Paperbound $3.00

THE RISE OF THE NEW PHYSICS, A. d'Abro. Most thorough explanation in print of central core of mathematical physics, both classical and modern; from Newton to Dirac and Heisenberg. Both history and exposition; philosophy of science, causality, explanations of higher mathematics, analytical mechanics, electromagnetism, thermodynamics, phase rule, special and general relativity, matrices. No higher mathematics needed to follow exposition, though treatment is elementary to intermediate in level. Recommended to serious student who wishes verbal understanding. 97 illustrations. xvii + 982pp.
20003-5, 20004-3 Two volumes, Paperbound$10.00

GREAT IDEAS OF OPERATIONS RESEARCH, Jagjit Singh. Easily followed non-technical explanation of mathematical tools, aims, results: statistics, linear programming, game theory, queueing theory, Monte Carlo simulation, etc. Uses only elementary mathematics. Many case studies, several analyzed in detail. Clarity, breadth make this excellent for specialist in another field who wishes background. 41 figures. x + 228pp.
21886-4 Paperbound $2.50

GREAT IDEAS OF MODERN MATHEMATICS: THEIR NATURE AND USE, Jagjit Singh. Internationally famous expositor, winner of Unesco's Kalinga Award for science popularization explains verbally such topics as differential equations, matrices, groups, sets, transformations, mathematical logic and other important modern mathematics, as well as use in physics, astrophysics, and similar fields. Superb exposition for layman, scientist in other areas. viii + 312pp.
20587-8 Paperbound $2.75

GREAT IDEAS IN INFORMATION THEORY, LANGUAGE AND CYBERNETICS, Jagjit Singh. The analog and digital computers, how they work, how they are like and unlike the human brain, the men who developed them, their future applications, computer terminology. An essential book for today, even for readers with little math. Some mathematical demonstrations included for more advanced readers. 118 figures. Tables. ix + 338pp.
21694-2 Paperbound $2.50

CHANCE, LUCK AND STATISTICS, Horace C. Levinson. Non-mathematical presentation of fundamentals of probability theory and science of statistics and their applications. Games of chance, betting odds, misuse of statistics, normal and skew distributions, birth rates, stock speculation, insurance. Enlarged edition. Formerly "The Science of Chance." xiii + 357pp.
21007-3 Paperbound $2.50

AMERICAN FOOD AND GAME FISHES, David S. Jordan and Barton W. Evermann. Definitive source of information, detailed and accurate enough to enable the sportsman and nature lover to identify conclusively some 1,000 species and sub-species of North American fish, sought for food or sport. Coverage of range, physiology, habits, life history, food value. Best methods of capture, interest to the angler, advice on bait, fly-fishing, etc. 338 drawings and photographs. 1 + 574pp. 6⅝ x 9⅜.
22196-2 Paperbound $5.00

THE FROG BOOK, Mary C. Dickerson. Complete with extensive finding keys, over 300 photographs, and an introduction to the general biology of frogs and toads, this is the classic non-technical study of Northeastern and Central species. 58 species; 290 photographs and 16 color plates. xvii + 253pp.
21973-9 Paperbound $4.00

THE MOTH BOOK: A GUIDE TO THE MOTHS OF NORTH AMERICA, William J. Holland. Classical study, eagerly sought after and used for the past 60 years. Clear identification manual to more than 2,000 different moths, largest manual in existence. General information about moths, capturing, mounting, classifying, etc., followed by species by species descriptions. 263 illustrations plus 48 color plates show almost every species, full size. 1968 edition, preface, nomenclature changes by A. E. Brower. xxiv + 479pp. of text. 6½ x 9¼.
21948-8 Paperbound $6.00

THE SEA-BEACH AT EBB-TIDE, Augusta Foote Arnold. Interested amateur can identify hundreds of marine plants and animals on coasts of North America; marine algae; seaweeds; squids; hermit crabs; horse shoe crabs; shrimps; corals; sea anemones; etc. Species descriptions cover: structure; food; reproductive cycle; size; shape; color; habitat; etc. Over 600 drawings. 85 plates. xii + 490pp.
21949-6 Paperbound $4.00

COMMON BIRD SONGS, Donald J. Borror. 33⅓ 12-inch record presents songs of 60 important birds of the eastern United States. A thorough, serious record which provides several examples for each bird, showing different types of song, individual variations, etc. Inestimable identification aid for birdwatcher. 32-page booklet gives text about birds and songs, with illustration for each bird.
21829-5 Record, book, album. Monaural. $3.50

FADS AND FALLACIES IN THE NAME OF SCIENCE, Martin Gardner. Fair, witty appraisal of cranks and quacks of science: Atlantis, Lemuria, hollow earth, flat earth, Velikovsky, orgone energy, Dianetics, flying saucers, Bridey Murphy, food fads, medical fads, perpetual motion, etc. Formerly "In the Name of Science." x + 363pp.
20394-8 Paperbound $3.00

HOAXES, Curtis D. MacDougall. Exhaustive, unbelievably rich account of great hoaxes: Locke's moon hoax, Shakespearean forgeries, sea serpents, Loch Ness monster, Cardiff giant, John Wilkes Booth's mummy, Disumbrationist school of art, dozens more; also journalism, psychology of hoaxing. 54 illustrations. xi + 338pp.
20465-0 Paperbound $3.50

CATALOGUE OF DOVER BOOKS

MATHEMATICAL PUZZLES FOR BEGINNERS AND ENTHUSIASTS, Geoffrey Mott-Smith. 189 puzzles from easy to difficult—involving arithmetic, logic, algebra, properties of digits, probability, etc.—for enjoyment and mental stimulus. Explanation of mathematical principles behind the puzzles. 135 illustrations. viii + 248pp.
20198-8 Paperbound $2.00

PAPER FOLDING FOR BEGINNERS, William D. Murray and Francis J. Rigney. Easiest book on the market, clearest instructions on making interesting, beautiful origami. Sail boats, cups, roosters, frogs that move legs, bonbon boxes, standing birds, etc. 40 projects; more than 275 diagrams and photographs. 94pp.
20713-7 Paperbound $1.00

TRICKS AND GAMES ON THE POOL TABLE, Fred Herrmann. 79 tricks and games—some solitaires, some for two or more players, some competitive games—to entertain you between formal games. Mystifying shots and throws, unusual caroms, tricks involving such props as cork, coins, a hat, etc. Formerly *Fun on the Pool Table*. 77 figures. 95pp.
21814-7 Paperbound $1.25

HAND SHADOWS TO BE THROWN UPON THE WALL: A SERIES OF NOVEL AND AMUSING FIGURES FORMED BY THE HAND, Henry Bursill. Delightful picturebook from great-grandfather's day shows how to make 18 different hand shadows: a bird that flies, duck that quacks, dog that wags his tail, camel, goose, deer, boy, turtle, etc. Only book of its sort. vi + 33pp. 6½ x 9¼.   21779-5 Paperbound $1.00

WHITTLING AND WOODCARVING, E. J. Tangerman. 18th printing of best book on market. "If you can cut a potato you can carve" toys and puzzles, chains, chessmen, caricatures, masks, frames, woodcut blocks, surface patterns, much more. Information on tools, woods, techniques. Also goes into serious wood sculpture from Middle Ages to present, East and West. 464 photos, figures. x + 293pp.
20965-2 Paperbound $2.50

HISTORY OF PHILOSOPHY, Julián Marias. Possibly the clearest, most easily followed, best planned, most useful one-volume history of philosophy on the market; neither skimpy nor overfull. Full details on system of every major philosopher and dozens of less important thinkers from pre-Socratics up to Existentialism and later. Strong on many European figures usually omitted. Has gone through dozens of editions in Europe. 1966 edition, translated by Stanley Appelbaum and Clarence Strowbridge. xviii + 505pp.
21739-6 Paperbound $3.50

YOGA: A SCIENTIFIC EVALUATION, Kovoor T. Behanan. Scientific but non-technical study of physiological results of yoga exercises; done under auspices of Yale U. Relations to Indian thought, to psychoanalysis, etc. 16 photos. xxiii + 270pp.
20505-3 Paperbound $2.50

Prices subject to change without notice.
Available at your book dealer or write for free catalogue to Dept. GI, Dover Publications, Inc., 180 Varick St., N. Y., N. Y. 10014. Dover publishes more than 150 books each year on science, elementary and advanced mathematics, biology, music, art, literary history, social sciences and other areas.

© THE BAKER & TAYLOR CO.